Fractal Geometry in Human Geography and Planning

SCIENCES

Geography and Demography, Field Director – Denise Pumain

Fractal Logics in Geography – Hierarchies and Centralities
Subject Head – Cécile Tannier

Fractal Geometry in Human Geography and Planning

Coordinated by
Cécile Tannier

iSTE

WILEY

First published 2024 in Great Britain and the United States by ISTE Ltd and John Wiley & Sons, Inc.

ISTE Ltd
27-37 St George's Road
London SW19 4EU
UK

www.iste.co.uk

John Wiley & Sons, Inc.
111 River Street
Hoboken, NJ 07030
USA

www.wiley.com

Library of Congress Control Number: 2023942096

British Library Cataloguing-in-Publication Data
A CIP record for this book is available from the British Library
ISBN 978-1-78945-159-7

ERC code:
SH2 Institutions, Values, Environment and Space
SH2_10 Land use and regional planning
SH2_11 Human, economic and social geography

Contents

Introduction

Cécile TANNIER
Laboratoire ThéMA, CNRS-Université de Franche-Comté, Besançon, France

Many natural objects such as coastlines, mountains, clouds, plants or cells have very irregular shapes. The shapes of the spatial distribution of human settlements are also very irregular, although logics of regularity often underlie their construction (architecture of buildings, continuity of infrastructures, urban planning and development rules, etc.). Indeed, irregularity and complexity of forms is the rule for many fields of both the natural (physics, biology, chemistry) and social sciences (finance, architecture, geography).

Historically, mathematicians have neglected the study of these complex and irregular shapes since they could not be analyzed with classical mathematical tools. Infinitesimal calculus, introduced by Newton and Leibniz in the 17th century, was perfectly adapted to the study of smooth and regular Euclidean shapes (circles, rectangles, ellipses), but not to that of complex and irregular shapes. However, since the late 1960s with the introduction of fractal geometry, mathematicians have focused on systematically studying these shapes. Since then, all fields of science have been considered from a fractal point of view, which has led to significant advances: in particular in medical imaging, in physics in the field of fluid turbulence, in biology for understanding the energy expenditure of organisms, and in cinematography for simulating virtual landscapes.

In geography, fractal geometry and analysis have been used since the 1980s, very shortly after the publication of Mandelbrot's (1975) founding work. Early work showed the value of fractals for measuring and modeling complex geographic forms,

Fractal Geometry in Human Geography and Planning,
coordinated by Cécile TANNIER.

and accounting for relationships between structures at different scales (Goodchild 1980; Arlinghaus 1985, 1987; Batty and Longley 1986, 1987; Goodchild and Mark 1987; Frankhauser 1988). Specifically in human geography, pioneering works dealt with the fractality of the outline of built forms (Longley and Batty 1989; Arlinghaus and Nystuen 1990; Frankhauser 1994), scaling relationships between the number of built clusters and their size (White and Engelen 1993), and fractal aspects of urban growth (Fotheringham et al. 1989; Frankhauser 1990; Batty and Xie 1996, 1999; Benguigui et al. 2000). Since then, geographers' interest in fractal analysis and modeling has not decreased, as evidenced by the regular publication of numerous articles discussing this approach in reference journals. Research carried out in this field over the last 40 years has contributed significantly to improving the characterization of the shapes of the spatial distribution of human settlements and understanding of their development (Tannier and Pumain 2005; Tannier 2018).

In this book, we wanted to review the concepts and methods of fractal analysis and modeling that allow us to study the spatial distribution of human settlements. The book has been constructed and written with the aim of providing a detailed understanding of the methods and to facilitate the use of existing software applications. The main concepts and mathematical tools are summarized, and emphasis is placed on the practical interests of their application for describing forms in human geography, as well as on pitfalls to avoid when using them. References from one chapter to another promote this learning by gradually familiarizing the reader with the manipulation of fractal objects and their practical interpretation, thanks to experiments conducted over a wide variety of empirical cases.

The book comprises three parts. In the first part (Chapters 1–3), we present the basic concepts and geometric reference figures that are useful for conducting informed and relevant fractal analysis in human geography. We show how fractal shapes have interesting properties for geographers, and in what way and how the formal framework of fractal geometry allows us to work on these shapes. The mathematical foundations are precisely described in language that we want to be accessible for the greatest number of readers. In the second part, we describe in detail the principles of four major families of analytical methods: fractal analyses of point patterns such as buildings or road intersections (Chapter 4), fractal methods of morphological delineation of urban agglomerations (Chapter 5), multifractal analyses of the spatial distribution of populations (Chapter 6) and cross-scale signatures for measuring deviations from scale invariance (Chapter 7). The modalities of application of these methods are described in detail and illustrated with numerous examples. The third and last part of this book is devoted to the applications of fractal geometry for urban and regional development and planning. We show in particular that introducing fractal geometric rules into models allows us to better understand how space is organized across scales, and how it is possible to modify this spatial organization by jointly taking into account the effects of distance and scale dependence (Chapter 8). An application of the *Fractalopolis* model to the Paris conurbation then gives a concrete example of multifractal prospective planning (Chapter 9).

References

Arlinghaus, S.L. (1985). Fractals take a central place. *Geografiska Annaler, Series B, Human Geography*, 67(2), 83–88.

Arlinghaus, S.L. (1987). *Essays on Mathematical Geography II*. Monographs Series. Institute of Mathematical Geography, Ann Arbor, MI.

Arlinghaus, S.L. and Nystuen, J.D. (1990). Geometry of boundary exchanges. *Geographical Review*, 80(1), 21–31.

Batty, M. and Longley, P.A. (1986). The fractal simulation of urban structure. *Environment and Planning A*, 18, 1143–1179.

Batty, M. and Longley, P.A. (1987). Urban shapes as fractals. *Area*, 19(3), 215–221.

Batty, M. and Xie, Y. (1996). Preliminary evidence for a theory of the fractal city. *Environment and Planning A*, 28, 1745–1762.

Batty, M. and Xie, Y. (1999). Self-organized criticality and urban development. *Discrete Dynamics in Nature and Society*, 3, 109–124.

Benguigui, L., Czamanski, D., Marinov, M., Portugali, Y. (2000). When and where is a city fractal? *Environment and Planning B: Planning and Design*, 27, 507–519.

Fotheringham, A.S., Batty, M., Longley, P.A. (1989). Diffusion-limited aggregation and the fractal nature of urban growth. *Papers of the Regional Science Association*, 67, 55–69.

Frankhauser, P. (1988). Fractal aspects of urban systems. *Sonderforschungsbereich 230 "Natürliche Konstruktionen"*, 1, 67–76.

Frankhauser, P. (1990). Aspects fractals des structures urbaines. *L'Espace géographique*, 19–20(1), 45–69.

Frankhauser, P. (1994). *La fractalité des structures urbaines*. Anthropos, Paris.

Goodchild, M. (1980). Fractals and the accuracy of geographical measures. *Mathematical Geology*, 12(2), 85–98.

Goodchild, M. and Mark, D. (1987). The fractal nature of geographic phenomena. *Annals of the Association of American Geographers*, 77(2), 265–278.

Longley, P.A. and Batty, M. (1989). On the fractal measurement of geographical boundaries. *Geographical Analysis*, 21(1), 47–67.

Mandelbrot, B. (1975). *Les objets fractals. Forme, hasard et dimension*. Flammarion, Paris.

Tannier, C. (2018). About fractal models in urban geography and planning: Refuting the aesthetics and the universal norm. *Cybergeo Conversation* (blog), 13 March [Online]. Available at: https://cybergeo.hypotheses.org/223.

Tannier, C. and Pumain, D. (2005). Fractals in urban geography: A theoretical outline and an empirical example. *Cybergeo: European Journal of Geography*, 307 [Online]. Available at: https://journals.openedition.org/cybergeo/3275.

White, R. and Engelen, G. (1993). Cellular automata and fractal urban form: A cellular modelling approach to the evolution of urban land-use patterns. *Environment and Planning A*, 25(8), 1175–1199.

Part 1

Basic Concepts and Reference Figures

1

Introduction to Fractal Analysis and Modeling in Human Geography

Cécile TANNIER and François SÉMÉCURBE
Laboratoire ThéMA, CNRS-Université de Franche-Comté, Besançon, France

> *(...) geography foregrounds the organization of space, its functioning; it seeks to explain its diversity or rather its diversification, at different scales; it prioritizes the question of location (why is it there and not elsewhere); starting from space, it questions reality, and especially society, in relation to its material basis, inherited both from the actions of past generations and from the forces of nature*
> (Durand-Dastès 1989).

Describing and understanding processes of concentration and dispersion of the spatial distribution of human settlements is at the heart of research in geography. A human settlement consists of the materialization in space, the physical inscription, of one or more activities. It is characterized by a certain level of durability. It may comprise a building, a group of a few buildings or a larger set of thousands of buildings. Each human settlement (a building or similar) is the location of one or more functions: residential, commercial, industrial, artisanal, religious, politics, logistics and so on. The durability of a building does not imply the temporal

---

The authors would like to thank Jean-Michel Floch, statistician at INSEE (French Institute of Statistics and Economic Studies), now retired, for his critical review of this chapter.

continuity of the activities occurring within it. For example, the transformation of a former farm into a second home creates a change in activity, which in this case becomes seasonal or more occasional. At the extreme, we may consider ruins no longer associated with an activity or a caravan parked in the same place and unoccupied for several years to be human settlements.

A human settlement is found within a more or less artificial unbuilt environment (parks, parking lots, orchards, fields). It is connected to other buildings, near or far, with the same or different functions, by means of transport and communication networks. All this – buildings, unbuilt environment, networks – constitutes a container for human activities. The contents associated with this container are individuals and groups performing activities, pursuing spatial practices and social relationships. In this work, we are only interested in the shapes of the content or the container, which can be geometrically assimilated to sets of points (centroids of buildings, nodes of road networks, built pixels and so on). Other types of spatial configurations, for example, flows or movement trajectories, will not be discussed.

Human settlements present varying degrees of concentration or dispersion in space and time: at the local scale, town centers or village centers, peri-central districts, peripheral extensions (business areas, housing estates); at the scale of an urban area, monocentric or polycentric agglomerations, compact or sprawling; at the regional scale, more or less diffuse and more or less hierarchical population systems. Concentrations and dispersions, considered as states or processes, vary according to the level of aggregation of the entities considered (Openshaw 1983; Fotheringham and Wong 1991; Madelin et al. 2009) and the spatial resolution selected (Müller 1978; Goodchild 1980; Toger et al. 2016). Spatial resolution corresponds to the fineness of the detail with which the elements of the container (buildings, unbuilt environment, networks) are represented. For example, the built-up fabric of an urban agglomeration may be mapped building by building, each building having its own spatial footprint delimited with high precision (metric or even infra-metric). In this case, the spatial resolution is fine. The same built-up fabric can be mapped in the form of urban patches roughly delimiting the envelope of different groups of buildings. In this case, the spatial resolution is coarser. A built-up fabric can also be mapped in the form of 10 m-sided square cells (fine spatial resolution) or 200 m-sided square cells (coarser spatial resolution). The level of aggregation of the analyzed or modeled entities is different from their spatial resolution. For example, we may study a district as a whole, without disaggregating it, or consider it as an aggregate of built blocks or even buildings. The fineness of the spatial resolution of the elementary entities under consideration partially determines the range of possible levels of aggregation.

When, in order to model a given phenomenon, we take into account several levels of aggregation (Grasland 2003; Sanders 2007) or a series of nested spatial resolutions (Goodchild and Mark 1987; Lam and Quattrochi 1992; Tate and Atkinson 2001), or even ranges (neighborhoods) of different sizes (White 2005; van Vliet et al. 2009; Mathian and Sanders 2014), the modeling is multi-scale[1]. When we represent variations of the phenomenon across levels of aggregation or spatial resolutions, the modeling is cross-scale. It is particularly interesting to study the concentration and dispersion of human settlements using multi- or cross-scale models, especially fractals, as this makes it possible to explicitly account for spatial hierarchies (e.g. the fact that an urban agglomeration has a large number of small built-up clusters, a smaller number of medium-sized aggregates and an even smaller number of large-sized clusters) and functional hierarchies (e.g. different levels of functional centrality). Multi- or cross-scale models allow us to represent and characterize variations in the concentration and dispersion of human settlements across scales; in other words, to determine in what way and how the concentration of human settlements varies from one level of analysis to another, and to identify possible invariances between scales, the degree of generality of which may vary in space and time.

1.1. Geometric models of the spatial distribution of human settlements: from the use of smooth and regular shapes to the adoption of fractal shapes

Many spatial models, which can be graphically represented, have been designed to describe and explain the shapes of the spatial distribution of human settlements in relation to the functioning of people and societies. As early as 1826, J. von Thünen proposed a model of radio-concentric organization of cultures around a center according to a principle of increasing transport costs from the center, where consumers are located, to the periphery. A century later, Burgess (1925) envisaged a model of city growth as the gradual addition of concentric rings of decreasing land rent from the center to the periphery. Subsequently, Christaller (1933) and Lösch (1940)[2] designed two models for the organization of hierarchically nested central areas, while Stewart (1947) and Clark (1951) proposed a mathematical model of decreasing density of the population of cities from the center to the periphery. Still a little later, Thünen's model was applied to the modeling of the spatial organization of urban spaces according to a principle of decreasing land rent from the center to the periphery by Alonso (1964), Mills (1967) and Muth (1969).

1. Not to be confused with so-called "multi-level" models, which are a category of statistical models (Mathian and Piron 2007).

2. English translation from 1954.

All these founding models have been the subject of numerous tests and applications, and have led to multiple developments. The geometric shapes of reference that underlie these models – circles, rings, triangles – are smooth and regular.

However, the shapes of the spatial distribution of human settlements are fundamentally non-uniform and very irregular. In the suburban fringes of contemporary European cities, for example, built-up areas have extremely variable spacings between buildings (Figure 1.1). The same is true when looking at cities as a whole: the size of the built-up clusters is highly variable, as are the distances between them (Figure 1.2).

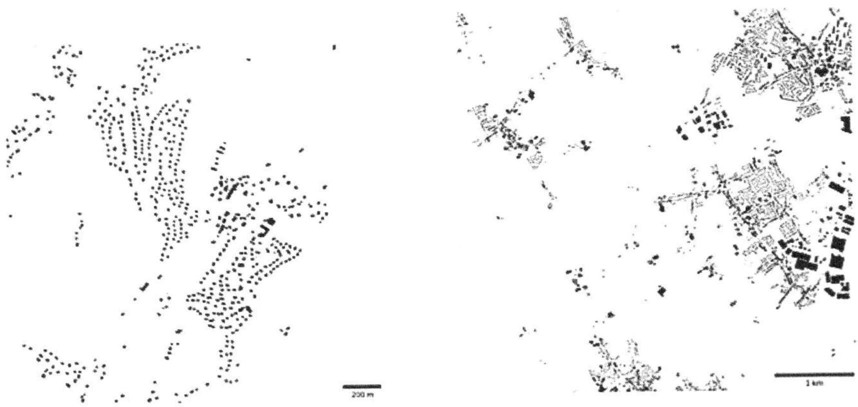

(a) Suburban fringe of the agglomeration of Besançon (eastern France)

(b) Suburban fringe of the agglomeration of Lille (northern France)

Figure 1.1. *Two examples of the built-up fabrics characteristic of peri-urban fringes: black outlines represent the footprint of buildings of different sizes and shapes*

This irregularity in the shapes observed is not unique to geography; it is found in many other fields of both natural (physics, biology, chemistry, etc.) and social science (finance, architecture, graphic arts, etc.). Natural objects in particular (coastlines, mountains, clouds, plants, cellular tissues, etc.) very often display such irregularities. Historically, mathematicians have neglected the study of shapes, whose irregularity and complexity meant they could not be described or analyzed using classical mathematical tools. Infinitesimal calculus, introduced by Newton and Leibniz in the second half of the 17th century, was indeed an ideal tool for the analysis of smooth objects forming part of the framework of classical Euclidean geometry, but not for objects displaying complex and irregular shapes.

Figure 1.2. *A built-up urban fabric: the agglomeration of Basel in 1957. Source: topographic maps of the Bundesamt für Landestopographie, 3084 Wabern, Germany. Manual digitization performed by the "Image, ville, environnement" laboratory, Strasbourg, France*

It was not until the end of the 1960s that the systematic study of such forms developed within mathematics, notably due to the efforts of Benoît Mandelbrot (1924–2010), often recognized as "The Father of Fractals" (Falconer 2013). Since the 1980s, all areas of science have been examined from a fractal perspective: biology, medicine, image compression, signal theory, finance, acoustics, film, music and geography. Fractal shapes have also gained mainstream interest through the widespread use of computers, which have allowed anyone with some knowledge of computer programming to generate elaborate fractal images by repeatedly applying a few simple computer commands.

Figure 1.3. *The initiator and the first three stages of construction of the Cantor set*

Figure 1.4. *The initiator and the first four stages of construction of the Koch snowflake*

"While fractal geometry dates from 1975, it is important in many ways to know that a number of shapes now counted as fractals have been known for a long time. But surprisingly few had actually been drawn before the computer era" (Mandelbrot 1988). In 1872, Weierstrass presented a continuous curve to the mathematical community that was non-differentiable at all points. In 1883, Cantor presented a set of points with the property of being self-similar (Figure 1.3). In 1890, Peano derived a "space-filling curve" from Cantor's work, whose sinuosities fill the entire space of a square. In 1904, von Koch proposed an elementary method of geometric construction of a continuous curve having no tangents (Figure 1.4).

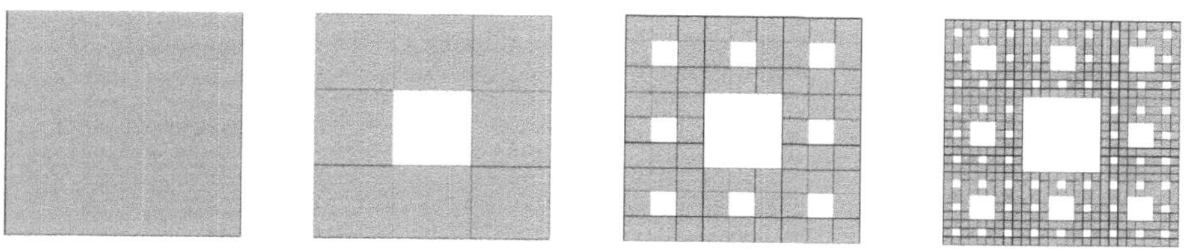

Figure 1.5. *The initiator and the first three stages of construction of a Sierpinski carpet (from Mandelbrot (1982))*

The principle of construction of fractal shapes explicitly incorporates a scale dependency from which derives, in the case of surface shapes, a hierarchy of free (or empty) spaces contained within the spatial area under consideration (Frankhauser 1994). Figure 1.5 illustrates this. The initiator of the figure is a square of side L. The generator consists of $N = 8$ copies of the initial square reduced by a factor of $r = 1/3$. The eight squares are placed so that an empty square is left in the middle of the initial square. In the second stage of construction, the generator is reproduced in each of the eight squares constituting it. The figure thus obtained consists of 64 squares. The same procedure is repeated successively. We observe that the number of elements increases according to a multiplicative logic (512 squares in the third construction step), that the empty spaces (gaps) inside the figure are spatially distributed in a non-uniform manner and that the relationship between the number and the size of gaps is nonlinear (small gaps are much more numerous than large gaps).

Graphical representations of geographical space displaying such an internal hierarchy were imagined long before the birth of fractal geometry. One example is the geometric division of France into departments proposed during the French Revolution by the select committee of the Constitutional Committee created in July 1789 (Figure 1.6).

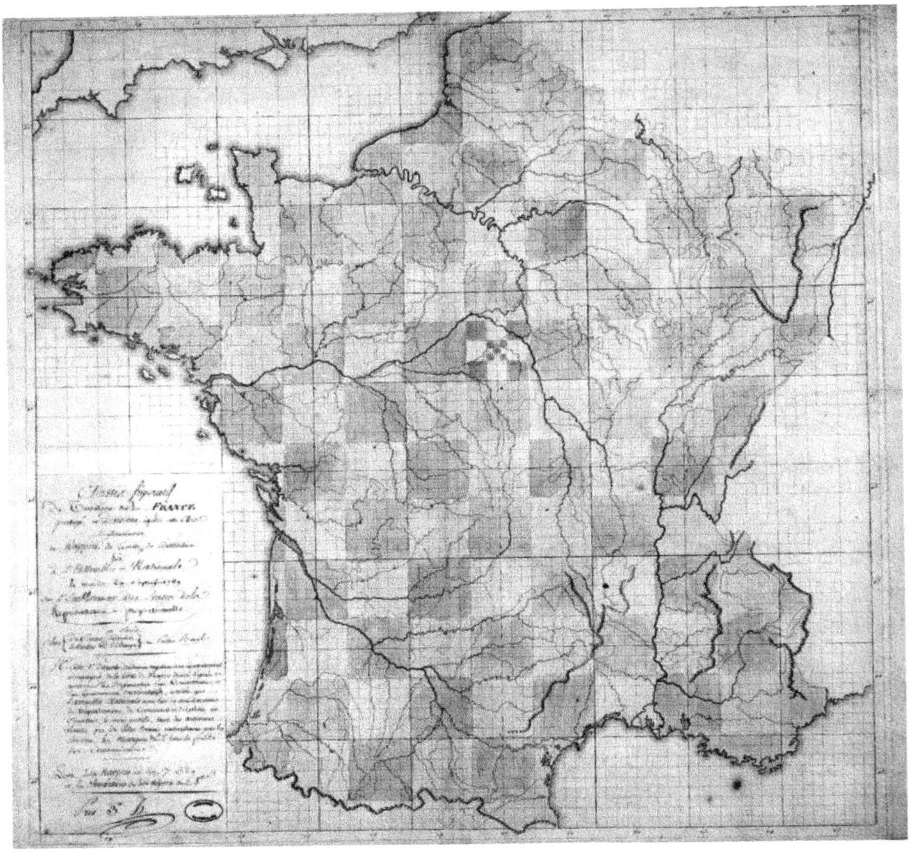

Figure 1.6. *Geometric division of France into departments proposed by the select committee of the Constitutional Committee created in July 1789. Source: figurative outline of the territory of France divided into equal divisions ©Center historique des Archives Nationales de France – Atelier de photographie. See: https://histoire-image.org/fr/etudes/formation-departements [Accessed November 23, 2020]. For a color version of this figure, see www.iste.co.uk/tannier/fractal.zip*

Smooth geometric figures, which fall within the scope of Euclidean geometry and classical mathematics (circles, ellipses, squares, triangles, etc.), have the particularity of having a tangent, that is, a sole and unique direction or slope, at each point. On this basis, various mathematical tools make it possible to establish whether or not two figures are similar, regardless of certain transformations (translation, change of scale, reflection and rotation) (Falconer 2013). It is therefore possible to determine whether a shape is a circle, a square or a triangle, regardless of its size and position. Fractal figures, on the other hand, have no tangent at any point. In other words, they are not differentiable. They also have the particularity of being self-similar; that is, their shape is more or less exactly the same whether we consider them as a whole or whether we zoom in on some of their parts. This explains why classical mathematical tools cannot help determine which of these shapes are similar and which are different. By contrast, the fractal geometry of Mandelbrot (1975) provides a formal mathematical framework for describing a wide variety of such self-similar shapes (Voss 1988).

1.2. Description and measurement of the shapes of the spatial distribution of human settlements

Describing the shape of patterns of human settlements presents a confusing aspect: it is very difficult to theoretically formalize in a general framework, while at the same time, any child would be able to categorize complex spatial forms. For example, the two images of bocage and the open-field system in Figure 1.7 visually display very different built configurations: in the open-field system, buildings are mainly concentrated in villages, unlike bocage where isolated buildings form a dense and homogeneous granular pattern. However, it is very difficult to differentiate these two types of configurations using quantitative indicators, a task which some geographers of the first half of the 20th century set themselves. At that time, a fertile field of research in geography concerned the study of the two predominant agrarian systems in Europe: bocage, featuring dispersed habitation, and the open-field system, featuring concentrated habitation (Lefèvre 1926; Demangeon 1927a, 1927b; Bloch 1931; Dion 1934). Quantitative analyses of the spatial distribution of the rural population emerged in this context. The resulting indicators of population concentration are among the first attempts at quantification in geography (Bretagnolle 1996). The proposed indicators were based on statistics aggregated by municipality from population censuses: both the population concentrated around the town hall, and the population dispersed in village hamlets and localities.

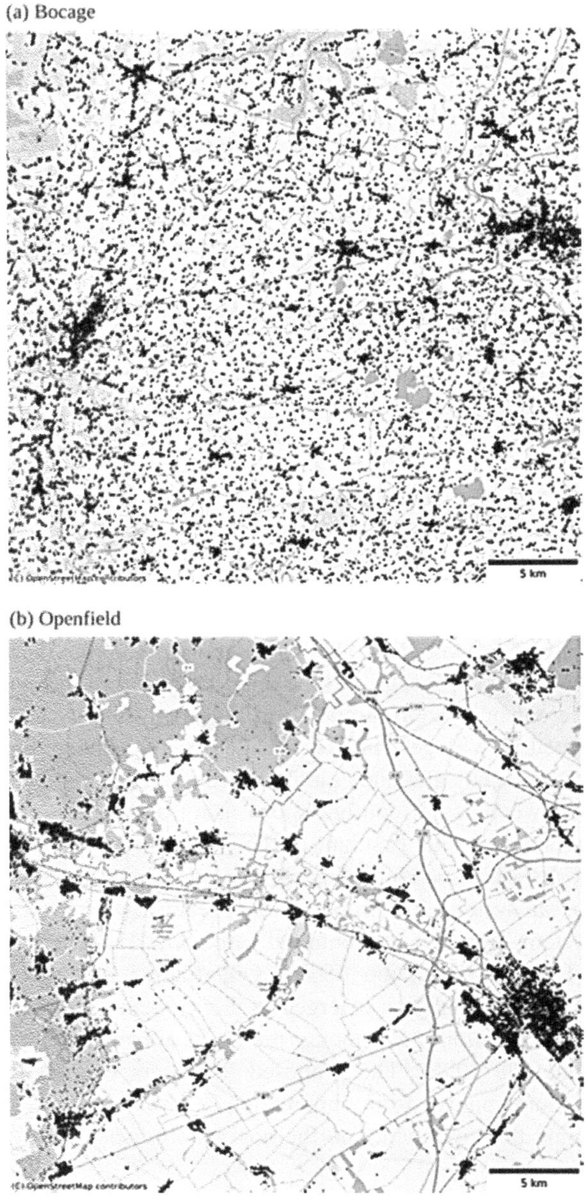

Figure 1.7. *Images of bocage (Western France, between Coutances and Saint-Lô) and the open-field system (Eastern France, near Châlons-en-Champagne). The black dots represent buildings. Data sources: OpenStreetMap and BD Topo IGN (2012). For a color version of this figure, see www.iste.co.uk/tannier/fractal.zip*

Demangeon (1933) was the first to give a synthetic formula for measuring the dispersion of rural habitation: $K = (E \times N)/T$, where N is the number of outlying areas (i.e. hamlets or housing units spatially unconnected to a village cluster), E is the overall population of the outlying areas, T is the total population of the municipality and K is the dispersion index. Demangeon's formula has received several criticisms (Colas 1945): very different values for the triplet (E, N, T) can produce similar values of K; the formula does not take into account the distance of the outlying areas from the main town, which is an essential factor of dispersion; it is difficult to interpret.

Many other formulas have subsequently been proposed to respond to the limitations of Demangeon's (Colas 1945; Chiffre 1969). Nevertheless, Colas (1945) concludes that: "we are convinced that it is illusory to try to represent rural dispersion by a number, in such a way that this number corresponds to one type of dispersion and only one". This has prompted some geographers to produce manually concentration maps of rural habitation, using only visual perception, rather than using quantitative indicators. However, the search for such indicators has not ceased. An overview of traditionally adopted approaches can be found in Duncan (1957) and Biraben and Duhourcau (1974). The Gini, Hoover and Duncan indices quantify unequal spatial distribution of a population based on a measure of deviation from a uniform distribution (Figure 1.8). Calculation of these indices is based on geographical divisions (most often administrative grids) to understand the positioning of the points relative to each other. Other methods based on quadrats (division of the space according to a homogeneous square grid) were also used in the 1970s (Haggett et al. 1977).

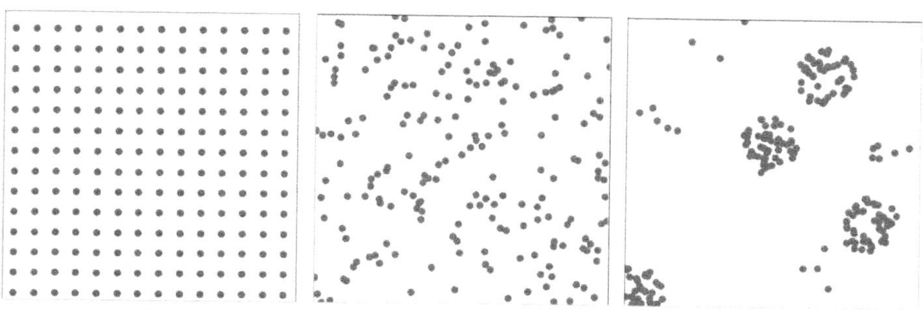

Figure 1.8. *Three different forms of distribution of points*

The common feature of all of these methods is that two points are considered to be close if they belong to the same spatial unit. However, this way of measuring proximity only very poorly reflects the mutual positioning of points in space. Moreover, it is particularly sensitive to the arbitrary nature of geographical division, in other words to the *modifiable areal unit problem* (MAUP) (Openshaw 1983), which is defined as follows: the states measured depend on the size of the elementary spatial units considered (scaling problem), on the shape and positioning of the elementary spatial units (zoning problem) and on the bounds of the area under study (Goodchild 2016). MAUP is the geographical avatar of an epistemological problem shared by all the humanities and natural sciences, namely the ecological error (Gehlke and Biehl 1934; Batty 2020). Thus, Bevan and Conolly (2006), in archaeology, find that "the scale of analysis can change what appears to be a nucleated or centralized pattern, to one better described as dispersed".

The MAUP is a consequence of scale dependence, which characterizes most geographic phenomena. Indeed, "remarkably few geographic variables are scale-less" (Goodchild 2001). Scale dependence means that the existence of a process of concentration with regard to a certain neighborhood implies a process of dispersion with regard to a wider or narrower neighborhood (Tannier 2017, 2019). Scale dependence exists only because there are significant heterogeneities or inequalities, and therefore socio-spatial differentiations within the geographical space. It has two consequences, which are described in the MAUP. The first is that states of concentration observed at a certain scale (a certain level of aggregation or a certain spatial resolution) imply states of dispersion at another scale, and vice versa. The second is that a correlation between two phenomena may be negative when calculated for certain spatial entities, and positive when calculated for others (Openshaw 1981).

Modern methods of geographical analysis based on regular grids and sliding neighborhoods make it possible to limit the effects of the choice of spatial division on the results of analysis. MAUP is then reduced to the arbitrariness induced by the choice of the scales of analysis and the encompassing study area (Brundson et al. 2002). Initially, Clark and Evans (1954) proposed calculating the average distance to the nearest neighbor of each individual in a population as a measure of its spatial distribution. A comparison of this average distance with that obtained if the population were randomly distributed makes it possible to assess its level of concentration. In practice, this method is rarely used as it is very arbitrary. For example, if individuals live in pairs but are very dispersed in space, the distance to the nearest neighbor is small, suggesting that the population is concentrated. But if instead of considering the distance to the nearest neighbor, we look at the distance to the second nearest neighbor, the distance becomes large and the population appears to be scattered (Biraben and Duhourcau 1974). This is why the neighborhood is, in most modern techniques of spatial statistics, a disk centered on each point of the set being studied. This disk acts as a mobile spatial unit and is used to count the points present near the point of interest. By varying the size of the disks, different scales of observation are introduced into the analyses. By construction, unlike geographic divisions, this notion of neighborhood has little sensitivity to MAUP. On this basis, relatively complex methods of spatial analysis have been proposed. The spatial distribution of reference is, in this case, random and of the same local density as that of the objects analyzed (Diggle 2013; Baddeley et al. 2015) (Figure 1.8).

Among advanced methods of spatial analysis, Ripley's function (1977) is used to describe the spatial structure of a set of points in many scientific fields (Bolibok 2008; Marcon and Puech 2017). As in the case of many statistical methods, it is built according to testing logic. The set of points studied is compared to a *Complete Spatial Randomness* distribution, obtained by means of a Poisson process. The way in which the set of points studied deviates from the reference distribution provides information on its more or less aggregated character. Poisson spatial processes are parameterized by a density value, called intensity in spatial statistics. For the comparison to be relevant, the Ripley function requires the intensity of the Poisson process to correspond exactly to that of the set of points studied. However, Ripley's function and its derivatives do not allow distinctions between certain forms which the human eye easily distinguishes. The fact is that human landscapes and forms of occupation of space result from a series of inheritances and actualizations by each generation, heritage being both a constraint limiting the fields of the possible, and the basis and framework for possible transformations related to technical and social developments. Thus, contrary to examples from natural sciences, such as the organization of cells in an organ or the distribution of trees in a forest, the hypothesis that the spatial distribution of human settlements is a realization of a random point process is scarcely relevant (Van Andel et al. 1986).

Medvedkov (1967) introduced into geography another way of measuring and detecting regularities within a spatial distribution, namely Shannon entropy. This indicator is very interesting from a theoretical point of view as it measures the quantity of information contained in a spatial distribution. Nevertheless, it has the disadvantage of also being particularly sensitive to the spatial division being used (Batty 1976). For example, the value of entropy increases mechanically if we divide a spatial unit into two segments of identical density. Batty (1976) proposed correcting this drawback by calculating a new indicator called "spatial entropy", which makes it possible to compare the spatial organization of a population at different levels of aggregation, as well as the quantity of information provided between one level and another. In this, we find ideas very similar to those of fractal analysis, which is no coincidence when we know that Prof. Batty was one of the researchers who first introduced this type of approach into geography (Batty and Longley 1994). Nevertheless, stimulating as it may be, the use of entropy in geography has remained mainly theoretical, probably because it does not directly measure the degree of dispersion or concentration of a spatial distribution, but rather its degree of complexity (Batty et al. 2014).

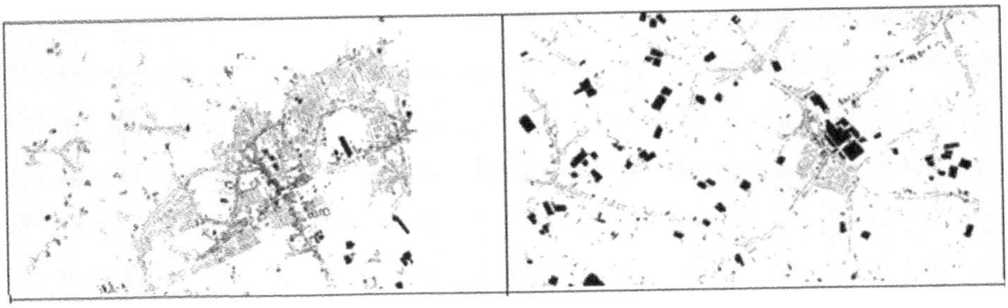

Figure 1.9. *Two Belgian municipalities with the same built-up density. Images kindly provided by I. Thomas (CORE, Louvain-la-Neuve, Belgium)*

As mentioned above, geographers very soon realized that using density was insufficient to describe a spatial configuration, as illustrated by the image of two Belgian municipalities with the same built-up density, but displaying very different built forms (Figure 1.9). When a density is measured, we assume a homogeneous distribution of elements in space as well as a linear, proportional relationship between population and surface area. The reference model of classical spatial analysis methods is therefore a uniform distribution. Each object is allocated the same amount of space for a given scale, regardless of its position relative to other objects. This very strong hypothesis explains the inadequacy of classical spatial analysis methods for characterizing very irregular spatial distributions. More complex methods of spatial analysis are based on random spatial distribution models. But even with this refinement, the results are not satisfactory because the reference distribution is defined for a given scale and is still too regular to adequately describe the relationships linking objects with space.

We indeed notice systematic variations in the spatial distribution of a given geographical fact across scales. These variations are non-proportional and nonlinear. They do not appear if we are satisfied with simpler measures of distributions, such as that of densities (François et al. 1995). Fractal dimensions, on the other hand, measure how the space occupied by a distribution of points changes across scales. This is why they enable the characterization of very heterogeneous spatial distributions. When measuring fractal dimensions, the basis of measurement is the shape of the measured object itself: a fractal dimension measures the change of the object with respect to itself when changing scale. While spatial analysis compares a spatial configuration with reference distributions exogenous to it, fractal analysis takes the configuration as its own reference distribution. In doing so, the information relevant to characterizing an object is not located at a given scale, but emerges from comparison of the properties of the configuration between two distinct scales. Spaces containing objects at a fine scale become contained spaces at a coarse scale, and the arbitrariness induced by assigning a portion of space to each object is partially addressed. The fractal dimension thus informs us about how spaces containing objects become contained spaces at coarser scales. In 2D, the higher the fractal dimension and the closer to 2, the more the container and content are of the same nature. When it is equal to 2, the spatial distribution is perfectly uniform. On the contrary, when the fractal dimensions have low values (close to 1 or less than 1), the container breaks up into contents isolated from each other across the scales.

1.3. References

Alonso, W. (1964). *Location and Land Use*. Harvard University Press, Cambridge, MA.

van Andel, T.H., Runnels, C.N., Pope, K.O. (1986). Five thousand years of land use and abuse in the southern Argolid, Greece. *Hesperia: The Journal of the American School of Classical Studies at Athens*, 55(1), 103–128.

Baddeley, A., Rubak, E., Turner, R. (2015). *Spatial Point Patterns: Methodology and Applications with R*. Chapman and Hall/CRC Press, New York.

Batty, M. (1976). Entropy in spatial aggregation. *Geographical Analysis*, 8(1), 1–21.

Batty, M. (2020). On scale and size – Editorial. *Environment and Planning B: Urban Analytics and City Science*, 47(3), 359–362.

Batty, M. and Longley, P.A. (1994). *Fractal Cities: A Geometry of Form and Function*. Academic Press, San Diego, CA and London.

Batty, M., Morphet, R., Masucci, P., Stanilov, K. (2014). Entropy, complexity, and spatial information. *Journal of Geographical Systems*, 16(4), 363–385.

Bevan, A. and Conolly, J. (2006). Multiscalar approaches to settlement pattern analysis. In *Confronting Scale in Archaeology – Issues of Theory and Practice*, Lock, G. and Molyneaux, B.L. (eds). Springer, 217–234.

Biraben, J.-N. and Duhourcau, F. (1974). La mesure de la population dans l'espace. *Population*, 29(1), 113–137.

Bloch, M. (1931). *Les caractères originaux de l'histoire rurale française*. Instituttet for sammenlignende kulturforskning, Oslo.

Bolibok, L. (2008). Limitations of Ripley K function use in the analysis of spatial patterns of tree stands with heterogeneous structure. *Acta Scientiarum Polonorum Silvarum Colendarum Ratio et Industria Lignaria*, 7(1), 5–18.

Bretagnolle, A. (1996). Étude des indices de concentration d'une population. *L'Espace Géographique*, 25(2), 145–157.

Brunsdon, C., Fotheringham, A.S., Charlton, M. (2002). Geographically weighted summary statistics – a framework for localised exploratory data analysis. *Computers, Environment and Urban Systems*, 26(6), 501–524.

Burgess, E.W. (1925). The growth of the city. An introduction to a research project. In *The City*, Park, R.E. and Burgess, E.W. (eds). Chicago University Press, 47.

Cantor, G. (1883). Über unendliche, lineare Punktmannigfaltigkeiten V. *Mathematische Annalen*, 21, 545–591.

Chiffre J. (1969). Une nouvelle formule de dispersion de l'habitat rural : son application au Nivernais. *Revue Géographique de l'Est*, 9(1–2), 149–175.

Christaller, W. (1980). *Die zentralen Orte in Süddeutschland. Eine ökonomisch-geographische Untersuchung über die Gesetzmäßigkeit der Verbreitung und Entwicklung der Siedlungen mit städtischer Funktion*. Wissenschaftliche Buchgesellschaft, Darmstadt (Reprint, originally published 1933, Fischer, Jena).

Clark, C. (1951). Urban population densities. *Journal of the Royal Statistical Society, Series A*, 114, 490–496.

Clark, P.J. and Evans, F.C. (1954). Distance to nearest neighbor as a measure of spatial relationships in populations. *Ecology*, 35(4), 445–453.

Colas, R. (1945). Répartition de l'habitat rural. *Bulletin de l'Association de Géographes Français*, 169–170, 51–56.

Demangeon, A. (1927a). La géographie de l'habitat rural – Premier article. *Annales de géographie*, 199, 1–23.

Demangeon, A. (1927b). La géographie de l'habitat rural – Deuxième article. *Annales de géographie*, 200, 97–114.

Demangeon, A. (1933). Une carte de l'habitat. *Annales de géographie*, 237, 225–232.

Diggle, P.J. (2013). *Statistical Analysis of Spatial and Spatio-Temporal Point Patterns*. Chapman and Hall/CRC Press, New York.

Dion, R. (1934). *Essai sur la formation du paysage rural français*. Arrault, Tours.

Duncan, O.D. (1957). The measurement of population distribution. *Population Studies*, 11(1), 27–45.

Durand-Dastès, F. (1989). Sur un espace intellectuel. *L'Espace géographique*, XVIII(2), 86–89.

Falconer, K.J. (2013). *Fractals. A Very Short Introduction*. Oxford University Press, Oxford, New York.

Fotheringham, A.S. and Wong, D.W.S. (1991). The modifiable areal unit problem in multivariate statistical analysis. *Environment and Planning A*, 23(7), 1025–1044.

François, N., Frankhauser, P., Pumain, D. (1995). Villes, densité et fractalité. Nouvelles représentations de la répartition de la population. *Les Annales de la recherche urbaine*, 67, 55–64.

Frankhauser, P. (1994). *La fractalité des structures urbaines*. Anthropos, Paris.

Gehlke, C.E. and Biehl, K. (1934). Certain effects of grouping upon the size of the correlation coefficient in census tract material. *Journal of the American Statistical Association*, 29, 169–170.

Goodchild, M. (1980). Fractals and the accuracy of geographical measures. *Mathematical Geology*, 12(2), 85–98.

Goodchild, M. (2001). Models of scale and scales of modelling. In *Modelling Scale in Geographical Information Science*, Tate, N. and Atkinson, P.M. (eds). John Wiley & Sons Ltd, Chichester, 3–10.

Goodchild, M. (2016). Laws in geography. *Conference of the Collège International des Sciences du Territoire*, Paris.

Goodchild, M. and Mark, D. (1987). The fractal nature of geographic phenomena. *Annals of the Association of American Geographers*, 77(2), 265–278.

Grasland, C. (2003). Richesse et population dans le monde : une représentation multiscalaire des inégalités. *Mappemonde*, 69, 20–25.

Haggett, P., Cliff, A.D., Frey, A.E. (1977). *Locational Analysis in Human Geography*, 2nd edition. Arnold, London.

von Koch, H. (1904). Sur une courbe continue sans tangente, obtenue par une construction géométrique élémentaire. *Arkiv för matematik, astronomi och fysik*, Band 1, 681–704.

Lam, N S. and Quattrochi, D. (1992). On the issues of scale, resolution, and fractal analysis in the mapping sciences. *Professional Geographer*, 44(1), 88–98.

Lefèvre, M. (1926). *L'habitat rural en Belgique. Etude de géographie humaine.* Vaillant-Carmanne, Liège.

Lösch, A. (1940). *Die räumliche Ordnung der Wirtschaft. Eine Untersuchung über Standort, Wirtschaftsgebiete und internationalem Handel.* Fischer, Jena.

Lösch, A. (1954). *The Economics of Location.* Yale University Press, New Haven, CT.

Madelin, M., Grasland, C., Mathian, H., Sanders, L., Vincent, J.-M. (2009). Das "MAUP": Modifiable areal Unit – Problem oder Fortschritt? *Informationen zur Raumentwicklung, Heft*, 645–660.

Mandelbrot, B. (1975). *Les objets fractals. Forme, hasard et dimension.* Flammarion, Paris.

Mandelbrot, B. (1982). *The Fractal Geometry of Nature.* W.H. Freeman and Company, New York.

Mandelbrot, B. (1988). Foreword: People and events behind the "Science of Fractal Images". In *The Science of Fractal Images*, Peitgen, H.O. and Saupe, D. (eds). Springer-Verlag, New York, Berlin, Heidelberg, 1–19.

Marcon, E. and Puech, F. (2017). A typology of distance-based measures of spatial concentration. *Regional Science and Urban Economics*, 62, 56–67.

Mathian, H. and Piron, M. (2007). Geographical scales and multidimensional statistical methods. In *Models in Spatial Analysis*, Sanders, L. (ed.). ISTE Ltd, London and John Wiley & Sons, New York.

Mathian, H. and Sanders, L. (2014). *Spatio-temporal Approaches: Geographic Objects and Change Process*. ISTE Ltd, London and John Wiley & Sons, New York.

Medvedkov, Y.V. (1967). The concept of entropy in settlement pattern analysis. *Papers in Regional Science*, 18(1), 165–168.

Mills, E.S. (1967). Transportation and patterns of urban development. An aggregative model of resource allocation in a metropolitan area. *Papers and Proceedings of the Seventy-Ninth Annual Meeting of the American Economic Association*, 57(2), 197–210.

Müller, J.-C. (1978). Map gridding and cartographic errors: A recurrent argument. *Canadian Cartographer*, 14, 152–167.

Muth, R.F. (1969). *Cities and Housing: The Spatial Pattern of Urban Residential Land Use*. University of Chicago Press, Chicago.

Openshaw, S. (1981). Le problème de l'agrégation spatiale en géographie. *L'Espace Géographique*, 10(1), 15–24.

Openshaw, S. (1983). *The Modifiable Areal Unit Problem. Concepts and Techniques in Modern Geography 38*, Geo Books, Norwich.

Peano, G. (1890). Sur une courbe, qui remplit toute une aire plane. *Mathematische Annalen*, 36, 157–160.

Ripley, B.D. (1977). Modelling spatial patterns. *Journal of the Royal Statistical Society: Series B (Methodological)*, 39(2), 172–192.

Sanders, L. (2007). Objets géographiques et simulation agent, entre thématique et méthodologie. *Revue internationale de géomatique*, 17, 135–160.

Stewart, J.Q. (1947). Empirical mathematical rules concerning the distribution and equilibrium of population. *Geographical Review*, 24, 461–485.

Tannier, C. (2017). Analyse et simulation de la concentration et de la dispersion des implantations humaines de l'échelle micro-locale à l'échelle régionale – Modèles multi-échelles et trans-échelles. HDR Thesis, Université de Franche-Comté, Besançon.

Tannier, C. (2019). Effects of distance and scale dependence in geographical models of cities and territories. In *Geographical Modeling: Cities and Territories*, Pumain, D. (ed.). ISTE Ltd, London, and John Wiley & Sons, New York, 53–94.

Tate, N. and Atkinson, P.M. (eds) (2001). *Modelling Scale in Geographical Information Science*. John Wiley & Sons Ltd, Chichester.

von Thünen, J.H. (1826). *Der isolierte Staatt in Beziehung auf Landwirtschaft und Nationalökonomie*. Friedrich Perthes, Hamburg.

Toger, M., Malkinson, D., Benenson, I., Czamanski, D. (2016). The connectivity of Haifa urban open space network. *Environment and Planning B: Planning and Design*, 43(5), 848–870.

van Vliet, J., White, R., Dragicevic, S. (2009). Modeling urban growth using a variable grid cellular automaton. *Computers, Environment and Urban Systems*, 33, 35–43.

Voss, R.F. (1988). Fractals in nature: From characterization to simulation. In *The Science of Fractal Images*, Peitgen, H.O. and Saupe, D. (eds). Springer-Verlag, New York, Berlin, Heidelberg, 21–70.

Weierstrass, K. (1872). Über continuirliche Functionen eines reellen Arguments, die für keinen Werth des letzteren einen bestimmten Differentialquotienten besitzen. *Gelesen in der Königl. Akademie der Wissenschaften am 18 Juli 1872, Mathematische Werke, Abhandlungen II (1895)*. Mayer & Müller, Berlin, 71–74.

White, R. (2005). Modelling multi-scale processes in a cellular automata framework. In *Complex Artificial Environments – Simulation, Cognition and VR in the Study and Planning of Cities*, Portugali, J. (ed.). Springer-Verlag, Berlin, Heidelberg, 165–177.

2

Basic Concepts for Fractal Analysis and Modeling in Human Geography

Cécile TANNIER
Laboratoire ThéMA, CNRS-Université de Franche-Comté, Besançon, France

Mandelbrot's fractal geometry provides both a description and a mathematical model for many of the seemingly complex forms found in nature. Shapes such as coastlines, mountains and clouds are not easily described by traditional Euclidean geometry. Nevertheless, they often possess a remarkable simplifying invariance under changes of magnification. [...] It may be quantified by a fractal dimension, a number that agrees with our intuitive vision of dimension, but need not be an integer
(Voss 1988).

Since the 1980s, geographers have recognized the value of introducing fractals in the form of geometric models and methods of measurement for analysis and modeling of the spatial distribution of human settlements. Fractal shapes, described as monstrous or pathological by the majority of mathematicians of the 19th century and the first half of the 20th century, do indeed have interesting properties that we will describe in this chapter. Fractal geometry, a formal mathematical framework proposed by Mandelbrot and relating to measure theory, makes it possible to work with such shapes. In doing so, it is possible to describe aspects of the real world that cannot be expressed using the traditional tools of Euclidean geometry (Lesmoir-Gordon et al. 1996). "[...] Between the domain of uncontrolled disorder

Fractal Geometry in Human Geography and Planning,
coordinated by Cécile TANNIER.

and the excessive order of Euclid, there is now a new zone of fractal order" (Mandelbrot 1975)[1].

By "shape", "object" or "figure", we mean a collection of points on the plane that we consider as an entity. Such an entity is a set in the mathematical sense (Falconer 2013). Studying fractal shapes, figures or geographical objects therefore consists of studying sets of fractal points. Mandelbrot (1975)[2] defines shapes as being mathematical idealizations, and objects as being the data of reality.

The figures of Euclidean geometry – triangles, circles, spheres, cylinders – have a characteristic scale of measurement; when one of their parts is enlarged, their structure is no longer recognizable. For example, if we zoom in on one side of the triangle in Figure 2.1(a), we no longer see a triangle but a straight line. In the same way, the perimeter of a circle becomes a featureless straight line when viewed on a large enough scale. People who think the Earth is flat do so because that's the way it looks to a tiny human (Stewart 1989). Conversely, fractal objects do not have a characteristic scale of measurement: each part of the magnified object resembles the object as a whole, as shown in Figure 2.1(b). The structure of such objects is visible over a large range of scales (Stewart 1989)[3]; they are said to have a fine structure (Falconer 2013). What is more, their structure is the same on all scales; they are self-similar[4]. The self-similarity of fractal objects is interesting because of their fine structure, which means that new details appear each time a portion of the object is enlarged. A line is also self-similar because each of its portions is similar to the whole. However, this self-similarity is trivial because no new detail appears when you enlarge a portion of the line.

In the case of smooth figures, circles or ellipses, for example, it is possible to trace a tangent at each point on their perimeter. The tangent at a point is a straight line which "perfectly touches" the curve considered at this point. The curve and its tangent form a zero angle at this point. Moreover, if a figure in Euclidean geometry is transformed by translation, scaling, reflection or rotation, its structure remains the same; that is, a circle is always a circle, a triangle is always a triangle, etc., regardless of their size and position. Fractal figures, in contrast, are not smooth but, on the contrary, show irregularities on all scales: they have no tangent at any point.

1. Fourth edition, 1995, p. 10.

2. Fourth edition, 1995, p. 13.

3. "Mandelbrot invented the term 'fractal' to describe a very different type of geometric object: one that continues to exhibit detailed structure over a large range of scales" (Stewart 1989).

4. "Not only do both the natural and mathematical fractals have structure on all scales: they have – within reason – the same structure on all scales" (Stewart 1989).

The irregularity and self-similarity of fractal figures mean that traditional mathematical tools cannot help determine which of them are similar and which are different.

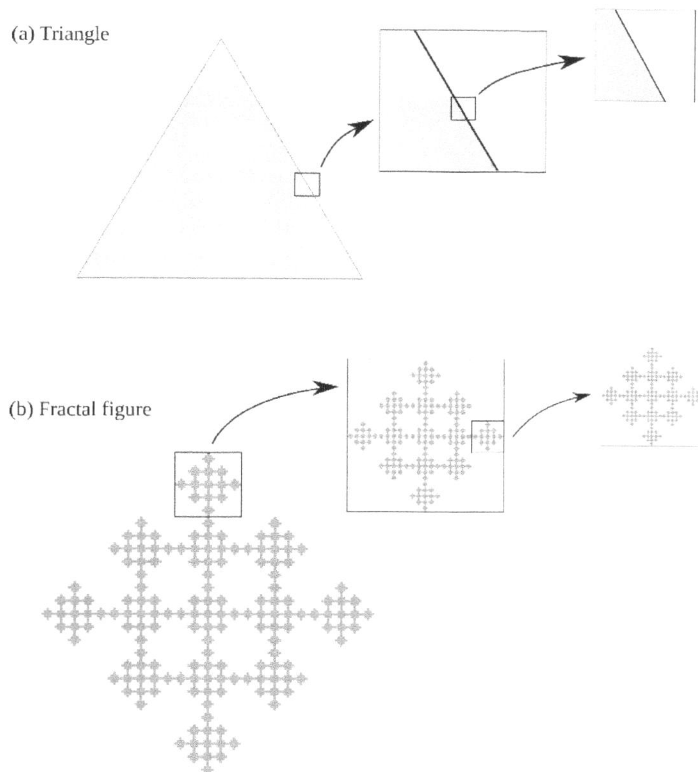

Figure 2.1. *A triangle and a fractal figure: full view and zoom*

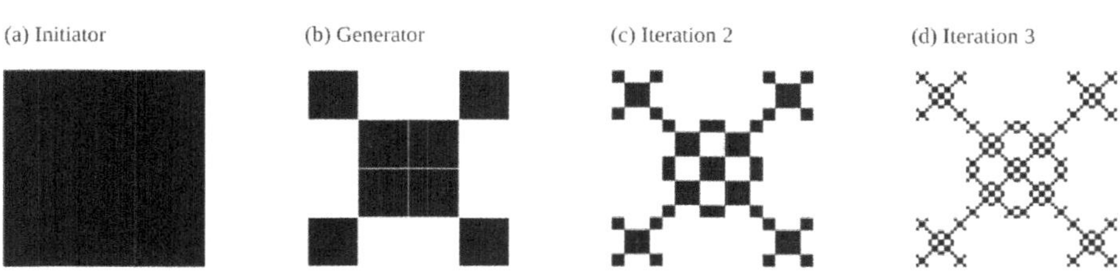

Figure 2.2. *First steps in the construction of a self-similar fractal figure. Figures created with GenFrac software (see: https://sourcesup.renater.fr/www/genfrac/)*

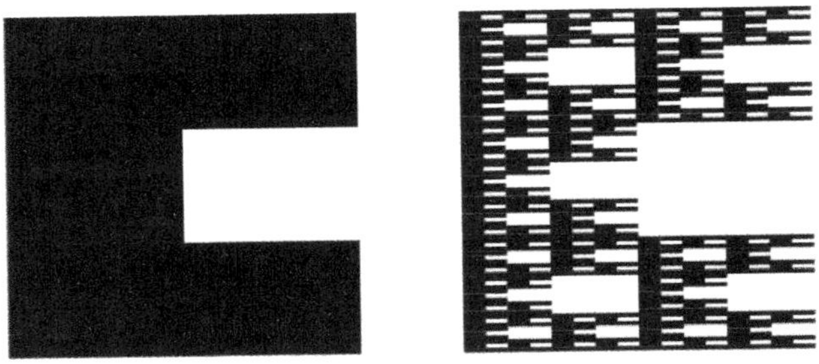

Figure 2.3. *The generator and the fourth iteration of a self-affine fractal form. Figures created with GenFrac software (see: https://sourcesup.renater.fr/www/genfrac/)*

2.1. Methods of construction of fractal figures

The method of construction of fractal figures differs from that of figures of Euclidean geometry. A *template* is defined at the start: a simple geometric figure (a square or a triangle, for example) is chosen to be the initiator of the shape (Figure 2.2(a)); then, a certain number of reduced copies of it are arranged in one way or another in the plane, which forms the generator (Figure 2.2(b)). Then, the elements of the generator are iteratively replaced by reduced copies of themselves (Figure 2.2(c) and (d)). The iteration continues in this way to infinity. The number of elements in a fractal figure increases with the iterations according to a multiplicative logic. Thus, the generator of Figure 2.2 comprises eight elements. At the next step (iteration 2), the figure has 8 × 8 = 64 elements. It then has 64 × 8 = 512 at iteration 3, and so on. When the construction of the figure stops after only a few iterations, the figures thus created are called "prefractals". In the case of some fractal figures, the generator consists of anisotropic geometric transformations of the initiator (Figure 2.3). These figures are self-affine and not self-similar: their parts are copies of the set reduced by different factors in both the x and y directions of the plane.

2.1.1. *Iterated function systems*

The template of a fractal figure often benefits from being written in the form of mathematical functions (Falconer 2013), that is, as applications defined over a domain containing the figure. From a geometric point of view, a function moves the set of points that make up a figure to a set of new positions so that the set of these new positions forms a new figure. In other words, the function transforms the initial figure into a new figure. The template of a fractal figure is a way of graphically expressing a set of functions.

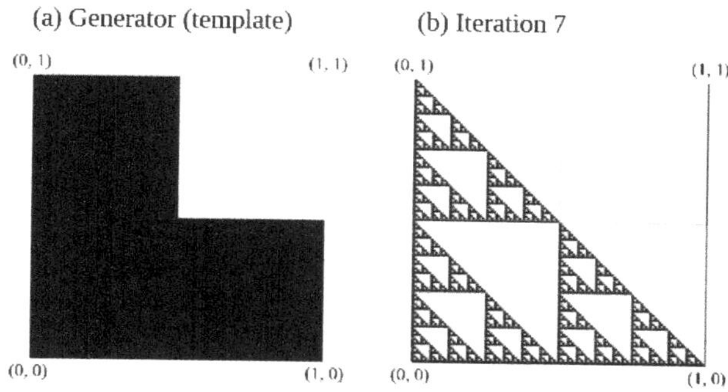

Figure 2.4. *Sierpinski triangle. Figures created with GenFrac software (see: https://sourcesup.renater.fr/www/genfrac/)*

Take the example of a figure called the Sierpinski triangle, as shown in Figure 2.4. The initiator of the shape is the unit square, of length 1 and with corners at coordinates (0, 0), (1, 0), (0, 1), (1, 1). The template, or generator, is composed of three smaller squares of length 1/2. The functions transforming the unit square into each of these three smallest squares are:

$$(x, y) \rightarrow (1/2x, 1/2y)$$

$$(x, y) \rightarrow (1/2x + 1/2, 1/2y)$$

$$(x, y) \rightarrow (1/2x, 1/2y + 1/2)$$

The Sierpinski triangle is completely defined by these three functions. Note that, although the initiator is a square, the fractal shape obtained after a certain number of iterations is a triangle. Indeed, the shape of the initiator has no influence on the fractal shape that is finally generated. Only the sizes and positions of the elements in the generator determine the shape that will ultimately be obtained. Modifying the generator leads to more or less significant modifications of the shape ultimately obtained, as shown by the two examples presented in Figure 2.5. The functions transforming the unit square into three smaller squares are, for variant A:

$$(x, y) \rightarrow (1/2x, 1/2y)$$
$$(x, y) \rightarrow (1/2x + 1, 1/2y)$$
$$(x, y) \rightarrow (1/2x, 1/2y + 1/2)$$

And for variant B:

$$(x, y) \rightarrow (1/2x, 1/2y)$$
$$(x, y) \rightarrow (-1/2x + 1, 1/2y)$$
$$(x, y) \rightarrow (1/2x, 1/2y + 1/2)$$

For both variants A and B, one of the generator elements is located outside the unit square, which causes the resulting fractal shape to lengthen as iterations progress.

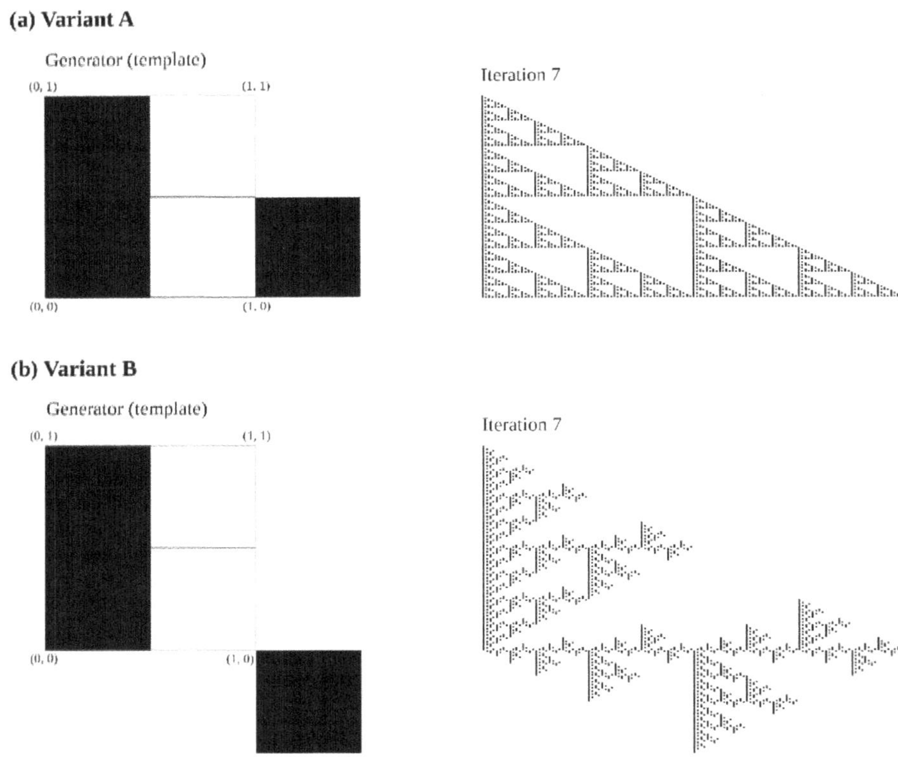

Figure 2.5. *Sierpinski triangle: two variants. Figures created with GenFrac software (see: https://sourcesup.renater.fr/www/genfrac/)*

A set (or family) of transformations of the form is called an *iterated function system* (IFS) (Figure 2.6). Considering a given system of iterated functions, there exists a unique set (called the attractor and usually a fractal) such that, taken together, the transformed copies of the set make up the entire set (Falconer 2013,

p. 27). An example of a system of iterated functions resulting in the reduction and rotation of a square is shown in Figure 2.7. Translations *Tx* and *Ty* starting from the lower left corner of the initiator (coordinate point (0, 0)) place the square in the center of the unit square. The *Sx* and *Sy* reductions mean that each square generated during a new iteration is half the size of the previously generated square. The *Shx* and *Shy* functions rotate the square from one iteration to another.

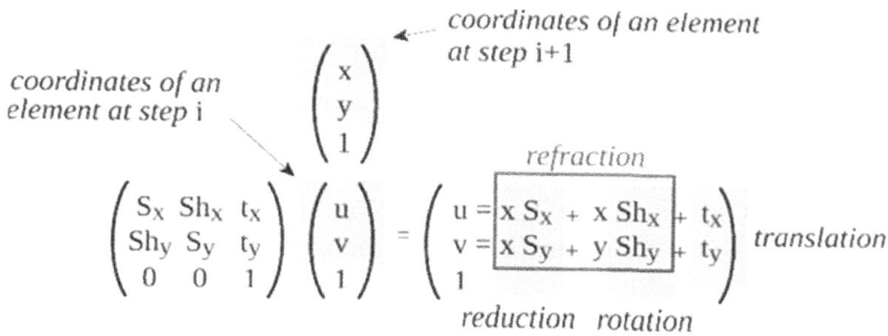

Figure 2.6. *Coding of a fractal shape generator (template) as an iterated function system*

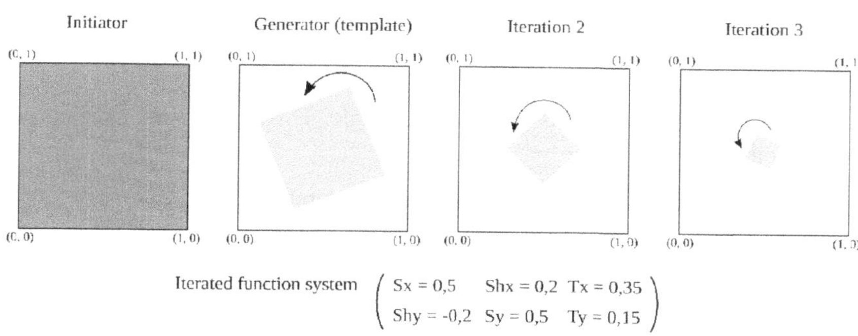

Figure 2.7. *Reduction and rotation of a square using an iterated function system (IFS)*

In addition to iterated function systems, there are a number of other methods for creating fractal figures.

2.1.2. *Fractal cascades*

The fractal cascade generator is a matrix composed of 0 for empty spaces and 1 for full spaces. In the first step of construction (iteration 1), the full spaces of the generator are replaced by a miniature copy of it. The fractal is obtained by repeating this operation, that is, by replacing the full spaces obtained in the previous step with a miniature copy of the generator. From an algorithmic point of view, this operation is very simple to perform with high-level programming languages such as Python, as shown in the example below (Sémécurbe 2020, p. 63–64):

```
def cascade(generateur, nb_iterations):
fractal = generateur
for i in range(nb_iterations):
fractal = np.kron(fractal, generateur)
return(fractal)
```

The command `kron` is the Kronecker delta. It allows replacement of the non-zero elements of the fractal matrix by the generator.

2.1.3. *Chaos game*

Barnsley (1988a, 1988b) devised a way to draw a fractal figure via a stochastic process comparable to rolling dice. Take the example of the Sierpinski triangle:

– Three points are positioned in the plane representing the three vertices of a triangle.

– A random point is selected inside this triangle. This point is the starting point for drawing the figure.

– One of the vertices of the triangle is chosen at random.

– A new point is drawn midway between the starting point and the selected vertex.

– Steps 3 and 4 are repeated from the newly drawn point.

After drawing about a hundred points, the outline of the Sierpinski triangle appears. As for the details of the figure, they appear after a few thousand points have been drawn (Figure 2.8). As Barnsley has shown, any fractal figure can be drawn following this principle, by choosing an appropriate position for the points forming the outline of the figure (rule 1) and appropriate rules for drawing the new points (rules 3 and 4), such as Barnsley's famous fern (Figure 2.9). Note, however, that if the chaos game starts with a square and with rules 2 and 3 explained above, no fractal figure appears: the inside of the square is uniformly filled with points. A fractal figure only appears if restrictions are placed on the choice of vertices of the square (rule 3), as shown in Figure 2.10. From a mathematical point of view, iterated function systems (IFS) are a generalization of the chaos game. Batty and Longley (1994, p. 87) point out that the real power of the chaos game "cannot be conveyed with strictly self-similar objects because these can be generated as quickly if not faster in a variety of more direct ways. However, where an object is composed of much less obvious self-affine copies of itself, then the method is truly magical".

Figure 2.8. *The chaos game starting from three points forming the vertices of an equilateral triangle. Result after 500, 1,000 and 2,000 iterations. Source: https://thatsmaths.com/2014/05/22/the-chaos-game/ [Accessed January 15, 2021]*

Figure 2.9. *Barnsley fern generated with the chaos game. From left to right: 1,000, 10,000 and 100,000 points. Source: https://thatsmaths.com/2014/05/22/the-chaos-game/ [Accessed January 15, 2021]. For a color version of this figure, see www.iste.co.uk/tannier/fractal.zip*

Figure 2.10. *Fractal figure generated with the chaos game starting from a square. The vertex of the square chosen in a given iteration cannot be chosen again in the next iteration. Source: MagistraMundi. 2016.CC BY-SA 4.0. See: https://en.wikipedia.org/wiki/Chaos_game [Accessed March 26, 2021]*

2.1.4. *L-systems (Lindenmayer systems)*

An *L-system* (Lindenmayer 1968) is composed of a set of symbols (also called an alphabet), production rules that define the creation of symbol strings and mechanisms that transcribe the symbol strings into geometric structures. Formally, an *L-system* is represented by a tuple (or uplet):

$$\mathbf{G} = (V, \omega, P)$$

where:

– **V** (the alphabet) is the set of symbols containing both elements that can be replaced (variables) and those which cannot be replaced (constants).

– **ω** (the start, the axiom, the initiator) is a string of symbols from **V** defining the initial state of the system.

– **P** is a set of production rules defining the way variables can be replaced with combinations of other variables and constants. Production rules are applied iteratively beginning from the initial state of the system.

As an example, Figure 2.11 shows the construction of a fractal curve using an *L-system*. The alphabet is as follows:

– "F" consists of drawing a line segment of a given fixed length.

– "+" consists of turning to the left by a certain angle.

– "–" consists of turning to the right by the same angle.

"F" is a variable and "+" and "–" are constants. The axiom of *L-system* is F, of length 10. The angle of the "+" and "–" symbols is 90°.

The production rule is F → F+F-F-F+F, which means:

– Draw a line segment of length 10.

– Turn to the left 90°.

– Draw a line segment of length 10.

– Turn to the right 90°.

– Draw a line segment of length 10.

– Turn to the right 90°.

– Draw a line segment of length 10.

– Turn to the left 90°.

– Draw a line segment of length 10.

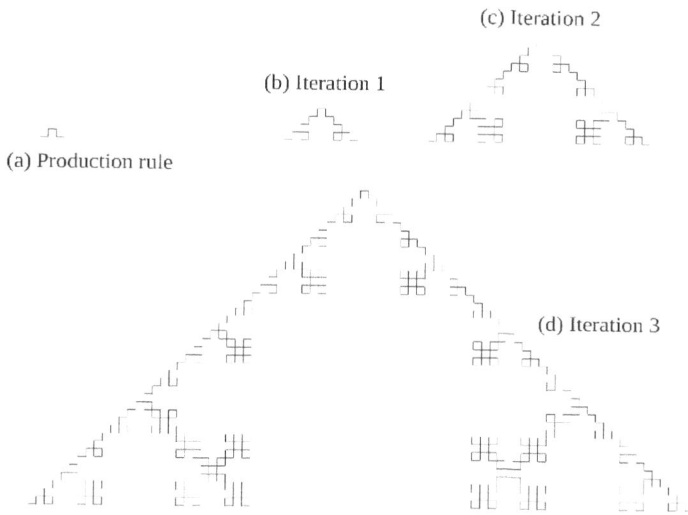

Figure 2.11. *First steps in the construction of a rectangular Koch curve by means of an L-system*

This production rule results in the drawing shown in Figure 2.11(a). In the second step of construction, each F, that is, each straight line segment of Figure 2.11(a), is replaced by the production rule, which gives F+F-F-F+F+F+F-F-F+F-F+F-F-F+F-F+F-F-F+F+F+F-F-F+F. This results in the design shown in Figure 2.11(b). Note that as the iterations progress (Figure 2.11(c) and (d)), the spatial footprint of the figure increases significantly, which is not the case (or very rarely) when using iterated function systems. This is understandable given that Lindenmayer initially designed *L-systems* to model plant growth.

2.1.5. *Other methods for creating fractal figures*

Four other categories of methods for creating fractal figures can be distinguished. The first consists of tiling the plane using smaller and smaller tiles. For this purpose, finite subdivision rules are used, which recursively divide a polygon (i.e. a tile) or other two-dimensional shape (i.e. the sides of a tile) into smaller and smaller pieces. To perform an elaborate tiling, we generally define different types of tiles, each having a specific method of being divided. It is also possible not to exactly repeat the same design from one subdivision to another, which allows the creation of figures with a richer geometric structure.

The second category of methods brings together a series of stochastic processes such as random walks at the origin of Brownian movements or Lévy flights, or stochastic aggregation processes of the *diffusion-limited aggregation* (DLA) type. These methods create so-called "random" fractal figures (see section 2.5).

The third category of methods relates to the study of dynamic systems in physics, whose behavior is represented in the form of systems of equations. The attractor of a dynamic system is a point or a portion of the phase space, an abstract space in which all trajectories of the system over time (i.e. possible successions of states) are represented. In the case of an elementary dynamic system, the attractor is a fixed point: an example is the simple pendulum which oscillates while losing energy and ends up stopping at an end point. Other systems never stop, and their evolution is cyclical and periodic, like the pendulum of a clock whose oscillations are maintained. When the behavior of a dynamic system is chaotic, the attractor of this system is fractal. This is known as a strange attractor.

The fourth category of methods allows the construction of so-called "*escape-time fractals*", also called "*orbital fractals*". The best known of these are the Julia sets and the Mandelbrot set. They are among the most frequently graphically depicted fractal figures and have been used for the creation of artwork and scenes in science fiction films, and as mystical symbols (Falconer 2013). The Julia sets and the Mandelbrot set are the result of the iterative application of functions, as in most of the cases mentioned above. In these particular cases, however, the coordinates of the points forming the path of these functions involve complex numbers. A complex number z is of the form $z = x + yi$, where x and y are real numbers and i is the square root of -1. In the set of real numbers $\mathcal{R}$, negative numbers have no square root, which is why y is the imaginary part of a complex number. The functions considered in the case of the Julia sets and the Mandelbrot set are of the form:

$$f(z) = z^2 + c \quad [2.1]$$

where z and c are complex numbers. Any starting point z_0 is associated with the sequence $z_n = f^n(z_0)$ of successive iterates of z_0 by f[5]. It turns out that, when the point z_0 takes certain values, the sequence z_n is bounded, while for other values of z_0 it

5. Source: https://mathcurve.com/fractals/julia/julia.shtml [Accessed December 18, 2020]. Website created by Robert Ferréol.

diverges towards infinity. The Julia set corresponding to a given sequence z_n is the demarcation line (the boundary) between the two sets of points z_0, those for which z_n is bounded (called the filled Julia set) and those for which z_n diverges to infinity (called the Fatou set). When $c = 0$, that is, when the real part and the imaginary part of c are both equal to 0, the Julia set is the circle of center 0 and radius 1.

Depending on the values of c, the Julia sets have very variable shapes (Figure 2.12). These shapes are fractal, some more compact and others more branching. Looking closer, we note that some forms are topologically related, that is, all in one piece, while others are not connected (in geometry, they are called dusts or Cantor sets). The Mandelbrot set is a complex-plane representation of the values of c for which the corresponding Julia sets are connected. Graphical representations of Julia sets or the Mandelbrot set are often colored so as to visualize the number of iterations required for a point outside one of these sets to leave its orbit, hence their name of *orbital fractals* or *escape-time fractals*.

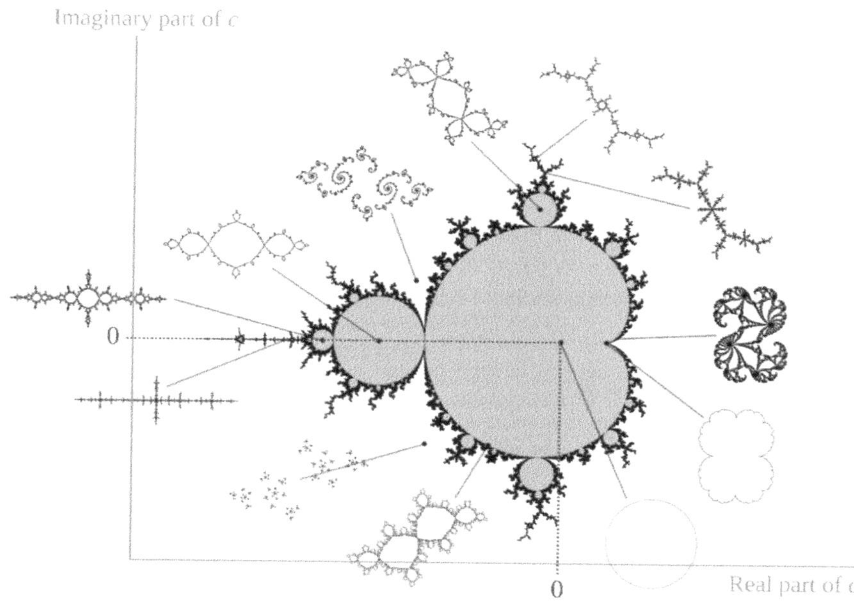

Figure 2.12. *In the complex plane, the Mandelbrot set (in gray) and Julia sets corresponding to different values of c and belonging or not belonging to the Mandelbrot set. Adapted from: Chéritat (2010). For a color version of this figure, see www.iste.co.uk/tannier/fractal.zip*

2.2. Measuring the size of fractal and non-fractal objects

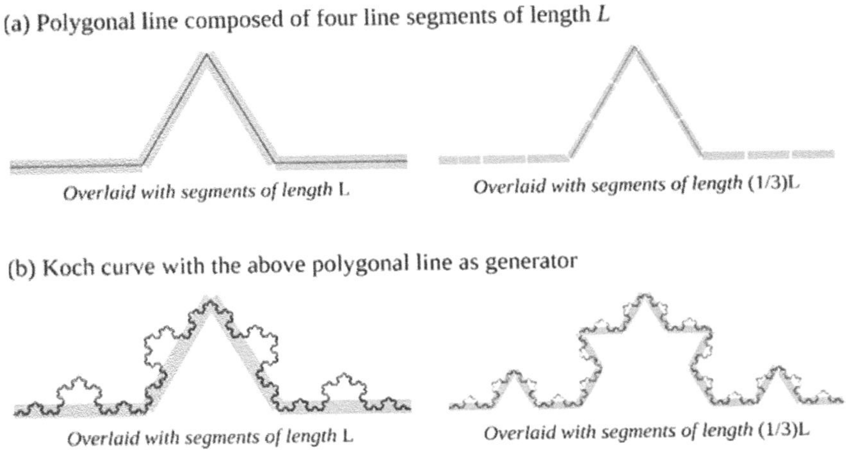

Figure 2.13. *First steps of overlaying a polygonal line and a Koch curve using 3^n measuring sticks of size $(1/3)^nL$*

To begin with, consider the example of a polygonal line made up of four line segments of size L (Figure 2.13(a)). The length L of this line is $\mathscr{L} = 4 \times L$. If each segment is overlaid with three measuring sticks of length $(1/3)L$, then the length of the line is $4 \times 3 \times (1/3)L = 4 \times L = \mathscr{L}$. Generalizing, we can say that if we overlay the line with $3 \times n$ measuring sticks of length $(1/3)^nL$, n being a positive integer, then the length of the line is:

$$4 \times 3^n \times (1/3)^n L = \mathscr{L} \quad [2.2]$$

If L has a length of 1 m, for example, then the length of $\mathscr{L}$ is in all cases equal to 4 m.

Let us now try to measure the length of the perimeter of a fractal curve, the von Koch curve (1904), by adopting this method (Figure 2.13(b)). Using measuring sticks of size L, the length of the curve is $4 \times L = \mathscr{L}$. But if we use sticks of size $(1/3)L$, the length of the curve is $L_1 = 4 \times 4(1/3)L$ (equal to about 5.33 m when L has a length of 1 m), which is greater than $\mathscr{L}$. And if we use sticks of size $(1/3)^2L$, the length of the curve is $l_2 = 4 \times 4^2\ (1/3)^2L$ (equal to approximately 7.11 m if L has a length of 1 m), which is greater than L_1. Generalizing, with measuring sticks of size $(1/3)^nL$, the length of the Koch curve is $L_n = 4 \times 4^n(1/3)^nL$. It therefore increases as the length of the measuring stick decreases. The size of a fractal object does indeed depend on the scale of measurement adopted (Falconer 2013).

Traditionally, a line is considered to be a one-dimensional object. A planar object (triangle, circle, square, etc.) has two dimensions, while a solid object has three dimensions. This dimension of which we speak is the topological dimension. It corresponds to the number of coordinates necessary to unambiguously locate a point on the object. In one dimension, lengths can be measured; in two dimensions, areas; in three dimensions, volumes. However, in the case of very irregular objects, the information provided by the topological dimension is not sufficient to measure their size; while an ordinary curve exactly fills a one-dimensional space, a fractal curve such as the Koch curve overflows into a two-dimensional space. But it is impossible to measure the size of complex geometric shapes, for example, a line unfolding in $\mathscr{R}^2$ and not in $\mathscr{R}$, with the methods of Euclidean geometry, as we have just seen previously.

A mathematicians' idea for a more satisfactory measurement of the shape of irregular objects is to consider only the local quantification of the occupied space as relevant. This is the idea adopted by Minkowski (1903) and later developed by Bouligand (1929). The principle used is that of step-by-step overlaying of the set of points with closed balls (or boxes) of ε decreasing size, and counting the number $N(\varepsilon)$ of boxes containing points at each stage of the overlaying. On this basis, the Minkowski–Bouligand dimension, also called the box dimension, is defined as:

$$D_B = \lim_{\epsilon \to 0} \frac{\log N(\epsilon)}{\log(1/\epsilon)} \qquad [2.3]$$

Note that D_B is not directly related to a coordinate system, but especially that it may be fractional while the topological dimension is, by definition, integer. This formula can use logarithms of base 10 or any other base (Lauwerier 1991).

This idea of measuring the space occupied by an object around its constituent points was also adopted by Hausdorff (1919) to measure the dimension of very irregular sets of points. Hausdorff's work on this issue has been extensively developed by Besicovitch. For a given shape, the Hausdorff–Besikovich dimension D_H, also simply called the Hausdorff dimension, corresponds to "the value of d for which the d-dimensional volume of a shape changes from infinity to zero" (Stewart 1989). This dimension D_H represents the "correct" dimension, the one at which the measurement of the object is perfectly accurate. It is more general than the box dimension, but mathematically much more complicated, particularly because it involves a measure, the Hausdorff measure, and not a simple counting of boxes[6]. In the case of self-similar fractal objects, the Minkowski–Bouligand dimension coincides with the Hausdorff dimension, and an object is considered fractal if the Hausdorff dimension is strictly greater than its topological dimension. Note that both Minkowski–Bouligand and Hausdorff dimensions involve limits to infinity. They are therefore not directly calculable for real-world objects. However, the first can be easily approximated (see Chapter 4).

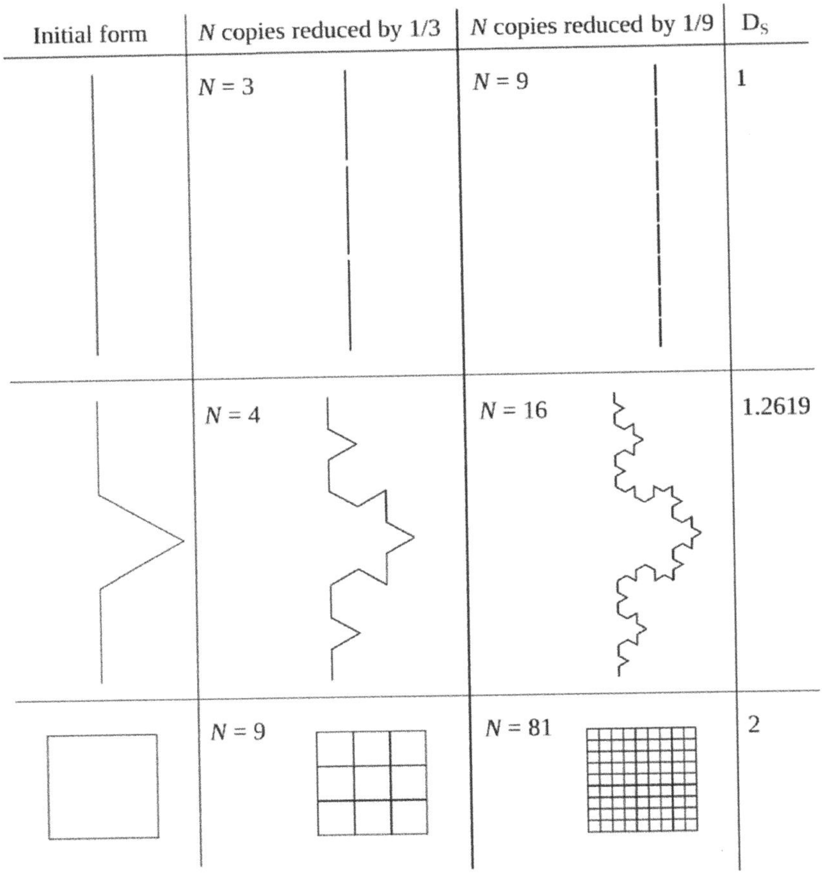

Figure 2.14. *Similarity dimension D_S of three figures decomposed into N copies of reduced size*

---

6. For a detailed mathematical definition of the Hausdorff measure and dimension, see Falconer (1990).

Another, much simpler, idea for measuring the dimension of an object is to decompose it into a certain number N of copies of itself reduced by a factor r (called the homothety ratio). For objects with internal homotheties of the same ratio, we have the following relation:

$$N = r^{-D_S} \quad [2.4]$$

where D_S is the similarity dimension. On this basis, D_S is obtained as follows:

$$D_S = - \log(N) / \log(r) = \log(N) / \log(1/r) \quad [2.5]$$

Take a reduction factor r equal to 1/3. A line segment may be decomposed into three copies of itself reduced by a factor of 1/3. Its dimension D_S is therefore equal to 1 (Figure 2.14). A square can be decomposed into nine reduced copies of itself. Its dimension D_S is therefore equal to 2. The Koch curve can be broken down into four reduced copies of itself. Its dimension D_S is therefore equal to 1.2619. Intuitively, observing Figure 2.14, we note that it is only possible to decompose the Koch curve into copies reduced by a factor of $(1/3)^n$, while it is possible to decompose the other figures into copies reduced by a factor of $(1/n)$. In the case of self-similar fractal objects, the similarity dimension D_S coincides with the Hausdorff dimension D_H.

2.3. Definition of a fractal object

The first definition of a fractal object proposed by Mandelbrot was as follows: "a set for which the Hausdorff-Besicovitch dimension strictly exceeds the topological dimension" (Mandelbrot 1982, p. 15). This definition is based on a theoretical mathematical notion, the Hausdorff dimension, and thus excludes a large proportion of natural phenomena for which this dimension is not calculable. Feder (1988, p. 11) notes that Mandelbrot later retracted this definition and proposed the following: "a fractal is a shape made of parts similar to the whole in some way" (Mandelbrot 1986). Other definitions have subsequently been proposed by other researchers without consensus being reached. This is why Falconer (1990) suggests that the definition of a fractal or of what is fractal should be regarded in the same way as the biologist regards the definition of "life": there is no hard and fast definition, but just a list of properties characteristic of a living thing. Thus, for Falconer (1990), fractals are figures or objects that possess some of the following characteristics:

1) They have a fine structure, that is, they display many details regardless of the scale of observation.

2) They are too irregular to be described effectively in traditional geometric language, both locally and globally.

3) They often show some form of self-similarity (exact, approximate or statistical).

4) Their fractal dimension (not necessarily Hausdorff dimension) is generally greater than their topological dimension.

5) In most cases, their principle of construction is very simple and may be recursive.

Apart from the fourth characteristic which is very technical, the other characteristics are commonly considered in geography for conceiving the forms of geographical space through a fractal lens, as found in the following statement by Batty and Longley (1994, Chapter 3): "any system in which the whole is composed of parts arranged hierarchically in some self-similar order is fractal".

2.4. Density and fractal dimension

The fact that the size of a fractal object depends on the scale of measurement adopted, as illustrated by the measurement of the length of the Koch curve in section 2.2, reiterates a well-known point in geography, that the level of detail of the representation of space, that is, the level of aggregation or spatial resolution, influences the measurement of geographical shapes. Tobler (2004), citing Arbia et al. (1996), specifies on this topic that everything is related to everything else, but things observed at a coarse spatial resolution are more related than things observed at a finer resolution. Indeed, aggregation has a smoothing effect (see Tobler (1969, 1990)), which is why it is impossible to define a "true" population density at a point. Any estimate of density depends on the level of detail at which it is measured, given that extremely few geographic phenomena are scale-independent (Goodchild 2001).

This appears very clearly for fractal figures such as those shown in Figure 2.15; the density of the number of elements occupied decreases as the spatial resolution becomes finer. This decrease in density is all the more marked when the fractal dimension is low (Table 2.1). At each spatial resolution i, the density $\rho(i)$ is calculated as:

$$\rho(i) = \mathrm{N}^i \times (r^2)^i = (N \times r^2)^i \qquad [2.6]$$

where N is the number of elements in the generator of the fractal figure and r is the reduction factor between the initiator and an element of the generator. The ratio between the density $\rho(i)$ measured at spatial resolution i and the density $\rho(i+1)$ at a finer spatial resolution $i+1$ is related to the fractal dimension D as follows:

$$\rho(i+1) / \rho(i) = N \times r^2 \qquad [2.7]$$

$$\rho(i+1) / \rho(i) = r^{(2-D)} \qquad [2.8]$$

The number 2 in the above equation is the dimension of the supporting space, namely here the 2D plane[7].

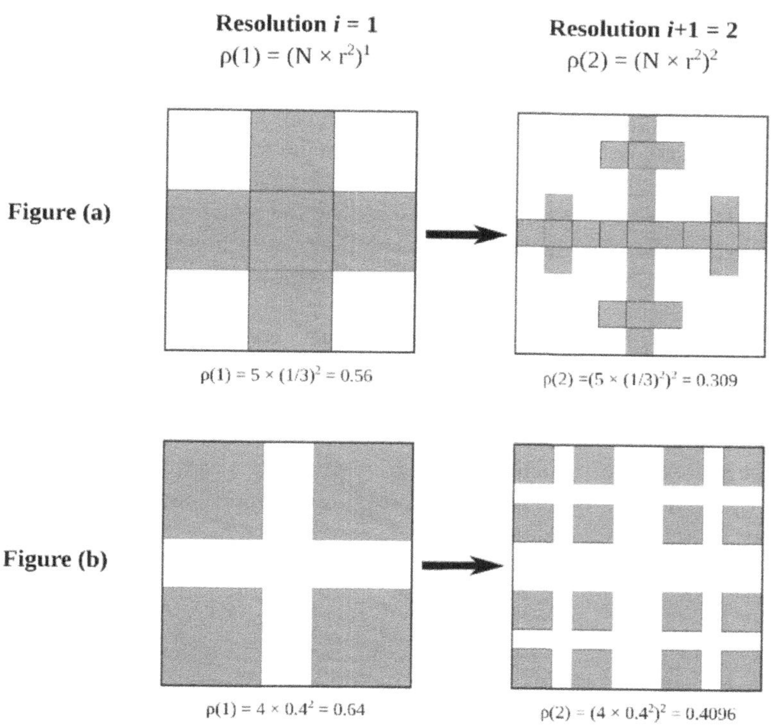

Figure 2.15. *Decrease in the density of elements in two fractal figures, depending on the spatial resolution adopted*

7. We note an error in the exponent of this formula, in the publications Tannier et al. (2012) – equation 2, Tannier et al. (2016) and Tannier (2019) – equation 1.

	$\rho(i+1) / \rho(i)$	
D	$r = 1/3$	$r = 1/4$
1.26	0.44	0.36
1.46	0.55	0.47
1.63	0.67	0.60
1.77	0.78	0.73

Table 2.1. *Ratio between the density calculated at a given spatial resolution i + 1 and at a coarser resolution i as a function of the fractal dimension D for two reduction factors r*

Now, consider the density of elements in a fractal figure, starting from the center of the figure and moving towards its periphery. In the fractal shown in Figure 2.16(a), the density of elements increases from the center of the figure to the periphery. In other words, the wider the zone of interest around the center of the figure, the higher the density. On the other hand, in the fractals shown in Figure 2.16(b) and (c), the density decreases or increases as a function of distance to the center.

Let us now consider the case of fractal clusters composed of particles clustered together according to a DLA (*diffusion-limited aggregation*)-type process, whose shape consists of a fractal nesting of digitations (*fractal fingering*) (Figure 2.17). In the 2D plane, the fractal dimension of such clusters is approximately equal to 1.71 when the path of the particles is not constrained by a lattice (Feder 1988). The density of particles in these clusters decreases from the center of the figure to the periphery. Within a radius *k*, the particle density is equal to:

$$\rho(k) \sim K_0^{-D}\, k^{D-2} \qquad [2.9]$$

where K_0 is the diameter of a particle and the exponent 2 is the dimension of the 2D plane. Batty and Kim (1992) and Chen (2009) have shown, in the case of cities, a similar relationship between their fractal dimension and the decrease in their population density or their built density from the city center to their outskirts[8].

8. Note: when the sets of points under consideration have this characteristic, it is not relevant to use spatial statistics methods because these assume that the average density of points is identical everywhere within the study area (Agterberg 2013); see Chapter 1.

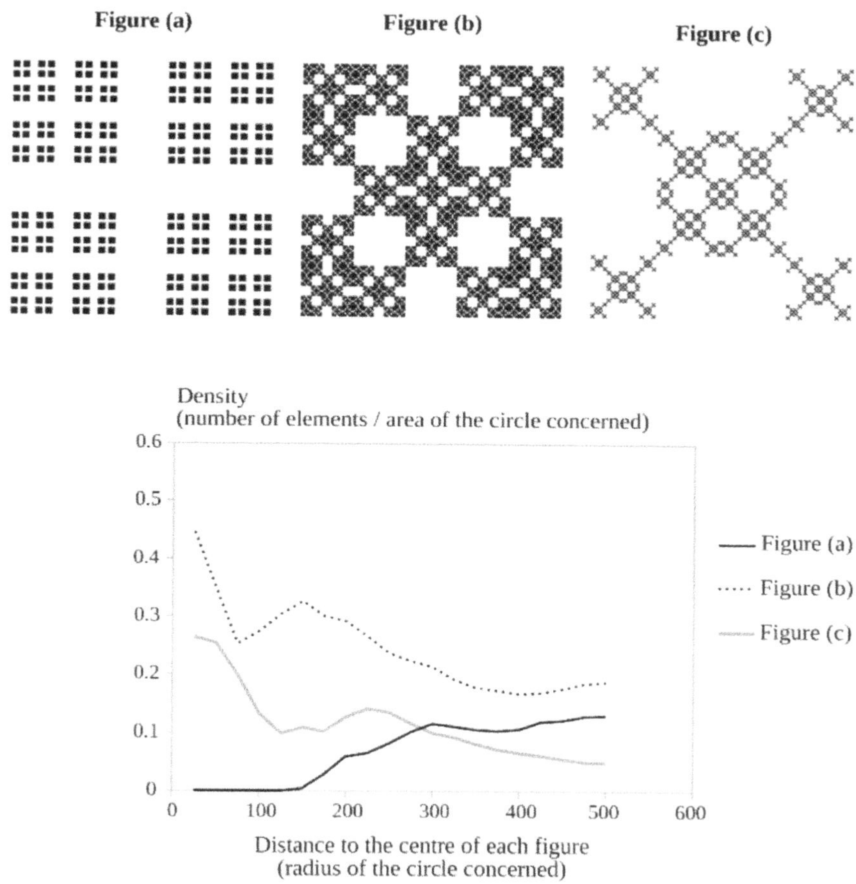

Figure 2.16. *Three fractal figures in the third iteration, created with GenFrac software (see: https://sourcesup.renater.fr/www/genfrac/). Change in the density of elements at increasing distances from the center of each figure. Density calculations performed by Gilles Vuidel, ThéMA laboratory (CNRS – Université Bourgogne Franche-Comté, Besançon, France)*

Figure 2.17. *Fractal cluster generated by a DLA (diffusion-limited aggregation) process. About 33,000 particles moving in a random walk to cluster to a seed placed in the center of the figure. The different colors show that the particles have clustered at different moments in time. Source: Wikimedia Commons by WingkLEE (see: https://en.wikipedia.org/wiki/Diffusion-limited_aggregation#/media/File:Of7_p0001_15h.jpg) [Accessed December 9, 2020]. For a color version of this figure, see www.iste.co.uk/tannier/fractal.zip*

2.5. Mathematical fractals versus fractals in the real world

> The most useful fractals involve chance and both their regularities and their irregularities are statistical. Also, the shapes described here tend to be scaling, implying that the degree of their irregularity and/or fragmentation is identical at all scales (Mandelbrot 1982).

In one of his first publications on fractals, and even before he introduced the term in 1975, Mandelbrot (1967) was interested in the self-similar character of the coast of Brittany. Visually, the coast of Brittany bears some resemblance to the Koch curve: just like it, a small portion of the coast of Brittany magnified several times resembles the coast seen as a whole. In the first case, however, the self-similarity of the figure is mathematically exact; in the second case, it is only the proportion of capes and bays that remains constant across the scales of measurement, while the arrangement of the elements in space may vary. "Statistically speaking, any portion of a coast [...] is homothetic to the whole, except for details that we choose not to deal with. Such a coast will be said to possess an internal homothety or to be self-similar" (Mandelbrot 1975)[9]. In fact, we distinguish three types of self-similarity. Strict self-similarity characterizes an object whose shape is exactly the same on any scale considered. A strictly self-similar set of points can be expressed as a union of sets, each of which is a reduced copy of (geometrically similar to) the full set. This is the case for fractal figures constructed using iterated function systems, as presented in section 2.1.1. Quasi-self-similarity characterizes figures constructed according to a non-random recursive mathematical rule, but whose component parts have a deformed or degenerated form in relation to the whole. The most well-known quasi-self-similar sets are the Julia sets and the Mandelbrot set: small parts of each of these sets can be enlarged and slightly deformed to correspond to a large part of the set considered (Figure 2.12). Objects in nature, such as coastlines, are generally neither strictly nor quasi-self-similar. They are statistically self-similar. Only numerical or statistical measurements characterizing their shape are identical across all scales. In this case, their shape is approximately, but not exactly, the same at different scales, as shown in Figure 2.18.

Figure 2.18. *Statistically self-similar hydrographic network (Egypt). Source: Google Earth fractals. See: http://paulbourke.net/fractals/googleearth/, Paul Bourke [Accessed March 26, 2021]. For a color version of this figure, see www.iste.co.uk/tannier/fractal.zip*

9. Fourth edition, 1995, p. 25.

Not all objects in nature are self-similar, even statistically. Stewart (1989) gives the example of the flea to illustrate this. A flea can make jumps of about 1 m. But if it were enlarged about a thousand times to be the size of an elephant, it could not make jumps of a thousand meters. On the contrary, its legs would be crushed under the weight of its body. This illustrates that the flea has a characteristic scale of length (or measurement), which is not the case for the coast of Brittany. The flea is a so-called *scalebound* object, after the neologism created by Mandelbrot (1981); that is, its structure is dominated by a very small number of distinct intrinsic scales. Fractal objects, on the other hand, are called *scaling*.

A crucial element intervenes in the creation of statistically self-similar objects: randomness. As noted by Feder (1988, p. 4), "randomness is an essential ingredient of most natural phenomena". Mandelbrot (1975, p. 6), also indicates that, to generate fractal irregularity, he uses constructions dominated by randomness. Randomness can be introduced very simply in the construction of a fractal figure. For example, it may be decided to randomly point the tips of the Koch snowflake up or down (Figure 2.19).

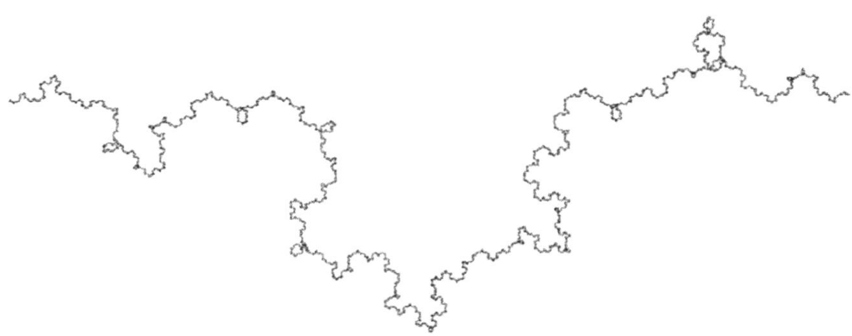

Figure 2.19. *Random Koch curve. Source: Wikimedia Commons. Author: Alexis Monnerot-Dumaine. See: https://commons.wikimedia.org/wiki/User:Prokofiev [Accessed March 26, 2021]*

By introducing chance in a more sophisticated way into the simulation of 3D fractal shapes, we can create spectacularly realistic landscapes. The algorithms traditionally used for this purpose are the *Random Midpoint Displacement* algorithm, the *Diamond-square* algorithm and various algorithms producing fractional Brownian motion (FBM) (Voss 1985). To increase the degree of realism of the simulated landscapes, textures (vegetation, snow, water, etc.) are applied to the fractal topographic shapes according to rules of slope and altitude (Figure 2.20). Imaginary landscapes, classics of science fiction films, have also been generated using fractal algorithms. The pioneers were the Genesis terraforming system in the film *Star Trek II: The Wrath of Khan* (1982) and the lunar landscape and contours of the Death Star in *Return of the Jedi*, episode VI of *Star Wars* (1983). Statistical (or random) fractals have thus become indispensable for generating realistic natural scenes where randomness of form exists within well-articulated structure (Batty and Longley 1994).

Figure 2.20. *Simulated fractal landscape. Source: Stevo-88, 2007. Fractal terrain texture. See: https://commons.wikimedia.org/wiki/File:Fractal_terrain_texture.jpg [Accessed March 26, 2021]. For a color version of this figure, see www.iste.co.uk/tannier/fractal.zip*

Concerning dynamics and generative mechanisms

As Mandelbrot points out (1975)[10], the "initial goal is to describe, from the outside, the shape of various objects. However, as soon as this first phase is successful, priority immediately shifts from description to explanation: from geometry to dynamics, to physics, and beyond". In physics in particular, the introduction of a new method of measurement is used to study an aspect of the structure in question and to verify existing theories by means of the results obtained. In the 1990s, for example, fractal analysis of the shape of clouds demonstrated the need to modify the theoretical explanation of their formation based on vortex theory (Frankhauser 1994). The deposition of gold on a surface also shows a fractal phase transition; as more gold is deposited on the surface, the branches of the different gold aggregates begin to join together, until they form a single aggregate. This transition is very significant and occurs in many physical systems. The process itself, which is a process of percolation, can be modeled with fractals. Indeed, in physical dynamic systems, either the shape of the objects themselves can be fractal, as in the case of clouds or gold deposits; or, when the dynamics of the system are chaotic, the attractor of the system represented in the phase space is fractal. A chaotic dynamic is the result of the operation of a deterministic system, that is, a system which is absolutely non-random, but which behaves in an apparently random manner (Stewart 1989).

Thus, fractals appear in science in two different ways: as a tool for describing irregular shapes and processes, or as a mathematical model of a chaotic dynamic (Stewart 1989). In human geography, fractals are used as geometric reference models and measurement methods, which represent two distinct but inseparable facets of the study of the spatial distribution of human settlements.

2.6. Fractality and scale invariance

In physics, biology, as well as in geography, economics and other sciences, we observe cases where a certain quantity or characteristic (size, income, number of elements) decreases or increases in terms of a measure (length, rank, intensity) in a non-proportional manner. For example, the relationship between the size of cities and their number is not proportional; in a given region of the globe, there are very few large cities compared to the number of medium-sized cities and even fewer compared to the number of small cities. Statistical distributions showing such non-proportional relationships are called leptokurtic, or *heavy-tailed distributions*; the tail of the distribution is long and has many more elements than the head of the distribution. The shape of these distributions is roughly a very stretched-out hyperbola on the x-axis (long tail resulting from the fact that the number of observations for small values of the analyzed variable is often high) and on the

10. Fourth edition, 1995, p. 6.

y-axis (the symmetric phenomenon of the long tail is a large initial peak resulting from the fact that the variable reaches very large values for a very small number of observations). The arithmetic mean is a very poor descriptor for such distributions, hence a joke commonly cited in economics: "If Bill Gates enters a bar, on average, the bar's customers are all millionaires."

Some of these heavy-tailed distributions vary regularly; when the unit of measurement of the samples in the distribution is changed, the shape of the distribution is not changed, nor are the properties resulting from it. In this case, the distribution has the property of scale invariance by change of units[11] (Dubrulle et al. 1997). Note that a line in the plane is invariant with respect to the change of scale. Indeed, if the measurement scale changes, all the points that constitute it remain unchanged. Similarly, if the points of a straight line are translated along the line, the same set of points is found. As for a line segment, it is scale-invariant when using a scale factor less than 1, but it is not invariant with respect to translation along its corresponding line, as the set of points constituting it is not the same once it has been translated. A circle is neither scale-invariant nor invariant with respect to translation in a direction, but it is invariant with respect to rotation around its center.

The invariance of a function or curve $f(x)$ considering change of scale by a factor λ of the variable x is expressed as follows:

$$f(\lambda x) = \lambda^{\Delta} f(x) \qquad [2.10]$$

If this equation holds, then $f(x)$ is scale-invariant of degree Δ. It is also said that the function is homogeneous. A line segment is scale-invariant of degree $\Delta = 1$ for all values of $\lambda = 1/n$, where n is an integer. The Koch curve (fractal figure presented above) is also scale-invariant of degree $\Delta = 1$, but only for values of $\lambda = 1/3n$. In general, fractal objects can be scale-invariant, but only for a discrete set of values of λ. And still, very often, additional operations of translation and rotation must be applied to the subsets of points constituting the object to find the object's global shape. This is why fractal objects are said to be self-similar and not scale-invariant in the strict sense. Conversely, many functions are scale-invariant without being fractal (Feder 1988).

The property of scale invariance necessarily implies a power law, otherwise called the scale law, which is a mathematical relationship of the form:

$$y = k x^{\alpha} \qquad [2.11]$$

11. There are other cases of scale invariance independent of the unit of measurement, when the quantities under consideration are dimensionless (Dubrulle et al. 1997).

where k is a proportionality constant and α is the scaling exponent. By definition, we have:

$$\alpha = dy / dx \qquad [2.12]$$

A power law may be expressed in linear form by switching to double logarithmic coordinates:

$$\log(y) = \alpha \log(x) + \log(k) \qquad [2.13]$$

Some authors equate scale-invariant functions with fractality. From their point of view, any set is fractal for which its corresponding statistical distribution takes the form of a power law. These scale-invariant sets are sometimes referred to as probabilistic fractals, to distinguish them from so-called geometric fractals (Crovelli and Barton 1995; Baveye and Boast 1998a). However, the existence of two different meanings of the term "fractal" leads to confusion, as Baveye and Boast (1998b) note. These two authors suggest using it only to designate geometrical constructs or physical systems that, within a range of observational scales, have geometrical properties similar to those of mathematical fractals.

Mandelbrot himself wondered about the fractality of certain scale-invariant functions. Notably, in a chapter entitled "Scaling and Power Laws Without Geometry" (Mandelbrot 1982, pp. 341–348), he studied the rank-size distribution of income within a population and showed that the scaling exponent of this distribution is formally a fractal dimension (p. 346). But two pages later, he notes that this scaling exponent, denoted D, does not seem to have any fractal interpretation worth writing down. In another chapter of his book "Islands, Clusters, and Percolation; Diameter-Number Relations" (pp. 116–130), Mandelbrot discusses the power law of the form $M(R) \propto R^Q$, where $M(R)$ represents the mass of elements present within a radius R around a point. According to him, the mere fact that it is a power law does not guarantee that Q is a fractal dimension. And to assert that Q is a fractal dimension is, according to him, meaningless and confusing.

In geography, Jiang has published a series of articles in which he supports the idea of describing as fractal any set with a much lower number of large elements than small elements, and for which the distinction between large and small elements can be made for both the whole and at least one subset of the set (Jiang and Yin 2014; Jiang 2015). Jiang thus claims a third definition of the term "fractal", the first being, according to him, strict self-similarity, and the second being statistical self-similarity. According to this third definition, any object is fractal when the statistical distribution of its component elements, classified in order of size, comprises many more elements of small size than elements of large size. This is ensured by dividing the distribution into two parts with the *head–tail breaks* method; the head of the distribution corresponds to the elements of larger than average size, and the tail of the distribution corresponds to the elements of smaller than average size. If a minority[12] of the elements belong to the head of the distribution, the distribution is again divided into two parts. The procedure is repeated until the elements larger than the average size of the sample are no longer a minority in number. An object is considered fractal if this procedure of iterative division of the distribution of its elements as a function of their size can be performed at least twice in succession. Thus, a fractal object is simply characterized by a series of data whose statistical distribution is heavy-tailed and displays a certain hierarchy (Jiang and Ma 2018). Fractality is therefore not "restricted" to power laws, but extends to exponential and log–normal functions. According to Jiang, this third definition of fractality implies that all geographic features may be fractal, if they are seen from the right perspective and scope. A country could be fractal, since seen from the perspective of cities there are far more small cities than large ones; a city could be fractal, since it consists of far more short streets than long ones[13]... However, this definition of fractality does not include the property of self-similarity (neither strict nor statistical), nor even those of scale invariance or scale dependence (Tannier 2019). It therefore has nothing in common, closely or remotely, with fractality as defined from a mathematical (geometric) point of view. It simply designates a form of hierarchical spatial organization, which may or may not be fractal. In fact, in more recent publications, Jiang no longer uses the term "fractal" to refer to objects whose structure is hierarchical; he now uses the expression "*living structure*" (Jiang 2019; Jiang and Huang 2021). "Living structure is defined as a mathematical structure, which consists of numerous recursively defined substructures with an inherent hierarchy. Across the hierarchy, there are far more small substructures than large ones, while on each level of the hierarchy the substructures are more or less similar in size" (Jiang and de Rijke 2022).

2.7. References

Agterberg, F.P. (2013). Fractals and spatial statistics of point patterns. *Journal of Earth Science*, 24(1), 1–11.

12. In general, less than 40%, but this threshold may be debated and varies from publication to publication. It is fixed at 25% in Jiang (2013a, 2013b).

13. Taken from: Jiang and Cattani (2018). Call for Papers Cartographica. *The International Journal for Geographic Information and Geovisualization, Special Issue: Fractal Geometry in Cartography and Geographic Information Science*. See: https://www.utpjournals.press/journals/cart/cfp [Accessed March 29, 2021].

Arbia, G., Benedetti, R., Espa, G. (1996). Effects of the MAUP on image classification. *Geographical Systems*, 3, 123–141.

Barnsley, M.F. (1988a). *Fractals Everywhere*. Academic Press, New York.

Barnsley, M.F. (1988b). Fractal modelling of real world images. In *The Science of Fractal Images*, Peitgen, H.O. and Saupe, D. (eds). Springer-Verlag, New York, 219–242.

Batty, M. and Kim, K. (1992). Form follows function: Reformulating urban population density function. *Urban Studies*, 29, 1043–1070.

Batty, M. and Longley, P.A. (1994). *Fractal Cities: A Geometry of Form and Function*. Academic Press, San Diego, CA and London.

Baveye, P. and Boast, C.W. (1998a). Fractal geometry, fragmentation processes and the physics of scale invariance: An introduction. In *Fractals in Soil Science*, Baveye, P., Parlange, J.-Y., Stewart, B.A. (eds). CRC Press, Taylor & Francis, 1–54.

Baveye, P. and Boast, C.W. (1998b). Concepts of "fractals" in soil science: Demixing apples and oranges (Comments & Letters to the editor). *Soil Science Society of America Journal*, 62(5), 1469–1470.

Bouligand, G. (1929). Sur la notion d'ordre de mesure d'un ensemble plan. *Bulletin des Sciences Mathématiques*, II(53), 185–192.

Chen, Y. (2009). A new model of urban population density indicating latent fractal structure. *International Journal of Urban Sustainable Development*, 1, 89–110.

Chéritat, A. (2010). L'ensemble de Mandelbrot. Images des mathématiques - La recherche mathématique en mots et en image. CNRS [Online]. Available at: http://images.math.cnrs.fr/L-ensemble-de-Mandelbrot.html.

Crovelli, R.A. and Barton, C.C. (1995). Fractals and the Pareto distribution applied to petroleum accumulation-size distributions. In *Fractals in Petroleum Geology and Earth Processes*, Barton, C.C. and La Pointe, P.R. (eds). Plenum Press, New York, 59–72.

Dubrulle, B., Graner, F., Sornette, D. (1997). Foreword: Scale invariance and beyond. In *Scale Invariance and Beyond (Les Houches Workshop, March 10–14, 1997)*, Dubrulle, B., Graner, F., Sornette, D. (eds). Springer Verlag, Berlin, Heidelberg, 1–21.

Falconer, K.J. (1990). *Fractal Geometry: Mathematical Foundations and Applications*. John Wiley & Sons, Chichester.

Falconer, K.J. (2013). *Fractals. A Very Short Introduction*. Oxford University Press, Oxford.

Feder, J. (1988). *Fractals*. Springer Science+Business Media, New York.

Frankhauser, P. (1994). *La fractalité des structures urbaines*. Anthropos, Paris.

Goodchild, M. (2001). Models of scale and scales of modelling. In *Modelling Scale in Geographical Information Science*, Tate, N. and Atkinson, P.M. (eds). John Wiley & Sons Ltd, Chichester, 3–10.

Hausdorff, F. (1919). Dimension und Äusseres Mass. *Mathematische Annalen*, 79, 157–179.

Jiang, B. (2013a). Head/tail breaks: A new classification scheme for data with a heavy-tailed distribution. *The Professional Geographer*, 65, 482–494.

Jiang, B. (2013b). The image of the city out of the underlying scaling of city artifacts or locations. *Annals of the Association of American Geographers*, 103(6), 1552–1566.

Jiang, B. (2015). Head/tail breaks for visualization of city structure and dynamics. *Cities*, 43, 69–77.

Jiang, B. (2019). Living structure down to earth and up to heaven: Christopher Alexander. *Urban Science*, 3, 96.

Jiang, B. and Huang, J.-T. (2021). A new approach to detecting and designing living structure of urban environments. *Computers, Environment and Urban Systems*, 88, 101646.

Jiang, B. and Ma, D. (2018). How complex is a fractal? Head/tail breaks and fractional hierarchy. *Journal of Geovisualization and Spatial Analysis*, 2, 6.

Jiang, B. and de Rijke, C. (2022). Representing geographic space as a hierarchy of recursively defined subspaces for computing the degree of order. *Computers, Environment and Urban Systems*, 92, 101750.

Jiang, B. and Yin, J. (2014). Ht-Index for quantifying the fractal or scaling structure of geographic features. *Annals of the Association of American Geographers*, 104(3), 530–540.

von Koch, H. (1904). Sur une courbe continue sans tangente, obtenue par une construction géométrique élémentaire. *Arkiv för matematik, astronomi och fysik*. Band 1, 681–704.

Lauwerier, H. (1991). *Fractals: Endlessly Repeated Geometric Figures*. Princeton University Press, Princeton, NJ.

Lesmoir-Gordon, N., Rood, W., Edney, R. (1996). *Introducing Fractals. A Graphic Guide*. Icon Books, London.

Lindenmayer, A. (1968). Mathematical models for cellular interactions in development II. Simple and branching filaments with two-sided inputs. *Journal of Theoretical Biology*, 18(3), 300–315.

Mandelbrot, B. (1967). How long is the coast of Britain? Statistical self-similarity and fractional dimension. *Science*, 156(3775), 636–638.

Mandelbrot, B. (1975). *Les objets fractals. Forme, hasard et dimension*. Flammarion, Paris.

Mandelbrot, B. (1981). Scalebound or scaling shapes: A useful distinction in the visual arts and in the natural sciences. *Leonardo*, 14, 45–47.

Mandelbrot, B. (1982). *The Fractal Geometry of Nature*. W.H. Freeman and Company, New York.

Mandelbrot, B. (1986). Self-affine fractal sets. In *Fractals in Physics*, Pietroner, L. and Tosatti, E. (eds). Elsevier, Amsterdam, 3–28.

Minkowski, H. (1903). Volumen und Oberfläche. *Mathematische Annalen*, 57, 447–495.

Sémécurbe, F. (2020). Analyse de la distribution spatiale des implantations humaines - Apports et limites d'indicateurs multi-échelles et trans-échelles. PhD Thesis, University of Franche-Comté, Besançon.

Stewart, I. (1989). *Does God Play Dice? The New Mathematics of Chaos*. Basil Blackwell, Oxford, UK/New York.

Tannier, C. (2019). Effects of distance and scale dependence in geographical models of cities and territories. In *Geographical Modeling: Cities and Territories*, Pumain, D. (ed.). ISTE Ltd, London, and John Wiley & Sons, Hoboken, 53–94.

Tannier, C., Vuidel, G., Houot, H., Frankhauser, P. (2012). Spatial accessibility to amenities in fractal and non fractal urban patterns. *Environment and Planning B: Planning and Design*, 39(5), 801–819.

Tannier, C., Bourgeois, M., Houot, H., Foltête, J.-C. (2016). Impact of urban developments on the functional connectivity of forested habitats: A joint contribution of advanced urban models and landscape graphs. *Land Use Policy*, 52, 76–91.

Tobler, W. (1969). Geographical filters and their inverses. *Geographical Analysis*, 1(3), 234–253.

Tobler, W. (1990). Frame independent spatial analysis. In *Accuracy of Spatial Databases*, Goodchild, M. and Gopal, S. (eds). Taylor & Francis, London, 115–222.

Tobler, W. (2004). On the first law of geography: A reply. *Annals of the Association of American Geographers*, 94(2), 304–310.

Voss, R.F. (1985). Random fractal forgeries. In *Fundamental Algorithms for Computer Graphics*, Earnshaw, R.A. (ed.). Springer-Verlag, 805–835.

Voss, R.F. (1988). Fractals in nature: From characterization to simulation. In *The Science of Fractal Images*, Peitgen, H.O. and Saupe, D. (eds). Springer-Verlag, New York, Berlin, Heidelberg, 21–70.

3

Fractal Reference Figures

Cécile TANNIER
Laboratoire ThéMA, CNRS-Université de Franche-Comté, Besançon, France

I coined fractal from the Latin adjective fractus (...) in addition to "fragmented" (as in fraction or refraction), fractus should also mean "irregular", both meanings being preserved in fragment
(Mandelbrot 1982, p. 4).

The appearance of self-similar figures (fractals) in the arts and sciences dates back several centuries (Mandelbrot 1988; Lesmoir-Gordon et al. 1996, pp. 154–159). Some are curves of topological dimension equal to 1; others are surfaces of topological dimension equal to 2; others are figures known as "dusts" of topological dimension equal to 0 (Mandelbrot 1982). Still, other figures are so oddly shaped that there are no good terms for them in either the sciences or the arts (Mandelbrot 1982, p. 1). Some fractal figures may be used as approximations to describe nature, in the same way as the figures of Euclidean geometry – triangles, circles, cylinder spheres, etc.– hence the interest of studying their properties in depth (Feder 1988). The Koch curve, for example, is a classic fractal model of a coastline. The fractal dimension of a coastline is generally between 1.15 and 1.25, while that of a Koch curve is close to 1.26 (Stewart 1989). Batty and Longley (1994, p. 91) point out that an urban boundary, that is, the curve that delimits the morphological envelope of a city, is quite similar to a coastline in terms of how it fills space. The Koch curve may therefore also be a model of an urban boundary. Meanwhile, Frankhauser and Sadler (1991) have observed that the ratio between the perimeter length and the area of large global cities tends to be proportional. This is the case for certain fractal figures, while the area of traditional figures of Euclidean geometry is proportional to the square of their perimeter.

Fractal objects in the real world result from generative processes in which chance plays a crucial role and are therefore statistically self-similar. Another important aspect that differentiates real fractal objects from mathematical fractal figures is the fact that the latter have a structure over an infinite range of scales, while the first have a structure over a quite small range of scales, sometimes referred to as a fractality interval (Mandelbrot 1975; Falconer 2013). It can be very useful to regard natural objects as fractals if they exhibit irregularities or self-similarity when viewed over a significant range of scales (Falconer 2013). For the geographic object "coastline", for example, the fractality interval extends from 1 cm, a scale of measurement below which the geographic object disappears in the detail of the interface between water, air and stone, to a few tens or hundreds of kilometers, a scale at which the "coastline" object is seen as a whole. Similarly, for a medium-sized city, the fractality range goes from 1 m or 10 m, a scale below which the outline of the footprint of buildings and streets is no longer discernible, to a few kilometers or tens of kilometers, the scale at which the city constitutes an identifiable whole and beyond which the shape of the city is no longer precisely distinguishable within a space of regional size.

The reference figures presented in this chapter are strictly self-similar. They are constructed according to a deterministic iterative process which makes it possible to precisely determine their geometric properties. The elementary entities used to create these figures are Euclidean shapes (squares, circles, triangles, etc.), given that only the sizes, numbers and positions of the elements in a generator determine the shape of the fractal figure that may be obtained (see Chapter 2, Figures 2.4 and 2.5). The properties of a fractal figure can only be fully seen at infinity. However, computer-generated images of fractal figures are used to build a visual intuition of these properties (Voss 1988).

3.1. Fractal curves and islands

3.1.1. *Fractal curves*

Fractal curves have a topological dimension equal to 1 and a fractal dimension $1 < D < 2$ when they exist in the 2D plane. Their length increases and tends towards infinity as the steps of their construction progress. The best known fractal curve is the Koch curve. Two versions of this curve were presented in Chapter 2. The most classic is the *triadic* Koch curve shown in Figure 2.13(b). Its fractal dimension is $D \approx 1.26$. A *quadratic* or *quadric* Koch curve of type 1 is shown in Figure 2.11. Its fractal dimension is higher, $D \approx 1.46$. In this chapter, a quadratic Koch curve of type 2, also called a "Minkowski sausage", is shown in Figure 3.1.

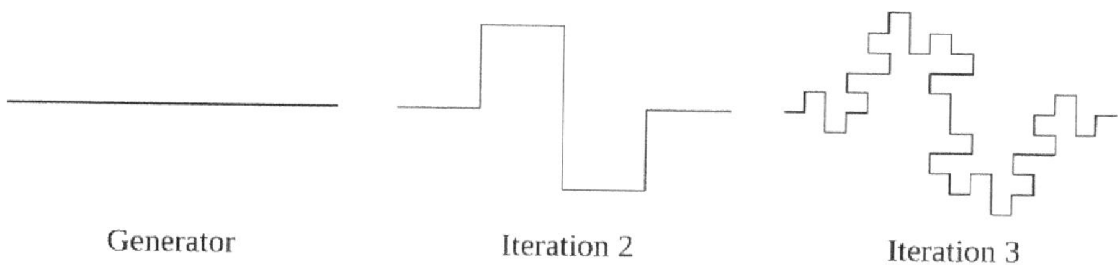

Figure 3.1. *Type 2 quadratic Koch curve*

Its generator is composed of $N = 8$ elements reduced by a factor $r = 1/4$. Its fractal dimension is therefore $D \approx 1.5$. Quadratic Koch curves lead directly to the Peano curves, the first of which was discovered in 1890 by the researcher of the same name (Mandelbrot 1982, p. 50). A Peano curve is a planar curve that passes through every point of the unit square (Figure 3.2). As it completely fills the space after an infinite number of iterations (i.e. of construction steps), it is said to be *plane filling*, and its fractal dimension D is equal to 2. In addition to Koch and Peano curves, there are a large number of other types of fractal curves (Levy curve and tapestry, dragon curve, etc.), some self-intersecting and others not (see Lauwerier (1991), Chapter 3 for examples with their methods of construction).

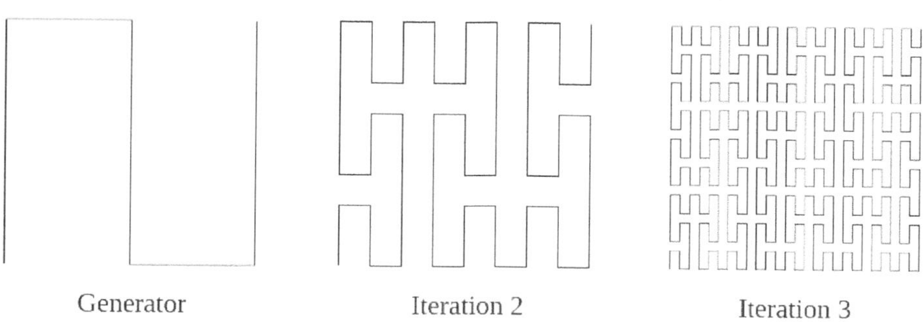

Figure 3.2. *First steps in the construction of a Peano curve*

3.1.2. *Teragons, fractal islands*

Teragons, or fractal islands, are surface figures without internal gaps and whose perimeter is a fractal curve. The length of their perimeter therefore lengthens indefinitely as the steps of construction progress. The best known of these figures is Koch's snowflake, shown in Figure 3.3 and also called the triadic Koch island (Mandelbrot 1982, p. 35). As the steps of its construction progress, the area of this figure converges to 8/5 of the area of the initial triangle. By construction, the fractal dimension of its perimeter is that of the triadic Koch curve $D \approx 1.26$, while that of its interior (i.e. of its surface) is equal to 2. Figure 3.4 is a rectangular Koch island (Mandelbrot 1982, pp. 50–51). Each side of the initial figure is replaced by a quadratic Koch curve of type 2. Its area remains constant from one step of construction to another, unlike the area of the previous teragon.

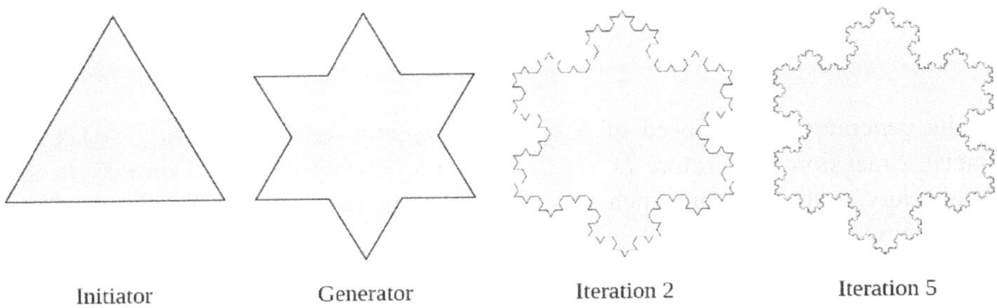

Figure 3.3. *Triadic Koch island (Koch snowflake)*

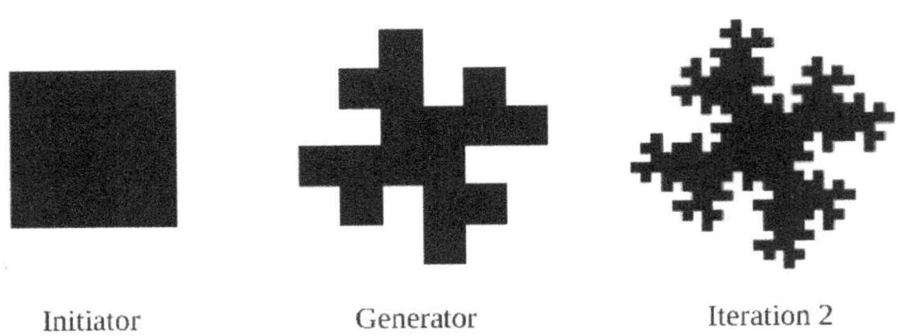

Figure 3.4. *Rectangular Koch island*

3.2. Fractal dusts, gaskets and carpets

3.2.1. *Fractal dusts*

Fractal dusts are formed of elements that are topologically unconnected across scales. In doing so, it is impossible to go from one point to another on the figure without leaving it (Falconer 2013). Their topological dimension is equal to zero. The elements of a dust may be spatially distributed on a line or in a plane. Cantor's dust (or set), studied by Henry Smith in 1875 and by Cantor in 1883 (Fleron 1994), was the first fractal dust to be imagined. It is constructed iteratively from the segment [0, 1] by removing the central third; then, the operation is repeated on the two remaining segments, and so on. Its fractal dimension is $D \approx 0.63$. It is very difficult to represent due to its thinness and fineness, which is why Mandelbrot (1982, p. 80) proposed enlarging it in the form of a bar (Figure 3.5(a)). This dust may also be represented in the 2D plane, which gives the Cantor square (Figure 3.5(b)), whose fractal dimension $D \approx 1.2619$ is very close to that of the Koch curve.

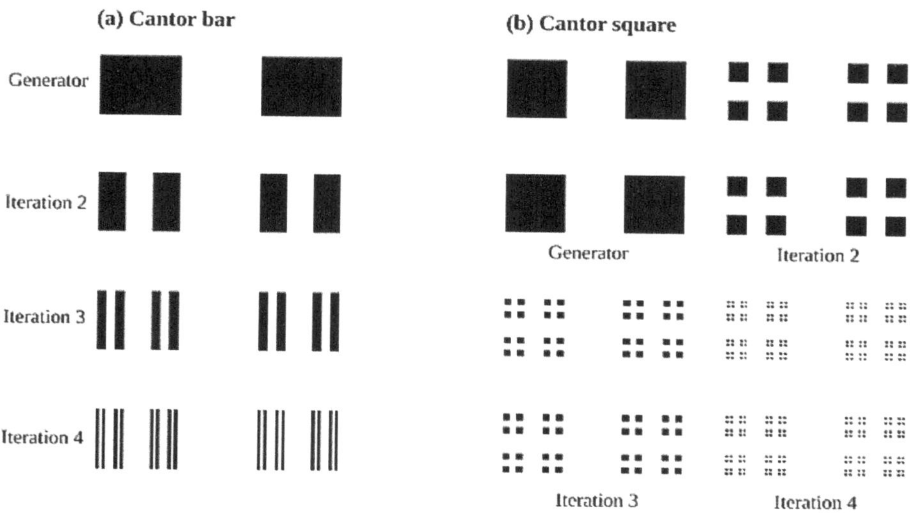

Figure 3.5. *Two Cantor sets*

Another category of fractal dusts is presented in the work of Mandelbrot (1982, p. 84) and those following: these are projections in the 2D plane of a descriptive model of the spatial distribution of stars and galaxies proposed by Fournier d'Albe (1907) (Figure 3.6). According to Mandelbrot, Fournier d'Albe unknowingly followed a parallel path to that of Cantor, the main difference between the two being that Fournier's construction extended through space, while that of Cantor extended along a line. In the example shown in Figure 3.6, the number of elements is $N = 5$ and the reduction factor is $r = 1/5$; the fractal dimension is therefore $D \approx 1$. Mandelbrot also presents another dust with a reduction factor of $r = 1/7$ and whose fractal dimension is $D \approx 0.8270$, therefore less than 1 (Mandelbrot 1982, p. 95).

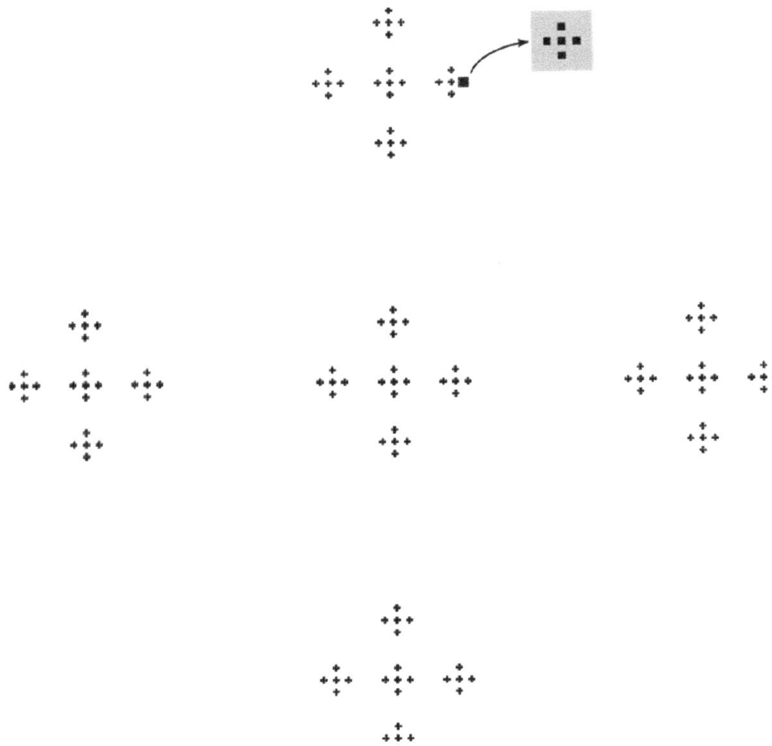

Figure 3.6. *A Fournier dust at the fourth step of iteration and a 600% zoom in on a group of elements in the figure. Fractal dimension D ≈ 1. After Mandelbrot (1982, p. 96)*

Fournier dusts are highly dispersed figures; most of the 2D plane is empty of points and their fractal dimension is less than or equal to 1. They are therefore of little interest as reference figures for the distribution of human settlements, unlike Cantor's square (Figure 3.5(b)) and other similar figures (Figure 3.7)[1]. As can be seen visually, the elements of Figures 3.5(b) and 3.7 form clusters separated from each other by free spaces of different sizes, which can be likened to traffic lanes (main streets, secondary streets, local service lanes) or to unbuilt spaces (parks, gardens, car parks). Very wide free spaces are few in number, while free spaces of

1. Fractal Figure 3.7 and those following in this chapter were drawn by means of iterated function systems (IFS) using the *GenFrac* software (https://sourcesup.renater.fr/www/genfrac/).

increasingly small width over the iterations are always greater in number. There also exist other types of fractal dust in which the shape and size of the gaps vary to a greater degree than in the dusts presented earlier (Figure 3.8).

(a) Dust D ≈ 1.58 (N = 9 and r = 1/5)

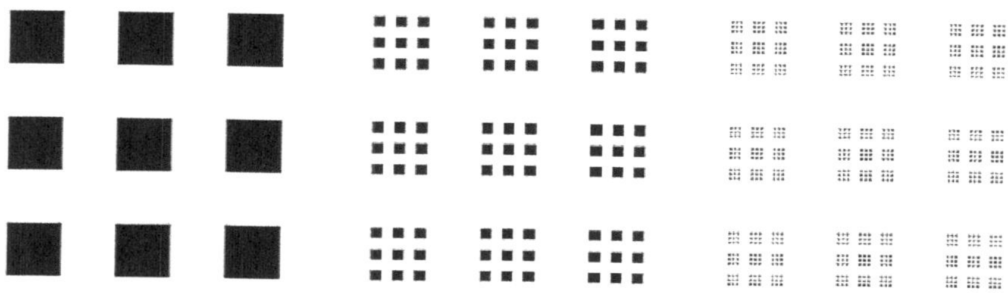

(b) Dust D ≈ 1.72 (N = 16 and r = 1/5)

Figure 3.7. *Two other fractal dusts related to Cantor's square*

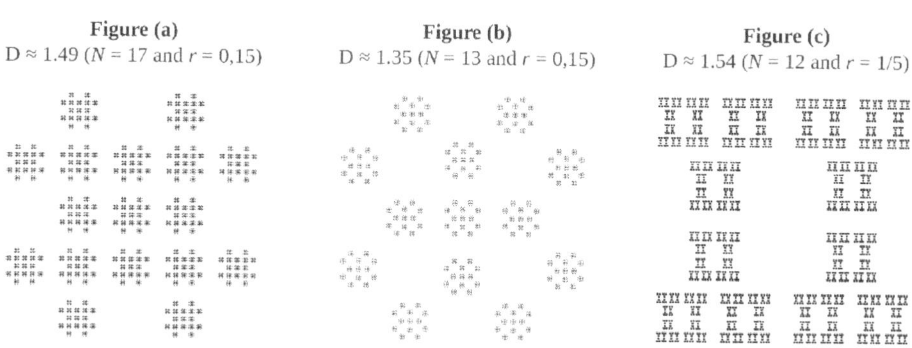

Figure 3.8. *Fractal dust showing a variety of shapes and sizes of spaces separating the elements from each other. Iteration 3*

3.2.2. *Sierpinski gaskets and carpets*

Some fractal figures are composed of related elements, thus forming a single and unique cluster. The best-known figures of this type are Sierpinski gaskets and carpets. Their topological dimension is equal to 1 (Mandelbrot 1982, p. 136) and their fractal dimension is $1 < D < 2$.

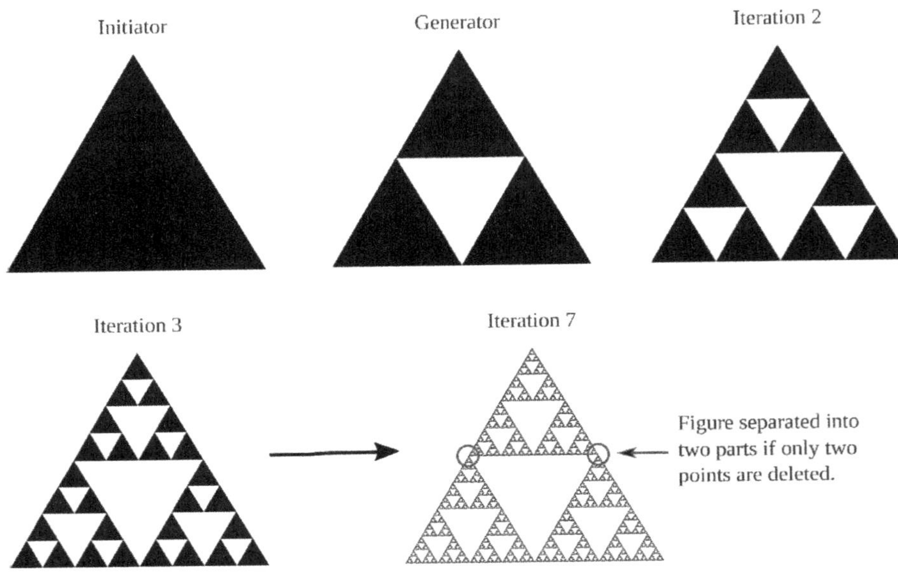

Figure 3.9. *Sierpinski gasket constructed from an equilateral triangle. For a color version of this figure, see www.iste.co.uk/tannier/fractal.zip*

The Sierpinski gasket was studied by the eponymous researcher in 1915, about 40 years after the Cantor set. It is traditionally constructed from an equilateral triangle by dividing it into four triangles and eliminating the central triangle (Figure 3.9). Its fractal dimension is $D \approx 1.5849$. An example of a Sierpinski gasket constructed from an isosceles triangle is shown in Figure 2.4 (see Chapter 2). The Eiffel Tower and Gothic cathedrals have structures similar to Sierpinski gaskets.

The Sierpinski carpet is also a fractal with a porous texture. The figure of a classic Sierpinski carpet is obtained by dividing a square into nine equal squares and then removing the central square (Figure 3.10(a)). Repeated indefinitely, this operation makes it possible to construct a figure whose area is zero and the total perimeter of whose internal gaps is infinite. The outer perimeter of the figure is smooth, of dimension $D = 1$. The fractal dimension of the figure itself is $D \approx 1.892$. The intersection of a Sierpinski carpet with a median or a diagonal is a Cantor set.

Figure (a)
$D \approx 1.892$ ($N = 8$ and $r = 1/3$)

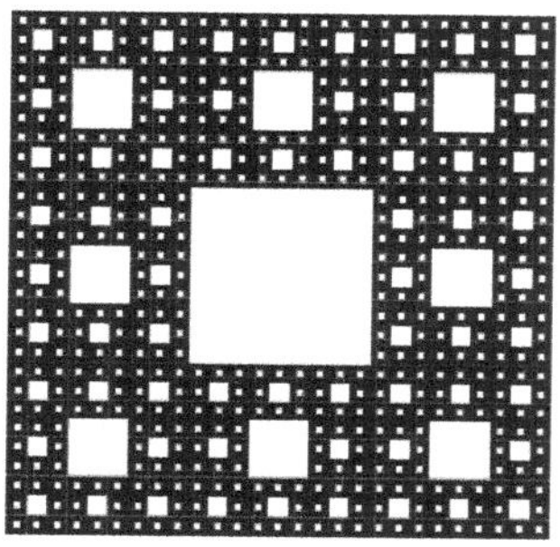

Figure (b)
$D \approx 1.861$ ($N = 20$ and $r = 1/5$)

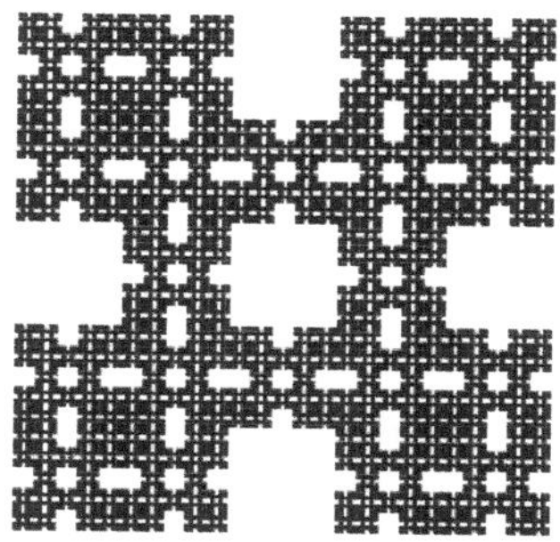

Figure 3.10. *Two Sierpinski carpets. Figure (a), iteration 4, after Mandelbrot (1982). Figure (b), iteration 3, after Frankhauser (1994)*

The interest of Sierpinski gaskets and carpets as reference models for the spatial distribution of human settlements is that they feature internal gaps of various sizes. They are therefore particularly well suited to modeling intra-urban built fabrics. Considering the connected nature of these figures, it is interesting to know that Sierpinski gaskets and carpets differ fundamentally in their branching, which is finite for the Sierpinski gasket and infinite for the Sierpinski carpet (Mandelbrot 1982, p. 138). Branching is the smallest number of links, that is, paths going from one point to another of the figure, which must be cut to disconnect a fairly large portion from the whole (Bunde and Havlin 1994). When a figure is of finite branching, cutting a finite number of points is enough to isolate a part of the figure. In the case of Figures 3.8 and 3.11, for example, it is sufficient to delete a few points to separate a fairly large part of the figure from the whole. Conversely, in the case of Figure 3.10, an infinity of points would need to be eliminated in order to disconnect a part of the figure.

A wide variety of figures can be constructed by generalizing the principle of construction of Sierpinski gaskets and carpets. For example, an initiator may be divided into nine squares ($r = 1/3$) (Figure 3.10(a)), into 25 squares ($r = 1/5$) (Figure 3.10(b)), or in some other way. The number and position of unoccupied squares may also vary (Figure 3.11), knowing that Sierpinski gaskets and carpets are not necessarily centered (Mandelbrot 1982, p. 133). If we consider the Sierpinski carpet presented in Figure 3.10(b), its fractal dimension is very slightly lower than that of Figure 3.10(a), but crucially, its outer perimeter is not smooth; it is a fractal curve composed of $N = 7$ elements and of fractal dimension $D \approx 1.2$. The examples shown in Figure 3.11 also have fractal outer perimeters. The one in Figure 3.11(b), for example, is of dimension $D \approx 1.365$.

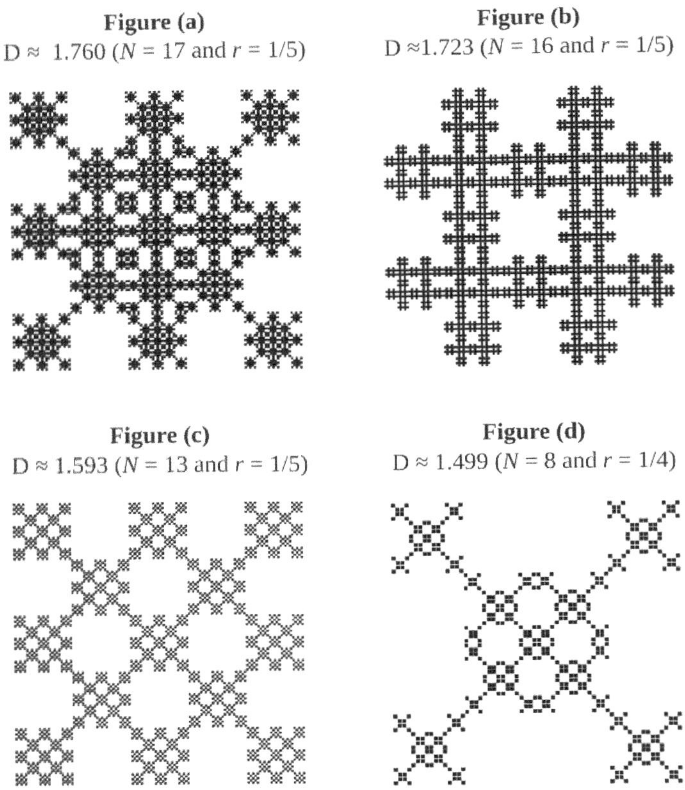

Figure 3.11. *Four figures in the family of Sierpinski gaskets and carpets built from a square, at iteration 3. Figure (b) after Bunde and Havlin (1994)*

By starting not from a square but from a pentagon, a hexagon or another regular polygon, it is possible to create other figures related to Sierpinski gaskets and carpets. These figures are called n-flakes, polyflakes or Sierpinski n-gons. Figure 3.12(a) is a pentagonal gasket. Its perimeter is a Koch curve. Figure 3.12(c) is a hexagonal gasket. The six sides of the figure are Koch curves, and the central recessed part is a Koch snowflake. Figures 3.12(b) and (d) show snowflakes. To construct them, an additional polygon is added to the center of the generator. The pentagonal snowflake (*pentaflake*) of Figure 3.12(b) is commonly called Dürer's pentagon after the name of the painter who imagined it, in 1525. It is constructed by iteratively substituting each pentagon by a snowflake of six pentagons, the central pentagon being rotated by 180° with respect to the others (Bunde and Havlin 1994, p. 8). Figure 3.12(d) shows a hexagonal snowflake (*hexaflake*). Its border is the Koch snowflake, and it contains an infinity of Koch snowflakes, both positive and negative. Note that among the four figures shown in Figure 3.12, the pentagonal snowflake shows a significantly higher degree of branching than the others, which may be important to consider for choosing a spatial distribution model of human settlements.

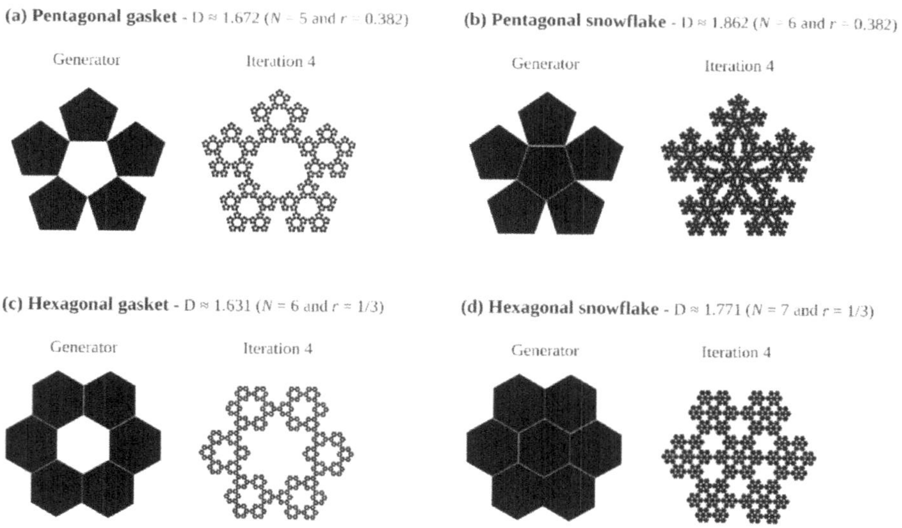

Figure 3.12. *Sierpinski polygons and polyflakes*

3.2.3. *Box fractal*

The process of constructing a box fractal, otherwise known as a Vicsek fractal, is analogous to that of a Sierpinski carpet. The fractal is obtained by dividing a square initiator into nine squares, then removing four of them while five remain (Frankhauser 1988, p. 68; Vicsek 1989, p. 12). Two versions of this fractal are shown in Figure 3.13. The envelope (outer perimeter) of the box fractal is a quadratic Koch curve. Like the Sierpinski carpet, the box fractal has an area of zero at infinity and an infinite perimeter. The big difference between these two families of shapes is the absence of internal gaps in the box fractal; all the free spaces are open and connected to each other. It can be noted that Figure 3.11(d) has a shape that is partially related to the box fractal, because of its rather linear and not very compact appearance, but like the Sierpinski carpets and gaskets, it has internal gaps.

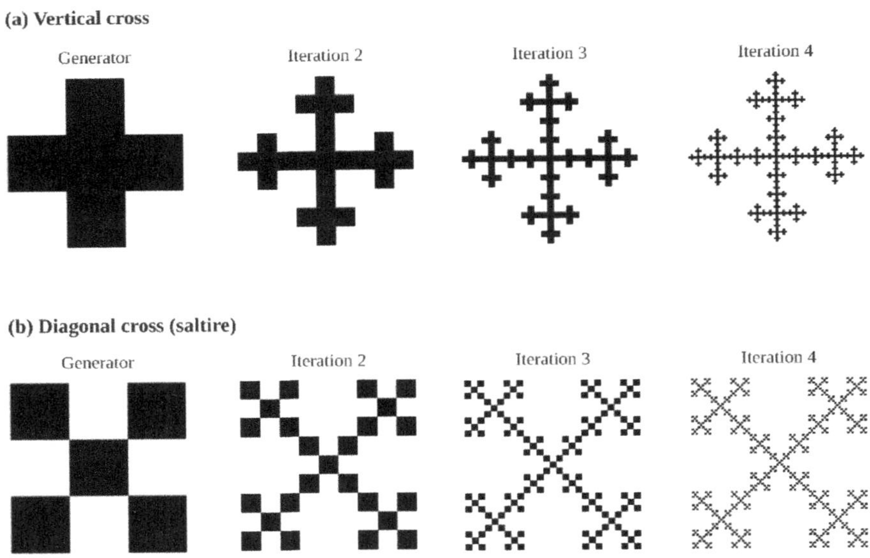

Figure 3.13. *Box fractal. D ≈ 1.464 (N = 5 and r = 1/3)*

3.3. Hybrid fractal figures and multifractal figures

3.3.1. *Hybrid fractal figures*

Hybrid fractal figures are particularly interesting for studies in urban geography because they have certain characteristics that make them resemble the built fabrics of cities (Tannier and Thomas 2013). Some hybrid figures are a combination of fractal dusts and Sierpinski gaskets or carpets, or box fractals; they include both clusters of connected elements and dispersed elements (Figure 3.14(a), (b) and (d)). Other hybrid figures only give rise, over the steps of their construction, to clusters of connected elements and no dispersed elements (Figure 3.14(c), (e) and (f)). Note that, in certain generators, the elements are connected but the fractal figure resulting from them comprises several clusters (Figure 3.14(c) and (e)). It can also be visually observed that the perimeter of the hybrid figures is not a single curve but a collection of "loops"[2], each one being more or less tortuous. The fractal dimension of the total perimeter of these figures (set of "loops") measures both their fragmentation and their irregularity (dendricity). In contrast, the fractal dimension of the perimeter of each cluster measures only its dendricity (tortuosity) (Mandelbrot 1982, p. 118).

Clarification should be made here as to the meaning of the term "cluster". Mandelbrot (1982, pp. 124–125) proposed distinguishing contact clusters, formed from topologically related elements, from visually identifiable clusters, as found in a fractal dust, but whose delimitation necessarily requires statistical rules or distance thresholds which are fixed a priori and therefore necessarily questionable. According to this definition, some among the figures shown in Figure 3.14 include a certain number of contact clusters composed of a single element. Depending on the figure, the number of contact clusters composed of several elements and clusters composed of a single element (dispersed elements) may vary greatly. Thus, at the fourth iteration, Figure 3.14(a) comprises 3,201 multi-element clusters and 2,576 scattered elements; Figure 3.14(b) has 308 multi-element clusters and 4,913 scattered elements; Figure 3.14(d) has 1,189 multi-element clusters and 1,632 scattered elements.

In Figure 3.15, we show the rank-size distribution of the clusters of the six hybrid figures from Figure 3.14 to iteration 4. The size of the largest cluster(s) varies between about 60,000 (Figures 3.15(b) and (c)) and 5 (Figure 3.15(d)) elements. The size ratio between the largest cluster(s) and the smaller cluster(s) also varies significantly, and the size of the clusters is more or less diversified depending on the figure (at iteration 4, three cluster sizes for Figure 3.15(e) vs. 11 cluster sizes for Figure 3.15(a)). In addition, the head of the rank-size distribution (i.e. the largest clusters) may consist of only one or two clusters (Figure 3.15(a)–(c)) or, on the contrary, a very large number of clusters (Figure 3.15(d)–(f)). There may be only one or two very large clusters in a figure for the reason that, within the generator, a relatively large number of elements may be related.

2. Hence the name *sigma-loop* (σ-loop) (or sum, Σ) proposed by Mandelbrot (1982).

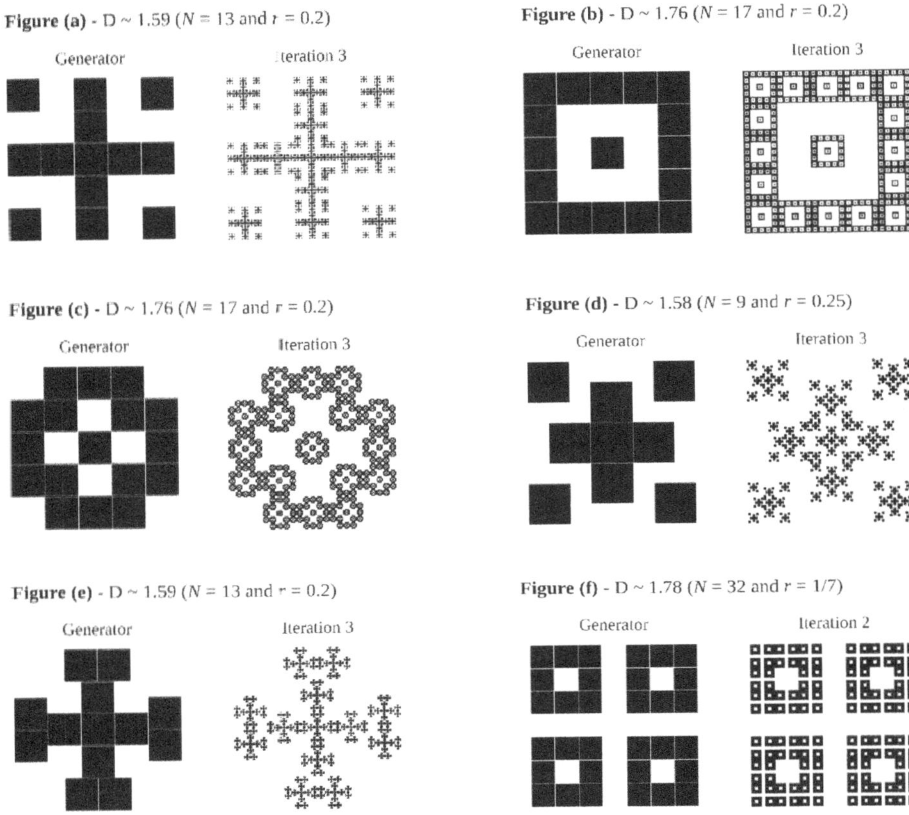

Figure 3.14. *Six hybrid fractal figures*

Table 3.1 shows the evolution of the number and size of the clusters in the six hybrid figures in Figure 3.14 as the steps of their construction progress. The number of clusters increases over the iterations in a more or less pronounced way according to the figures. For example, between iteration 2 and iteration 6, the number of clusters increases from 41 to 965,945 for (a) and from 5 to 122,045 for (e). In the case of figures with dispersed elements (size of the smallest clusters equal to 1), their number increases from 17 (figure (b)) or 24 (figures (a) and (d)) at iteration 2 to 14,564, 32,628 or 83,521 at iteration 6 (figures (d), (a) and (b), respectively). Similarly, the diversity of cluster sizes increases more or less according to the figures: in the case of (d) and (e), the size of the smallest and largest clusters does not change at all from one iteration to another; however, in the case of (c), the respective sizes of the smallest and largest clusters are equal to 17 and 272 at iteration 2 and then equal to 17 and 1,114,112 at iteration 5.

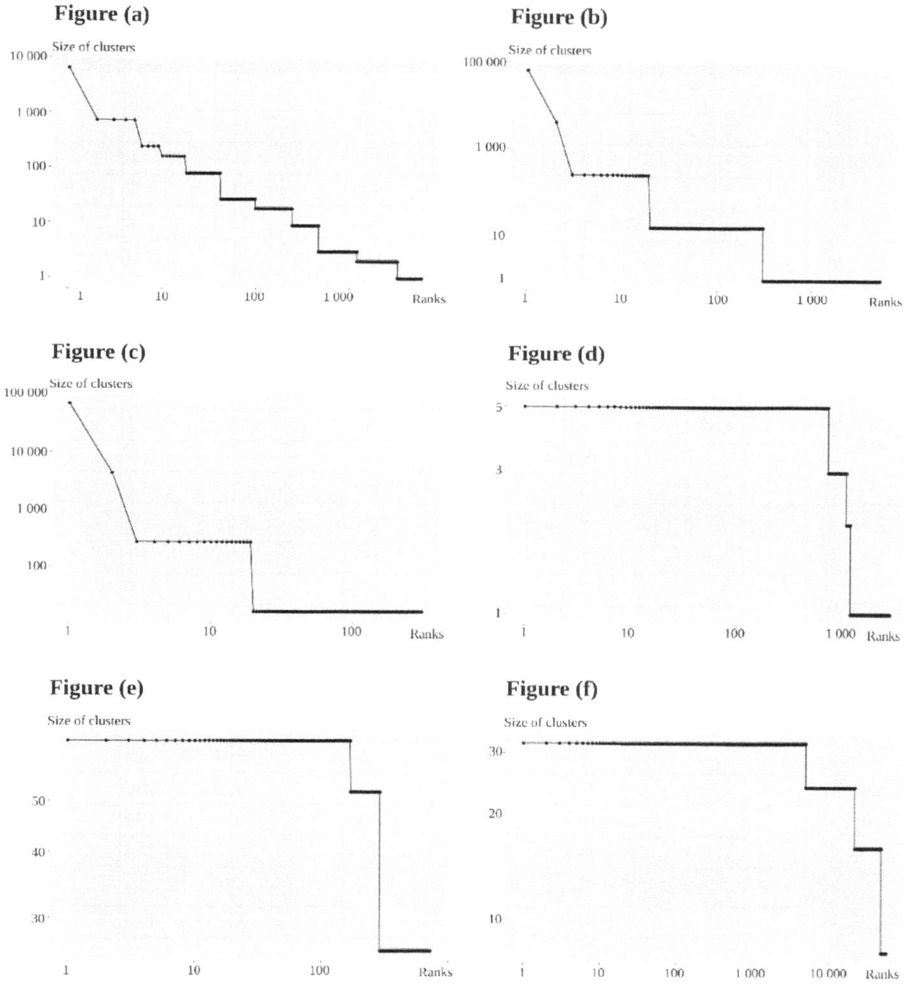

Figure 3.15. *Rank-size distributions of the clusters of each fractal in Figure 3.14, at iteration 4. The measure of the size of each cluster is the number of topologically related elements making up the cluster. Calculations performed by Gilles Vuidel, ThéMA laboratory (CNRS – Université de Franche-Comté, Besançon, France). Note: the scales of the x-axis and y-axis differ greatly from one graph to another*

	Step of construction (iteration)	Number of clusters	Number of classes of cluster size	Number of cluster(s) of the smallest size	Size of the smallest cluster(s)	Number of cluster(s) of the largest size	Size of the largest cluster(s)
Figure (a)	2	41	5	24	1	1	81
	3	461	8	220	1	1	729
	4	5,777	11	2,576	1	1	6,561
	5	74,453	14	32,628	1	1	59,049
	6	965,945	17	421,576	1	1	531,441
Figure (b)	2	19	3	17	1	1	256
	3	308	4	289	1	1	4096
	4	5,221	5	4,913	1	1	65,536
	5	88,742	6	83,521	1	1	1,048,576
Figure (c)	2	2	2	1	17	1	272
	3	19	3	17	17	1	4,352
	4	308	4	289	17	1	69,632
	5	5,221	5	4,913	17	1	1,114,112
Figure (d)	2	37	3	24	1	9	5
	3	317	4	188	1	81	5
	4	2,821	4	1,632	1	729	5
	5	25,325	4	14,564	1	6,561	5
	6	227,797	4	130,824	1	59,049	5
	7	2,049,917	4	1,176,908	1	531,441	5
Figure (e)	2	5	2	4	26	1	65
	3	57	3	36	26	13	65
	4	725	3	436	26	169	65
	5	9,393	3	5,604	26	2,197	65
	6	122,045	3	72,724	26	28,561	65
Figure (f)	2	64	3	16	8	16	24
	3	1,792	4	272	8	128	32
	4	55,936	4	7,696	8	4,992	32

Table 3.1. *Evolution of the number and size of clusters in the six hybrid figures of Figure 3.14 considering the first steps of their construction*[3]

3. Calculations performed on supercomputer facilities of the "Mésocentre de calcul de Franche-Comté" by Gilles Vuidel, ThéMA laboratory (CNRS – Franche-Comté University, Besançon, France).

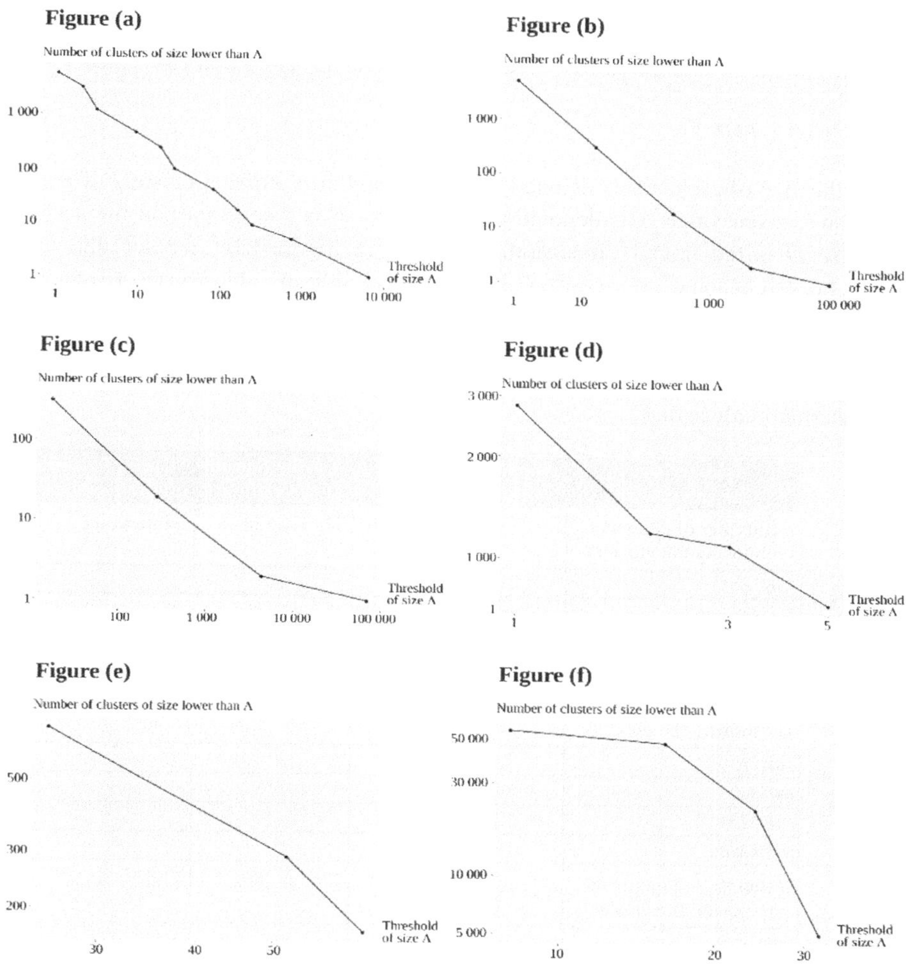

Figure 3.16. *Cumulative distributions, by size class, of the clusters of each fractal in Figure 3.14, at iteration 4. The measure of the size of each cluster is the number of topologically related elements that make up the cluster[4]. Note: the scales of the x-axis and y-axis differ greatly from one graph to another*

4. Calculations performed by Gilles Vuidel, ThéMA laboratory (CNRS – Franche-Comté University, Besançon, France).

In theory, in fractal figures, the number of clusters of different size classes follows a power law of the form (Mandelbrot 1982, p. 117, 118 and 134):

$$Nr(\Lambda > \lambda) \propto F\lambda^{-D} \quad [3.1]$$

where the size of a cluster is denoted as λ, the number of contact clusters of a size less than a given value Λ is denoted as Nr(Λ > λ), *F* is a constant and the scaling exponent *D* is the fractal dimension. However, in Figure 3.16, we see that only curves (a), (b), (c) and (e) are fairly close to a straight line, while curves (d) and (f) have very different shapes. According to Mandelbrot, some fractal figures show a massive cross lumping, that is, large clusters emerge over iterations by the fusion of pre-existing clusters. In this case, the proportionality relation $Nr(\Lambda > \lambda) \propto \lambda^{-D}$ is valid only for small values of λ.

	Cluster size measurement	Approximate value of *D*	D_S	Lower bound of the estimate	Upper bound of the estimate
Figure (a)	*a* (number of elements composing the cluster)	1.113	1.59	Third point	Ninth point
	sqrt_a	2.226			
	extMax	1.315			
	rectMax	1.515			
	diamMax	1.564			
Figure (b)	*a* (number of elements composing the cluster)	0.908	1.76	Second point	Fourth point
	sqrt_a	1.816			
	extMax	1.564			
	rectMax	1.564			
	diamMax	1.564			
Figure (c)	*a* (number of elements composing the cluster)		1.76	First point	Third point
	sqrt_a	1.816			
	extMax	1.564			
	rectMax	1.564			
	diamMax	1.584			
Figure (e)	*a* (number of elements composing the cluster)	1.326	1.59	First point	Second point
	sqrt_a	2.653			
	extMax	3.591			
	rectMax	7.912			
	diamMax	5.115			

Table 3.2. *Slope values (exponent D) of Figure 3.16(a)–(c) and (e). Estimates performed by Gilles Vuidel, ThéMA laboratory (CNRS – Franche-Comté University, Besançon, France)*

In Figure 3.16(a)–(c), we estimate the slope of the curve, that is, the value of the scaling exponent D, for the smallest values of λ, that is, for the portion of the curve that is a straight line or near to it (Table 3.2) – unit of measurement a of the size of the clusters, which is the number of elements composing the cluster. These values are much lower than the fractal dimension values (similarity dimension D_S) of these four figures. One possible explanation is that these figures are prefractals, just like the geographical objects we are interested in, and not fractals constructed for a number of infinite iterations as studied by Mandelbrot. However, the value of the scaling exponent D of these same figures, when estimated for some additional iterations, remains very different from the fractal dimension D_S (results not shown here).

We then considered the measures of the size of a cluster other than the number of topologically related elements comprising it (measure a in Table 3.2): the square root of the area of the cluster (*sqrt_a*), the extent (width and height) of the rectangle enclosing as closely as possible the unrotated cluster (*extMax* and *extMin*, respectively), the diameter of the smallest circle circumscribing the cluster (*diamMax*), etc. It turns out that if we adopt a measure of the size of the clusters other than a, the shape of the curves of Figure 3.16 is not modified, or only modified a very little, but the slope values (exponent D) vary significantly (Table 3.2). In all cases except one (*diamMax* of figure (a)), the values of D are quite different from the similarity dimension D_S of each figure.

3.3.2. *Multifractal figures*

Multifractal figures have the particularity of combining at least two reduction factors in their generator, which causes the size of the elements of the figure to vary. Apart from this, they may fall into one or other of the categories mentioned above. Figure 3.17 gives four examples. Figure 3.17(a) is a multifractal dust. Figure 3.17(b) is a multifractal Vicsek cross. Figure 3.17(c) and (d) are hybrid figures, multifractal variants of Figure 3.14(d) and (b), respectively.

Since the generator of each multifractal figure has several reduction factors (r_1, r_2, …), the relation:

$$N = r^{-D_S} \qquad [3.2]$$

used to calculate the similarity dimension D_S of a fractal figure must be generalized in the following way (Gouyet 1992, p. 14):

$$\sum_{i=1}^{N} r_i^D = 1 \qquad [3.3]$$

where i represents each element in the generator, r_i is their reduction factor and N is the total number of elements in the generator.

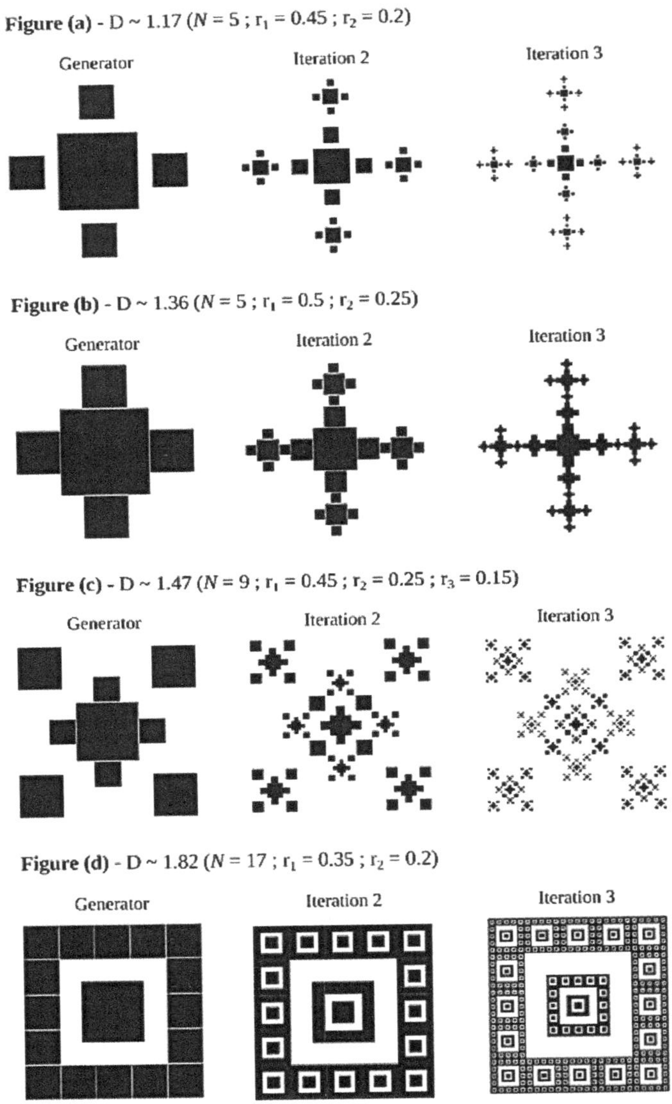

Figure 3.17. *Four multifractal figures. (a) After Yamu and Frankhauser (2015). (b) After Frankhauser (1998). (c) Adapted from Frankhauser (2015). In the GenFrac software, the approximate value of D for multifractal figures is found using a binary search algorithm*

3.3.3. *Composite figures*

Other types of fractal figures can be created by alternately applying two different generators an infinite number of times (Semécurbe 2020); after n_1 iterations with the first generator, n_2 iterations are performed with the second, then again n_1 iterations with the first generator and n_2 iterations with the second, etc. After k repetitions of this process, the number of elements is $N_k = \delta^{-k\,(n1\,D1\,+\,n2\,D2)}$, where D_1 and D_2 are the fractal dimensions associated with the two generators. The fractal dimension D_{tot} of the set of points obtained with this process is an average of the two fractal dimensions weighted by the frequency of appearance of the two generators:

$$D_{tot} = [\, n_1 D_1 + n_2 D_2] \div [\, n_1 + n_2] \qquad [3.4]$$

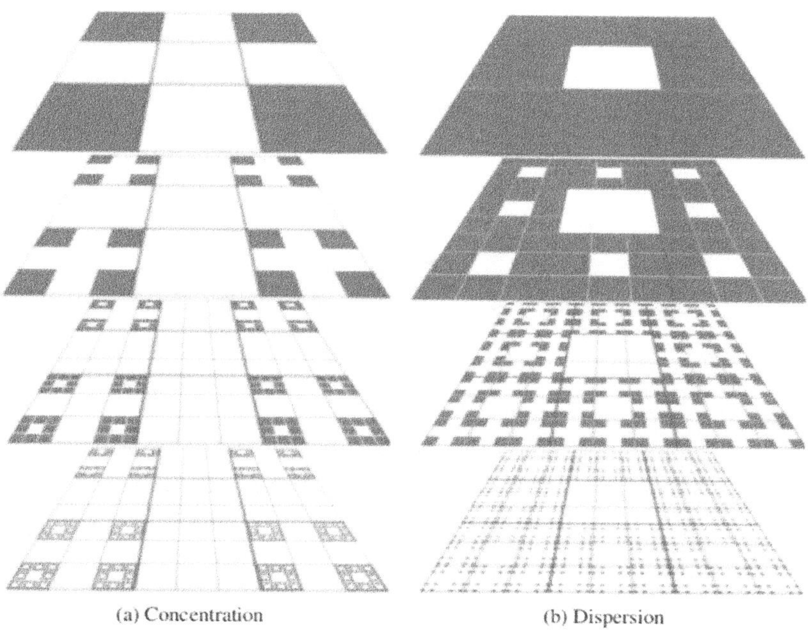

Figure 3.18. *Two examples of composite fractal figures, taken from Sémécurbe (2020, p. 145). The two configurations (a) and (b) are constructed by successively applying two different generators. The first is associated with the fractal dimension log(4)/log(3) and the second with the dimension log(8)/log(3). The change of generator takes place between the second and the third steps in the construction of each figure. For a color version of this figure, see www.iste.co.uk/tannier/fractal.zip*

NOTE.– It is interesting to note that, despite the use of two generators, the set of points obtained converges at infinity towards a theoretical fractal that is perfectly scale-invariant. It is possible to break this scale invariance by greatly modifying the frequency of appearance of the generators across the scales – for example, by using the factorial function. We start by iterating once with the first generator, then 2! = 2 with the second generator, then we start again 3! = 6 with the first generator and 4! = 24 with the second, and so on. In this case, the limit at infinity which allows determination of the fractal dimension is not defined, and D_{tot} oscillates between the values of the two dimensions D_1 and D_2. Through this example, we see that the scale invariance property used to characterize real-world objects, considering by definition a limited range of scales, differs from the scale invariance used to characterize mathematical figures considering an infinite range of scales (Sémécurbe 2020).

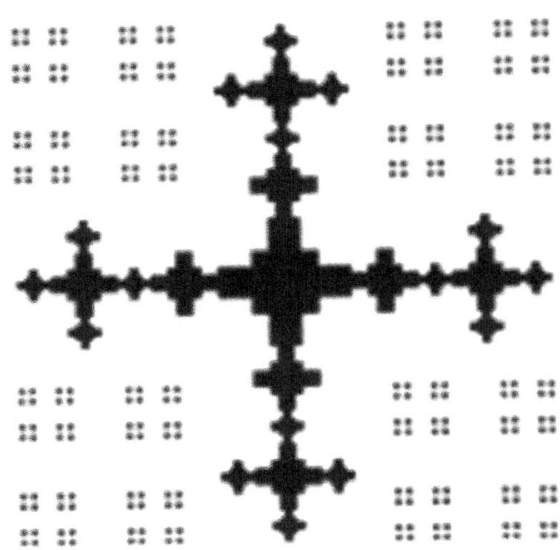

Figure 3.19. *Juxtaposition of a multifractal cross (D ≈ 1.36) and a fractal dust (D ≈ 1.26), taken from Frankhauser (2017a)*

Meanwhile, Frankhauser (2017a) suggested representing two very different fractal figures together, for example, a multifractal cross and a fractal dust, ensuring that no element of one of the figures intersects an element of the other figure. The example shown in Figure 3.19 is an idealized representation of an axial city whose periphery (rural space) is formed by a discontinuous pattern of buildings. Since the two fractal figures are completely independent of each other, the fractal dimension

D_{tot} of the set cannot be found by means of a mathematical formula; it can only be estimated. In all cases, the estimated value of D_{tot} will be greater than the highest fractal dimension of the two juxtaposed figures.

3.4. Fractal geometric models of central places

A central place system is a set of centers (cities, towns, villages) of different sizes that presents a hierarchy from a functional point of view, that is, according to the types of functions (shops and services in particular) that they house. Geometric models of central place systems define the number and spatial distribution of centers; centers are located relative to each other according to their hierarchical level. The first formalized central place systems were proposed by Christaller (1933) and Lösch (1940). The assumptions and functional implications of these two models will not be discussed here (see Woldenberg and Berry 1967; Saey 1973; Beguin 1979; Robic 1982; Pumain 2004, 2006; Banos et al. 2011): we are interested only in the geometric pattern of the central places. In the graphic representations proposed by Christaller (1933) and Lösch (1940), the basic geometric figure is the hexagon, which allows a uniform tiling of the plane in equal surfaces (Figure 3.20).

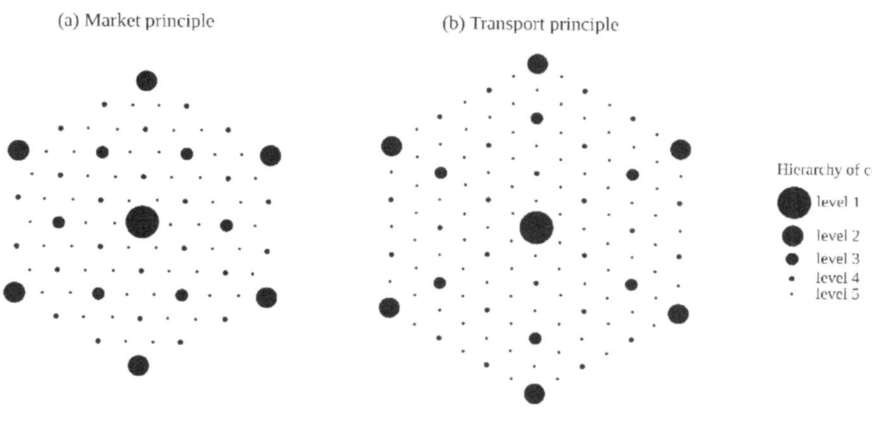

Figure 3.20. *Two central place systems proposed by Christaller (1933)*

However, the spatial distribution of human settlements, particularly that of towns and villages, is not uniform but rather very irregular. This is why some researchers have imagined distributing the central places of a given theoretical region according to fractal logic. Chen (2011) proposed a distribution on a hexagonal Koch snowflake (hexaflake), which exhibits a classic fractal nesting of smaller hexaflakes within larger ones (self-similarity) as well as empty spaces (gaps) of different sizes (see Figure 3.12(d) shown earlier). On this tiling, we have represented three central place systems which each follow different rules for positioning the centers with respect to each other according to their hierarchical level (Figure 3.21). The rules of positioning from which Figure 3.21(a) results are comparable to those of Christaller's model based on the market principle (Figure 3.20(a)). The centers of the highest hierarchical levels are positioned at the center of the hexaflakes, while the smallest centers (of level 5) are positioned on a boundary line between two hexaflakes. The rules of positioning for Figure 3.21(b) are identical, except that the level 2 centers are this time positioned close to the level 1 center. Finally, the rules for positioning centers in Figure 3.21(c) are identical to those in Figure 3.21(b), except that all centers are placed at the center of a hexaflake, including the smallest level 5 centers.

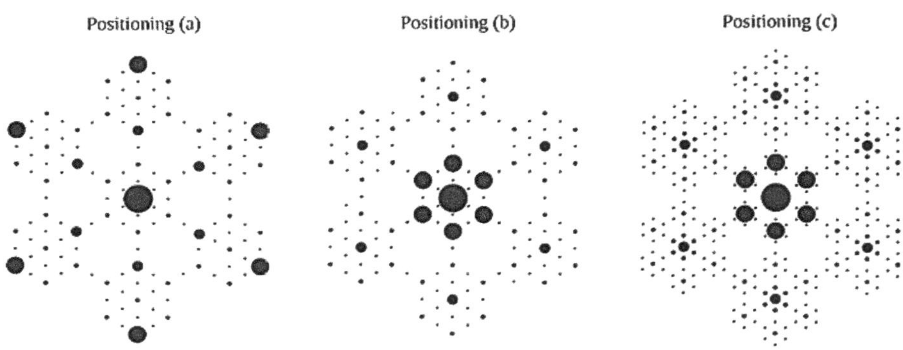

Figure 3.21. *Positioning of three different central place systems on a fractal hexagonal tiling of the plane (hexaflake). Each size of the circle, from the smallest to the largest, corresponds to a hierarchical level, from the lowest to the highest*

Beyond the positioning of centers on a fractal tiling of the plane, other authors have proposed multifractal central place systems based on an iterative principle of construction. In Figure 3.22, we have shown a hexagonal multifractal central place system. This system was first presented and discussed by Benguigui and Czamanski (2004) in a pentagonal version and then by Frankhauser et al. (2007) and Chen (2014) in a hexagonal version. Depending on the version, the size ratio between the elements of the generator is slightly different, as are the distances separating the elements from each other. Note that the spatial distribution of the centers of different hierarchical levels is very different from the systems of Figure 3.21, as is the shape of the gaps. The system of Figure 3.22 has an interesting feature, namely that the centers of lesser importance are only located in the periphery of the centers of medium importance furthest from the main center, thus following a logic of decentralized concentration.

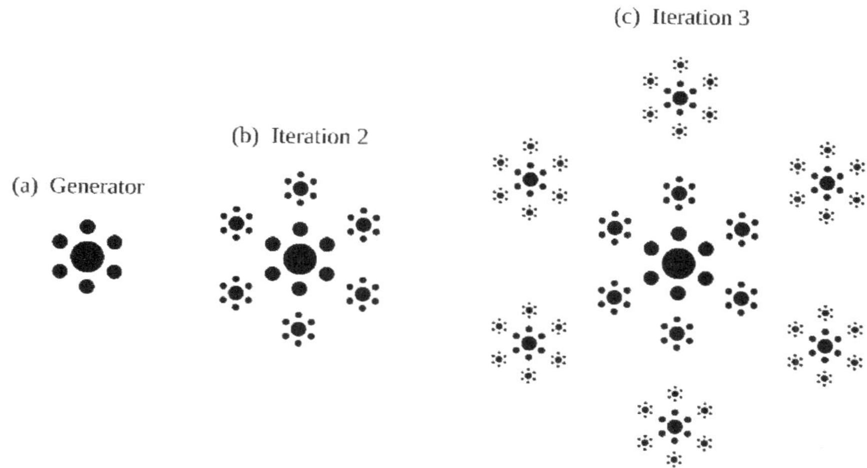

Figure 3.22. *Generation of a hexagonal multifractal central place system according to a logic of deconcentrated concentration*

Frankhauser et al. (2007) give a functional interpretation, oriented towards urban planning, of this multifractal central place model. According to them, due to its radio-concentric organization, the model allows the concentration of centers in areas that can be served well by a public transport network, the location of centers of different hierarchical levels at the nodes of this public transport network and the existence of large contiguous unbuilt spaces (green protrusions from the outside towards the heart of the system). As for the interpretations of the model made by Benguigui and Czamanski (2004) and Chen (2014), they are dynamic and not functional. For Benguigui and Czamanski (2004), the model represents the creation of a central place system following a process of growth by “leapfrogging”: new centers of the lower hierarchical level gradually appear at a certain distance from existing centers of the higher hierarchical level. For Chen (2014), the modeled growth process combines an “incomplete” and “intermittent” filling of space – one that preserves free spaces where no new center can appear – with the fact that the centers grow more or less over time according to their hierarchical level and their location relative to the other centers. However, the growth of centers over time, which according to Chen (2014) is different depending on their hierarchical level and their relative location with respect to other centers, is not integrated into the process of creating the system. As shown in Figure 3.22, the centers do not grow during the process of creating the system; only new centers appear. In addition, the process exclusively advances by adding new centers at the periphery of the figure, but no new sub-centers appear in the immediate vicinity of the existing centers.

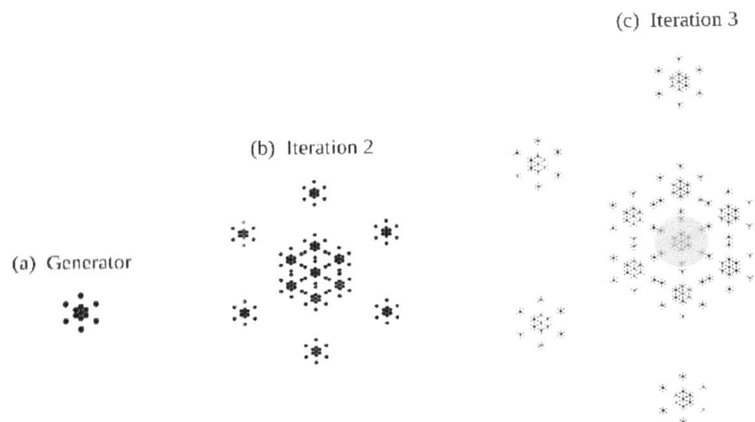

Figure 3.23. *Generation of a hexagonal monofractal central place system according to a logic of deconcentrated concentration. For reasons of space, (c) has been reduced and is only 28% of the size that it would have been if kept at the same scale of representation as (a) and (b). The relative spatial extent of (b) is represented by a gray disk in (c)*

It is possible to imagine a monofractal central place system, that is, one where all the elements of the generator have the same size (Figure 3.23), which is similar to the multifractal system of Figure 3.22. At the third iteration of the system, we observe the existence of a central cluster with a certain morphological continuity and which is clearly distinguished, in terms of distances, from the secondary centers located on the periphery. However, the hierarchy of centers and sub-centers of different hierarchical levels does not appear clearly in this monofractal system, whereas it is present in the multifractal system previously presented.

Different authors have tried to demonstrate the fractality of Lösch and Christaller's central place systems. Interested in the Lösch model, Arlinghaus (1985) developed a method of fractal generation of the perimeters of hexagons representing the nested market areas of the centers of different hierarchical levels. Arlinghaus and Arlinghaus (1989) then demonstrated mathematically how to choose the appropriate fractal generator to simulate the perimeters of market areas whose size and orientation is defined by a positive integer called the Löschian market-area number (or more simply the Loschian number). Later, Chen and Zhou (2006) showed that the number of central places at a given hierarchical level is related to the distance between the adjacent central places of this same level according to a power law whose exponent is the fractal dimension of the system considered. This is defined as the ratio of the logarithms of the ratio of the number of centers to the ratio of the distances considering two successive hierarchical levels. For Woldenberg and Berry (1967), self-similarity is directly invoked in the central place theory, since modeling self-similarity at different scales involves finding an appropriate generating function which can be applied to each scale in a recursive manner (Batty and Longley 1994). However, it is impossible to generate a central place system comparable to those of Christaller or Lösch from a method of generating fractal figures, for example, a system of iterated functions (Barnsley 1988). Indeed, the patterns of Christaller or Lösch's central places are uniform, therefore intrinsically non-fractal. Starting from a classic fractal figure, the Sierpinski gasket, Frankhauser (1994) was able to generate a geometric model of central places comparable to that of Christaller based on the market principle. However, its model does not correspond to the fractal figure itself but to its complementary set, that of gaps (or free spaces), which is not fractal.

3.5. Concerning lacunarity

An important point to note is that the fractal dimension of the figures presented in this chapter does not, on its own, make it possible to differentiate them. Indeed, many visually very different figures have the same fractal dimension. Additional properties must therefore be considered to characterize them: their topological dimension, their degree of branching, their distribution of clusters by size classes (or their rank-size distribution of clusters) and the fractal dimension of their total perimeter, which measures both its fragmentation and its irregularity. Another important property of fractal figures that we have not mentioned so far is their lacunarity. In a very lacunar fractal, the "lacunae" (gaps, empty spaces) are very wide; it is the opposite in a fractal with little lacunarity. According to Mandelbrot (1982, p. 140), a fractal remains incompletely specified for many purposes without considering its lacunarity, which is a particularly important property.

To address this notion of lacunarity, let us take the example of a Cantor set constructed from a line segment, therefore in 1 dimension (1D) (Figure 3.5(a)). All the elements of the generator are of the same length; all the gaps are also of the same length:

– let L be the length of the initiator;

– let N be the number of elements in the generator;

– let Λ be the number of gaps in the generator;

– let r be the reduction factor of the generator elements. The size (length) of an element in the generator is therefore $r \times L$;

– let s be the reduction factor of the gaps in the generator. The size (length) of a gap in the generator is therefore $s \times L$.

We have:

$$NrL + \Lambda sL = L \quad [3.5]$$

$$Nr + \Lambda s = L \quad [3.6]$$

The total length of gaps at each iteration can be expressed as follows:

– at iteration 1:

$$\Lambda sL \quad [3.7]$$

– at iteration 2:

$$\Lambda sL + NrL \times s \times \Lambda \quad [3.8]$$

– at iteration 3:

$$\Lambda sL + NrL \times s \times \Lambda + N^2r^2L \times s \times \Lambda \quad [3.9]$$

– at iteration n:

$$\sum_{i=1}^{n} N^{i-1} r^{i-1} \times Ls\Lambda = Ls\Lambda \sum_{i=0}^{n-1} N^i r^i \quad [3.10]$$

$$= Ls\Lambda \frac{1-(Nr)^n}{1-Nr} = Ls\Lambda \frac{1-(Nr)^n}{\Lambda s} = L(1-(Nr)^n))$$

At infinity, the total length of the gaps is equal to the length L of the initiator (Frankhauser 1994, pp. 53–54). In fact, we saw at the beginning of this chapter that the Cantor set is, at infinity, made up of isolated points having no length by definition. These points do exist (the Cantor set is not empty), but they do not contribute to L. Another particularity of the set of gaps is that it is not fractal. While the occupied elements of the figure are subdivided from one step of the construction to another, via the application of reduced copies of the generator, the gaps are not; those which pre-exist at a given iteration remain as they are at the next iteration.

As the iterations progress, the gaps become more numerous. What is more, those added to the pre-existing gaps are always smaller and more numerous. The double-logarithmic representation of the number of gaps (on the y-axis) at each iteration (on the x-axis) is a straight line. In the case of regular fractal figures whose generator is simple, such as the Cantor set considered here, the exponent of the hyperbolic distribution of the gaps is equal to D_S. This is not the case for more complicated fractals: the distribution remains hyperbolic, but the exponent is no longer equal to D_S (Frankhauser 1994, p. 64).

Now consider 2D figures. We have seen previously that Sierpinski carpets have internal gaps of varying sizes. Their lacunarity has been defined mathematically by Gefen et al. (1984) and Lin and Yang (1986). This lacunarity of Sierpinski carpets is described very clearly in Frankhauser (1994, pp. 60–64). In general, and by simplifying the writings of Mandelbrot (1982, p. 316) and Voss (1988, p. 67 et seq.), we can define lacunarity as the variance of the local mass of a fractal for different scales of analysis considered separately.

Based on these pioneering works, different methods of measurement, implemented in the form of different algorithms, have been proposed (see in particular the algorithm of Allain and Cloitre (1991) and the application made of it by Plotnick et al. (1993)). However, it is a priori only possible to compare the lacunarity of figures having the same fractal dimension[5], which represents an important limit to the analyses that can be performed. In addition, in the case of fractal dusts and hybrid figures, lacunarity is much more difficult to measure than for Sierpinski carpets as the gaps consist not only of internal "holes" in the figure, but also of free spaces that have no border with its outside.

5. "(...) *the notion of lacunarity has nothing to do with topology, and* [that] *it concerns comparisons at given* D*; its possible use for inter-*D *comparisons remains unexplored*" (Mandelbrot 1982, p. 316).

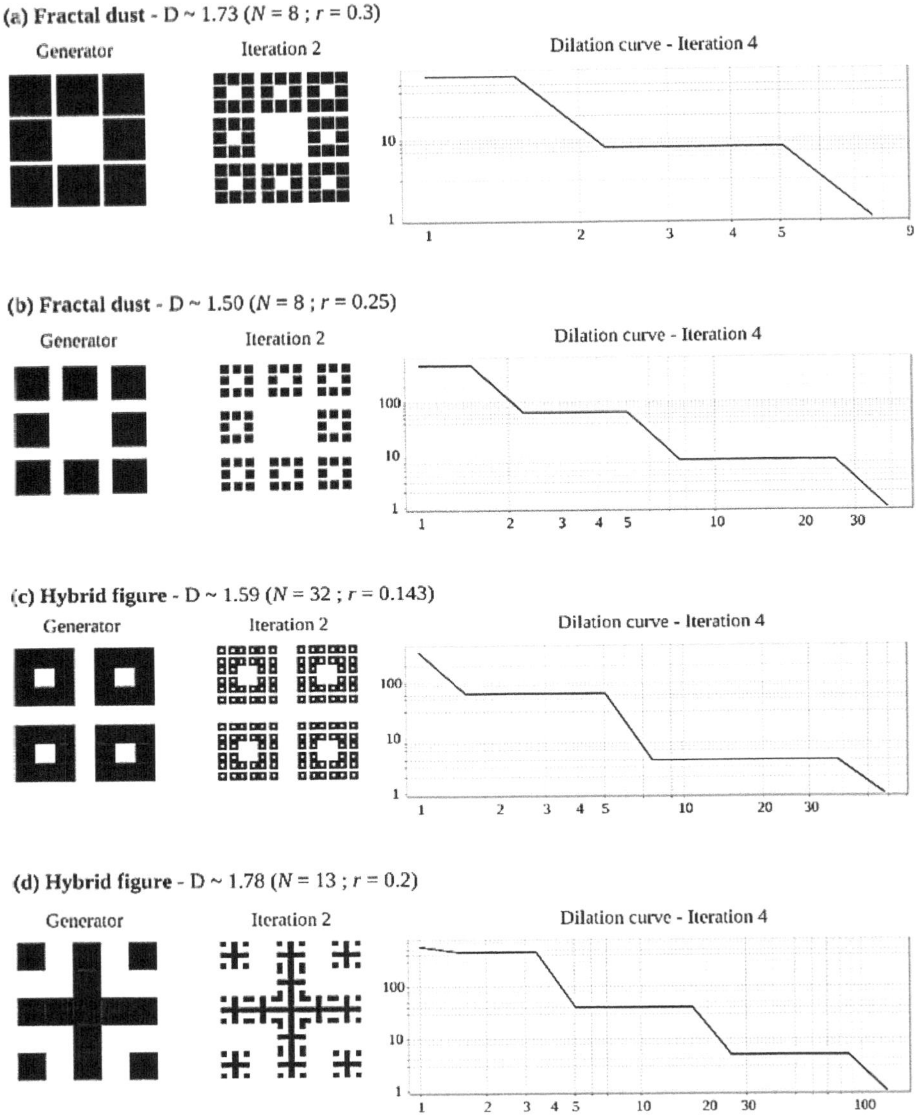

Figure 3.24. *Dilation curves of four fractal figures. x-axes: steps of dilation. y-axes: number of clusters identified at each step. Note: the scales of the axes differ greatly from one graph to another*

An alternative to the measurement of lacunarity is to consider not the gaps themselves, but the distances separating the clusters of elements from each other. This alternative is particularly interesting for the study of fractal dusts or hybrid figures. A step-by-step dilation process is used for this purpose (Tannier et al. 2011). Each element is surrounded by a strip (buffer zone) whose thickness increases with each step of dilation. As dilation progresses, elements merge that were initially separated, due to widening produced by the buffer zone, and form clusters. These clusters grow gradually and their number decreases. The dilation operation stops when all elements are contained in a single cluster. At each step of dilation, the number of clusters is counted. This makes it possible to graphically construct a curve, where the x-axis represents the steps of dilation and the y-axis gives the number of clusters. Four curves resulting from the dilation of four fractal figures are shown in Figure 3.24. The magnitude of the values on the x-axis varies according to the spacing of the elements within each figure. When all the elements are located, step by step, close to each other, few dilations are necessary for all to be brought together into a single cluster. This is the case in Figure 3.24(a). On the other hand, in the case of Figure 3.24(b) and (c) and especially (d), the elements gradually become further apart from each other. Thus, many dilations are necessary for them to be all gathered into a single cluster. The amplitude of the values on the y-axis reveals, for its part, the greater or lesser diversity of the distances separating the clusters from each other. Thus, the dilation curves inform us about the amplitude and diversity of the distances separating the elements from each other.

It is interesting to compare the fractal dusts of Figures 3.24(a) and (b) and the Sierpinski carpet of Figure 3.10(a). The generators of these three figures have the same number of elements N arranged in the same way, but they differ in their reduction factor r (Thomas et al. 2008). In the case of the carpet, the elements of the generator are connected and form a single and unique contact-cluster at all iterations. In the case of the dusts, free spaces separate the elements from each other. These spaces are wider and their widths vary more in Figure 3.24(b) than in Figure 3.24(a), which is reflected in the two corresponding dilation curves.

3.6. Conclusion

We have chosen to present only 2D figures in this chapter since, to our knowledge, no research in human geography has used 3D fractal figures. It should be noted, however, that many 3D fractal figures have been imagined, the best known among them surely being the Menger sponge, originally described by the author of the same name in 1926. It is also known as the Menger cube, or the Sierpinski cube or sponge. It is a three-dimensional generalization of the one-dimensional Cantor set and the two-dimensional Sierpinski carpet.

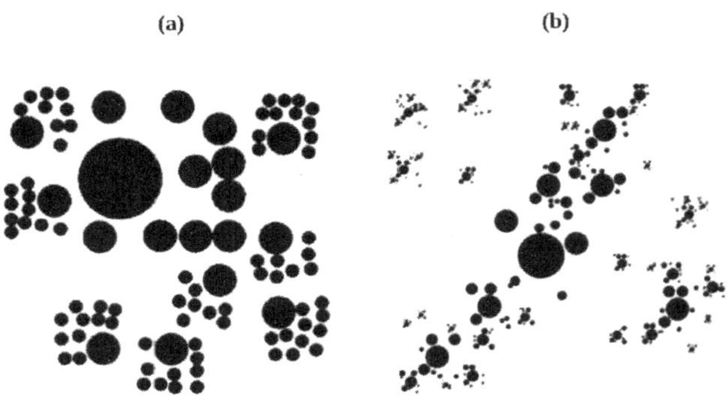

Figure 3.25. *Two examples of random multifractal dusts. (a) Taken from Frankhauser (2017a). (b) Taken from Frankhauser (2017b)*

All the figures we have presented are perfectly regular, apart from the fractal irregularity inherent in them. In this respect, their shape is quite dissimilar to that of real geographical objects. On the other hand, by their strict self-similarity, we know their similarity dimension D_S with accuracy as well as some of their morphological properties (connectivity of elements, number of clusters, length of the border). These figures may therefore serve as references with which to compare the real forms of cities and systems of settlement. Moreover, it is easy to create more realistic fractal figures by randomly varying the position of the elements created at each iteration while respecting the number N and the reduction factor r fixed in the generator (Figure 3.25, see also Chapter 2)[6].

3.7. References

Allain, C. and Cloitre, M. (1991). Characterizing the lacunarity of random and deterministic fractal sets. *Physical Review A*, 44(6), 3552–3558.

Arlinghaus, S.L. (1985). Fractals take a central place. *Geografiska Annaler, Series B, Human Geography*, 67(2), 83–88.

6. Mandelbrot (1982, p. 88) further points out that Fournier d'Albe's model, representing the spatial distribution of stars and galaxies in the universe and discussed in section 3.2.1, is so regular as to be grotesque, and mentions another possible way to introduce more irregularity: "As a corrective, Charlier 1908, 1922 suggests that one allow N and r to vary from one hierarchical level to another, taking on the values N_m and r_m.".

Arlinghaus, S.L. and Arlinghaus, W.C. (1989). Fractal theory of central place geometry: A Diophantine analysis of fractal generators for arbitrary Loschian numbers. *Geographical Analysis*, 21, 103–121.

Banos, A., Moreno, D., Pivano, C., Taillandier, P. (2011). Christaller, toujours vivant ! *Cybergeo: European Journal of Geography* [Online]. Available at: http://journals.openedition.org/cybergeo/24877.

Barnsley, M. (1988). *Fractals Everywhere*. Academic Press, New York.

Batty, M. and Longley, P.A. (1994). *Fractal Cities: A Geometry of Form and Function*. Academic Press, San Diego, CA and London.

Beguin, H. (1979). Urban hierarchy and the rank-size distribution. *Geographical Analysis*, 11(2), 149–164.

Benguigui, L. and Czamanski, D. (2004). Simulation analysis of the fractality of cities. *Geographical Analysis*, 36(1), 69–84.

Bunde, A. and Havlin, S. (1994). A brief introduction to fractal geometry. In *Fractals in Science*, Bunde, A. and Havlin, S. (eds). Springer-Verlag, Berlin Heidelberg, 1–25.

Chen, Y. (2011). Fractal systems of central places based on intermittency of space-filling. *Chaos Solitons Fractals*, 44(8), 619–632.

Chen, Y. (2014). Multifractals of central place systems: Models, dimension spectrums, and empirical analysis. *Physica A*, 402, 266–282.

Chen, Y. and Zhou, Y.X. (2006). Reinterpreting central place networks using ideas from fractals and self-organized criticality. *Environment and Planning B: Planning and Design*, 33(3), 345–364.

Christaller, W. (1933). *Die zentralen Orte in Süddeutschland. Eine ökonomisch-geographische Untersuchung über die Gesetzmäßigkeit der Verbreitung und Entwicklung der Siedlungen mit städtischer Funktion*. Gustav Fischer, Jena.

Falconer, K.J. (2013). *Fractals. A Very Short Introduction*. Oxford University Press.

Feder, J. (1988). *Fractals*. Springer Science+Business Media, New York.

Fleron, J.F. (1994). A note on the history of the Cantor set and Cantor function. *Mathematics Magazine*, 67(2), 136–140.

Fournier d'Albe, E.E. (1907). *Two New Worlds: 1. The Infra-World II. The Supra-World*. Longmans, Green, and Co., London, New York, Bombay, Calcutta.

Frankhauser, P. (1988). Fractal aspects of urban systems. *Sonderforschungsbereich 230 "Natürliche Konstruktionen"*, 1, 67–76.

Frankhauser, P. (1994). *La fractalité des structures urbaines*. Anthropos, Paris.

Frankhauser P. (1998). The fractal approach: A new tool for the spatial analysis of urban agglomerations. *Population*, 4, 205–240.

Frankhauser, P. (2015). From fractal urban pattern analysis to fractal urban planning concepts. In *Computational Approaches for Urban Environments, Geotechnologies and the Environment*, Helbich, M., Jokar Arsanjani, J., Leitner, M. (eds). Springer International Publishing, Switzerland, 13–48.

Frankhauser, P. (2017a). Between self-organization and planning: Cities and the fractal order principle. In *The Virtual and the Real in Planning and Urban Design*, Yamu, C., Poplin, A., Devisch, O., de Roo, G. (eds). Taylor & Francis (Routledge), 112–135.

Frankhauser, P. (2017b). La modélisation fractale des tissus urbains De l'analyse morphologique au concept d'aménagement. In *Villes, réseaux et transport. Le défi fractal*, Dupuy, G. (ed.). Economica, Paris, 73–111.

Frankhauser, P. and Sadler, R. (1991). Fractal analysis of agglomerations. In *Natural Structures: Principles, Strategies, and Models in Architecture and Nature*, Hilliges, M. (ed.). University of Stuttgart, 57–65.

Frankhauser, P., Tannier, C., Vuidel, G., Houot, H. (2007). Approche fractale de l'urbanisation – Méthodes d'analyse d'accessibilité et simulations multi-échelles. *11th World Conference on Transportation Research*, Berkeley, 24–28 June 2007.

Gefen, Y., Aharony, A., Mandelbrot, B.B. (1984). Phase transitions on fractals: III. Infinitely ramified lattices. *Journal Physics A: Mathematical and General*, 17, 177–1289.

Gouyet, J.-F. (1992). *Physique et structures fractales*. Masson, Paris-Milan-Barcelona.

Lauwerier, H. (1991). *Fractals: Endlessly Repeated Geometric Figures*. Princeton University Press, New Jersey.

Lesmoir-Gordon, N., Rood, W., Edney, R. (1996). *Introducing Fractals. A Graphic Guide*. Icon Books, London.

Lin, B. and Yang, Z.R. (1986). A suggested lacunarity expression for Sierpinski carpets. *Journal Physics A: Mathematical and General*, 19, 49–52.

Lösch, A. (1940). *Die räumliche Ordnung der Wirtschaft. Eine Untersuchung über Standort.* Wirtschaftsgebiete und internationalem Handel, Fischer, Jena.

Mandelbrot B. (1975). *Les objets fractals. Forme, hasard et dimension*. Flammarion, Paris.

Mandelbrot, B. (1982). *The Fractal Geometry of Nature*. W.H. Freeman and Company, New York.

Mandelbrot, B. (1988). Foreword: People and events behind the "Science of fractal images". In *The Science of Fractal Images*, Peitgen, H.O. and Saupe, D. (eds). Springer-Verlag, New York, Berlin, Heidelberg, 1–19.

Plotnick, R.E., Gardner, R.H., O'Neill, R.V. (1993). Lacunarity indices as measures of landscape texture. *Landscape Ecology*, 8(3), 201–211.

Pumain, D. (2004). L'invention permanente : la hiérarchie urbaine. In *Mélanges en l'honneur de Giovanni Busino*, Bridel, P. (ed.). Labor et Fides, Geneva, 66–86.

Pumain, D. (2006). Systèmes de villes et niveaux d'organisation. In *Morphogenèse. L'origine des formes*, Bourgine, P. and Lesne, A. (eds). Belin, Paris, 239–263.

Robic, M.-C. (1982). Cent ans avant Christaller… Une théorie des lieux centraux. *L'Espace géographique*, 11(1), 5–12.

Saey, P. (1973). Three fallacies in the literature on central place theory. *Tijdschrift voor Economische en Sociale Geografie*, 64 (3), 181–194.

Sémécurbe, F. (2020). Analyse de la distribution spatiale des implantations humaines – Apports et limites d'indicateurs multi-échelles et trans-échelles. Geography PhD Thesis, Université de Franche-Comté, Besançon, France.

Stewart, I. (1989). *Does God Play Dice? The New Mathematics of Chaos*. Basil Blackwell, Oxford, and New York.

Tannier, C. and Thomas, I. (2013). Defining and characterizing urban boundaries: A fractal analysis of theoretical cities and Belgian cities. *Computers, Environment and Urban Systems*, 41, 234–248.

Tannier, C., Thomas, I., Vuidel, G., Frankhauser, P. (2011), A fractal approach to identifying urban boundaries, *Geographical Analysis*, 43(2), 211–227.

Thomas, I., Frankhauser, P., Biernacki, C. (2008). The morphology of built-up landscapes in Wallonia (Belgium): A classification using fractal indices. *Landscape and Urban Planning*, 84, 99–115.

Vicsek, T. (1989). *Fractal Growth Phenomena*. World Scientific Publishers, Singapore.

Voss, R.F. (1988). Fractals in nature: From characterization to simulation. In *The Science of Fractal Images*, Peitgen, H.O. and Saupe, D. (eds). Springer-Verlag, New York, Berlin, Heidelberg, 21–70.

Woldenberg, M.J. and Berry, B.J.L. (1967). Rivers and central places: Analogous systems? *Journal of Regional Science*, 7(2), 129–139.

Yamu, C. and Frankhauser, P. (2015). Spatial accessibility to amenities, natural areas and urban green spaces: Using a multiscale, multifractal simulation model for managing urban sprawl. *Environment and Planning B: Planning & Design*, 42(6), 1054–1078.

PART 2

Fractal and Multifractal Analyses of the Spatial Distribution of Human Settlements

4

Fractal Analysis Methods for Characterizing the Spatial Distribution of Human Settlements

Cécile TANNIER[1], Gaëtan MONTERO[2], François SÉMÉCURBE[1] and Isabelle THOMAS[2]

[1]Laboratoire ThéMA, CNRS-Université de Franche-Comté, Besançon, France

[2]Centre for Operational Research and Econometrics (CORE), Université catholique de Louvain-la Neuve, Belgium

Fractals reveal a new regime of nature susceptible to mathematical modelling. They open our eyes to patterns that might otherwise be considered formless. They raise new questions and provide new kinds of answers. "Fractals", says the science writer Jeanne McDermott, "capture the texture of reality" (Stewart 1989).

Fractal analysis has been widely used in geography for almost 40 years. Pioneering works have highlighted the fact that local and global spatial configurations are interrelated and present certain scale-invariant characteristics (Goodchild 1980; Arlinghaus 1985, 1987; Batty and Longley 1986, 1987; Goodchild and Mark 1987; Frankhauser 1988). Specifically in urban geography, researchers have explored the hypothesis that the processes driving urban growth lead to fractal spatial configurations (Fotheringham et al. 1989; Frankhauser 1990, 1994; Batty and Longley 1994; Batty and Xie 1996, 1999; Makse et al. 1998; Benguigui et al. 2000;

Fractal Geometry in Human Geography and Planning,
coordinated by Cécile TANNIER.

Shen 2002). Subsequently, a large number of publications have shown an interest in using fractal analyses to characterize urban fabrics (e.g. Longley and Mesev 1997, 2000; De Keersmaecker et al. 2003; Tannier and Pumain 2005; Thomas et al. 2007, 2008; Feng and Chen 2010; Chen 2011).

Originally, the spatial resolution of the digital cartographic data accessible to researchers (images from remote sensing or paper maps, digitized then rasterized) was quite coarse, the pixels of the images most often having a width of 100–200 m from side to side. From around the 2000s, researchers had access to digital data with finer spatial resolution. This, combined with the increased computational capabilities of computers, has made it possible to estimate fractal dimensions much more accurately than before (Longley and Mesev 2000; Frankhauser 2003).

In this chapter, the shapes (spatial configurations) of human settlements in 2D, which may be assimilated geometrically to sets of points, are studied by means of fractal analysis. We mostly consider the footprint of buildings, their centroids or the segments and nodes of road or public transport networks.

4.1. Estimating the fractal dimension of a collection of human settlements (point, linear or surface objects)[1]

In mathematics, dimension measures the size of an object. Topological dimension is the number of coordinates necessary to unambiguously locate a point on an object and thus measure its size, that is, its length, its surface or its volume, for objects that are, respectively, one-dimensional (lines), two-dimensional (plane objects: triangles, circles, squares, etc.) or three-dimensional (volume objects). Topological dimensions are, by construction, positive integers. In the case of smooth and regular objects (circles, ellipses, rectangles, etc.), the topological dimension determines the exact number of coordinates needed to measure their size. In contrast, the information provided by the topological dimension is not sufficient to measure the size of objects of very irregular shapes, for example, a set of points corresponding to a spatial distribution of human settlements. The fractal dimension allows this in a more satisfactory way, as it is a measure of the space occupied by an object around the points that constitute it (Falconer 1990). It is, by definition, not an integer. It measures the rate at which an ever-increasing number of details of an object become apparent when the object is examined more and more finely (Falconer 2013), as if using magnifying glasses of gradually increasing magnification power. The fractal dimension is therefore a measure of the amount of

1. The concepts used to understand the methods and results of analysis presented in this chapter are defined in Chapter 2.

space occupied as the spatial resolution adopted becomes finer. It is an indicator of the complexity of the shape of an object. Since it measures how the space occupied evolves across scales, the way that it characterizes an object is not located at a given scale but emerges from comparing the properties of the object between several scales in succession.

4.1.1. *General principle governing the estimation of fractal dimensions*

The mathematicians' idea for measuring the dimension of irregular objects is to consider that only the local quantification of the occupied space is relevant. The principle used is that of step-by-step overlaying of the set of points with closed balls (or boxes) of ε decreasing size, and counting the number $N(\varepsilon)$ of boxes containing points at each stage of the overlaying (Minkowski 1903; Bouligand 1929). On this basis, the Minkowski–Bouligand dimension, also called the box-counting dimension, is defined as:

$$D_B = \lim_{\varepsilon \to 0} \frac{\log(N(\varepsilon))}{\log\left(\frac{1}{\varepsilon}\right)} \qquad [4.1]$$

The Minkowski–Bouligand dimension involves a limit to infinity. It corresponds to the asymptotic behavior of N as ε tends towards 0. It is therefore not directly calculable for real-world objects, but it may be estimated considering sufficiently wide ranges of values of ε, for which the curve representing N as a function of ε corresponds to a power law:

$$N(\varepsilon) \sim a\,\varepsilon^{D} + c \qquad [4.2]$$

Mandelbrot (1975) notes that:

> [...] the physicist is right to treat the passage to the mathematical limit with caution. The fractal dimension implies such a passage, so is suspect. [...] It is perhaps because of this suspicion that the physical role of the fractal dimension was not discovered until my own work. But we see that [...] the application of the infinitesimal to the finite should cause no fear if it proceeds with caution[2].

In practice, to estimate a fractal dimension, we begin by counting the constituent points of an object (centroids of buildings, black pixels in a binary image, intersections of roads) that are present in neighborhoods of increasing or decreasing size. Two variables change as a function of the count step: the number of

2. Fourth edition, 1995, p. 32.

enumerated elements N and the size of the neighborhood ε. On this basis, a curve is drawn which relates N and ε in a graph using logarithmic coordinates (Figure 4.1). If it is possible to fit this curve to a line, the slope of this line is the fractal dimension D:

$$\log N(\varepsilon) = D \log \varepsilon + \log a \quad [4.3]$$

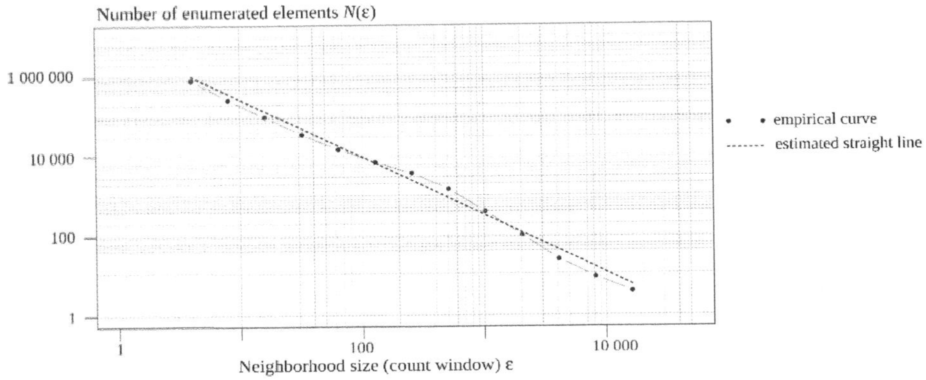

Figure 4.1. *Bi-logarithmic graph representing the number of enumerated elements N(ε) as a function of the size of the neighborhood ε*

According to Falconer (2013, p. 50), the coefficient a of equation [4.2] can be interpreted as a D-dimensional measure of the object. In theory, it may be neglected if sufficiently small ε neighborhoods are considered. In practice, the estimate of D corresponds to the slope of the line (log ε; log N(ε)) for a certain range of values of ε. Since the slope of such a line is independent of the value of the coefficient a, it may be neglected.

As we have just mentioned, the fractal dimension is calculated only over a limited range of values for ε, corresponding to visible or significant values of ε with respect to the properties of the object we wish to study. In doing so, the equalities or inequalities between dimensions, mathematically proven by considering the asymptotic behavior of N when ε tends towards 0, may no longer be valid. The smallest neighborhood size (respectively the largest) below (respectively beyond) which the estimation of a fractal dimension loses all its meaning is that after which the measured object is no longer statistically self-similar. As Pumain (2010) notes,

> [...] processes do not unfold ad infinitum on the surface of the earth, hence the appearance of thresholds between forms of structuring (or

> filling) of geographical space, some of which present systematic variations according to the scale, according to a given fractal dimension, and others which are independent of the scale. Any empirical measurement of fractal dimension thus highlights a domain of validity of the identified structure, generally between a lower threshold and an upper threshold below or beyond which the fractal dimension changes or can no longer be established.

Mandelbrot (1975) discusses these lower and upper thresholds in a section of his book entitled "Physical meaning of fractal dimensions when we refuse to pass to the limit. Internal and external ruptures"[3]. He adds that "[...] dimensions should not be multiplied beyond necessity, and [...] some fractal areas may be too narrow to be worth distinguishing"[4]. Gonzato et al. (1998) specify that the range of distances (or sizes) separating the lower threshold from the upper threshold must be at least two or three orders of magnitude wide. Self-similarity identified within a smaller interval, half an order of magnitude for example, is not indicative of important physical phenomena and should therefore be disregarded.

The level of resolution of the analyzed data also interferes with the determination of the domain of validity of the objects' fractality: "[...] the physical dimension inevitably has a pragmatic and therefore subjective basis. It is a matter of resolution" (Mandelbrot 1975)[5].

> Indeed, since the measure of fractal dimension consists in counting the sites occupied by a given type of object in boxes of increasingly smaller size, this counting becomes impossible when the scale of the supporting cartographic document does not allow discernment of full and empty boxes or the network segments at a very large scale (Pumain 2010).

The statistical self-similarity of geographical forms is therefore spatially circumscribed and concerns restricted ranges of scales.

By definition, the fractal dimension of 2D sets of points is between 0 and 2. The fractal dimension of a pattern of highly dispersed and non-uniformly spatially distributed points is greater than 0 but less than 1. A smooth line (linear object) has a fractal dimension equal to its topological dimension which is equal to 1. A very tortuous line has a fractal dimension greater than 1. Surface objects are also

3. Fourth edition, 1995, pp. 31–32.
4. Ibid. p. 14.
5. Ibid. p. 13.

characterized by a fractal dimension between 1 and 2. The more a surface object features many gaps of different sizes, the lower the fractal dimension. Finally, a fractal dimension equal to 2 characterizes a set of points that are either uniformly concentrated in space or uniformly dispersed.

> Set of points in 2D: $0 \leq D \leq 2$.
>
> Pattern of highly dispersed and non-uniformly spatially distributed points: $0 < D \leq 1$.
>
> Smooth line: $D = 1$.
>
> Very tortuous line: $1 < D \leq 2$.
>
> Surface object with more or fewer gaps: $1 \leq D < 2$. The more gaps of different sizes a surface object has, the lower its fractal dimension.
>
> Set of uniformly concentrated or uniformly dispersed points: $D = 2$.

There are different methods of fractal analysis, all of which are a measure of the space occupied by an object around its constituent points, considering neighborhoods of increasing or decreasing sizes. Depending on the definition of the neighborhood ε in question and according to the method chosen for obtaining $N(\varepsilon)$, different fractal dimensions may be estimated. In human geography, the most commonly used methods are box-counting analysis and correlation analysis, presented below, in addition to dilation analysis, radial analysis and multifractal analysis. Other methods in common use outside the field of human geography are lacunarity analysis, on the one hand, as well as rescaled range analysis and detrended fluctuation analysis dedicated to the study of temporal data series (self-affine fractals), on the other hand. For a description of the traditional methods of calculating fractal dimensions, the reader may consult Theiler (1990). For the practical implementation of these analyses, the reader may refer to the manual for the *Fractalyse* fractal analysis software[6]. Not all fractal dimensions estimated using these different methods are sensitive to the same types of spatial irregularities and thus characterize different aspects of a spatial configuration. The choice of one dimension or another must depend on both the properties of the object studied and those that the researcher wishes to analyze.

The fractal dimension alone does not make it possible to unequivocally distinguish between different objects. As Batty and Longley (1994, Chapter 3) point

6. See: https://sourcesup.renater.fr/www/fractalyse/.

out, the fractal dimension says little about the orientation and overall shape of a figure. Indeed, many visually very different figures have the same fractal dimension. For example, totally unconnected fractal dusts or totally connected Sierpinski carpets may have the same fractal dimension. Additional properties must therefore be considered to characterize each figure: topological dimension, degree of branching, distribution of clusters by size classes (or rank-size distribution of clusters), fractal dimension of the perimeter, lacunarity or distribution of distances separating the clusters of elements from each other, etc.

Another important point to note is that even though it is possible to estimate the fractal dimension of an object, in other words, if the empirical curve relating log ε and log $N(\varepsilon)$ sufficiently approaches a straight line for a given range ε of values, this does not necessarily prove that the object considered is fractal. Nonetheless, measures of fractal dimensions are interesting for the study of the spatial distribution of human settlements because they make it possible to characterize very heterogeneous spatial distributions, with the differences or similarities identified being visually coherent.

4.1.2. *Box-counting dimension*

The simplest fractal dimension to estimate is the box-counting dimension D_B. The method involves arranging in space the minimum number of boxes of a given size that can cover all the elements of the set of points studied. Most often, the method is simplified by covering the space with a regular grid (square or rectangular) and counting the number of squares of the grid (or boxes) that include at least one element. The size of the grid squares is increased (or decreased) step by step by a certain factor and the number of occupied squares decreases (or increases). Ideally, a large number of boxes should contain one or more elements of the object of study at each step of the count. In other words, the measured object must cover a large part of the study area to prevent the results from being biased by the existence of empty areas artificially introduced into the analysis (Gonzato et al. 1998).

By considering each size of the grid square successively, we may position the grid in several different ways and select the grid position for which the number of occupied squares is minimal. On this basis, a double-logarithmic coordinate graph is constructed comprising the number of occupied squares on the y-axis and the size of the squares on the x-axis. This empirical curve is then adjusted by means of a straight line whose slope is the fractal dimension D_B. It is verified that the estimated curve includes a sufficient number of points, each corresponding to a square size and a fairly wide interval of square sizes (according to the authors, at least three orders of magnitude or a ratio between the largest and smallest box size in excess of 10). Finally, the statistical quality of the fit is checked. If this is not satisfactory, we try to fit the empirical curve to a straight line for a narrower range of square sizes.

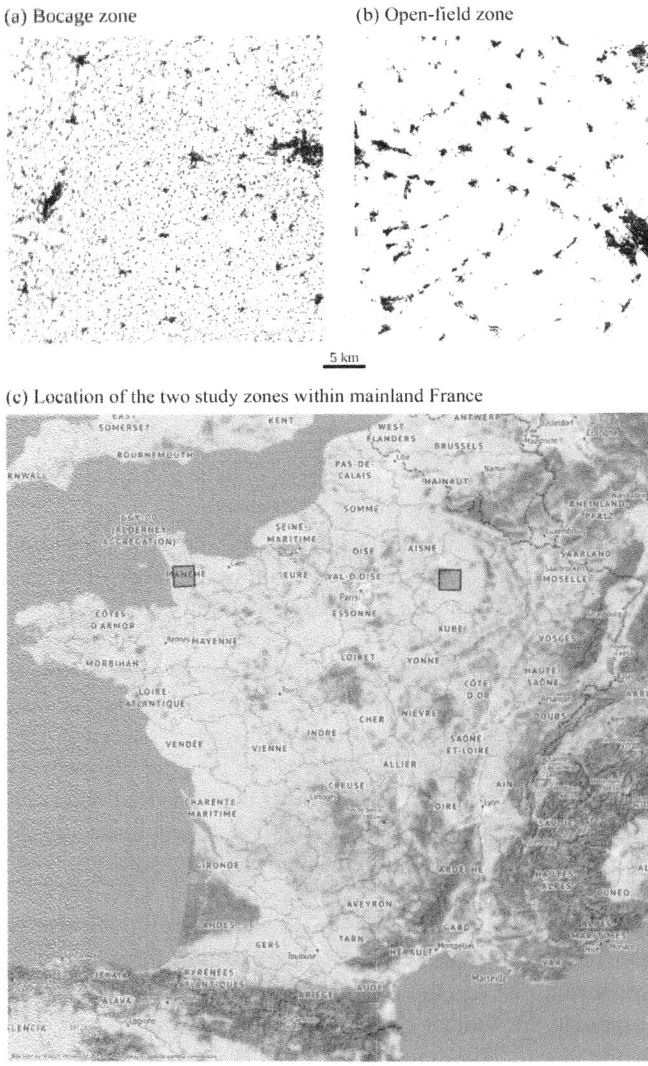

Figure 4.2. *Spatial distribution of buildings in a bocage area (west) and an open-field area (east) in mainland France. Source: IGN BD Topo (2012). Taken from Sémécurbe (2020). For a color version of this figure, see www.iste.co.uk/tannier/fractal.zip*

As an example, we present the analysis of two ideal-typical built configurations in bocage and open-field areas (Figure 4.2). Data from the IGN BD Topo (2012) (polygons representing the footprint of buildings) have been rasterized to 4 m of resolution. The fractal box-counting analysis was performed with *Fractalyse* software[7]. The analyzed fabric is covered iteratively with a square mesh of increasing size. The number of boxes containing at least one pixel is counted at each iteration. This counting method is comparable to other methods, in particular the so-called grid counting of Feng and Chen (2010) based on a rectangular (rather than square) mesh of decreasing size. The empirical counting curves of the two built-up configurations are shown in Figure 4.3. It is visually clear that the curve of the built-up area in the open-field zone is closer to a straight line than that in the bocage zone. Indicators of the statistical goodness-of-fit of these curves with a straight line are:

– for the open-field area: $R^2 = 0.996303$; p-value: 0.00000; estimated fractal dimension (D_B, slope of the fitted line): 1.358; 95% confidence interval: [1.317–1.399]; bootstrap confidence interval[8]: [1.303–1.397];

– or the bocage area: $R^2 = 0.979324$; p-value: 2.22045×10^{-16}; estimated fractal dimension (D_B, slope of the fitted line): 1.397; 95% confidence interval: [1.297–1.498]; bootstrap confidence interval: [1.281–1.515].

For the estimation of a fractal dimension to be valid, conditions are generally set to $R^2 > 0.99$ and p-value < 0.001. The amplitude of the estimated confidence intervals of D must be less than 0.10. Given these conditions, the quality of the statistical estimation of the fractal dimension of the built area in the bocage zone is unsatisfactory. The hypothesis of self-similarity is therefore invalidated. When the empirical counting curve of the bocage area is segmented in order to improve the quality of the estimates for narrower scale intervals, the following results are obtained:

– box size range [8; 344 m]: R^2: 0.999841; p-value: 2.22045×10^{-16}; $D_B \sim 1.047$; 95% confidence interval: [1.036–1.057]; bootstrap confidence interval: [1.037–1.060];

7. See: https://sourcesup.renater.fr/www/fractalyse/.

8. The bootstrap involves creating new statistical samples from an initial sample, or extending the latter, by drawing with replacement. The method is based on stochastic simulations, such as Monte Carlo methods and Bayesian numerical methods, the difference being that the bootstrap requires no other information than that available for the individuals of the original sample.

– box size range [344; 8,868 m]: R^2: 0.999118; p-value: 5.94080×10^{-12}; D_B ~1.867; 95% confidence interval: [1.818–1.917]; bootstrap confidence interval: [1.812–1.917].

Given these results, it appears that the spatial distribution of buildings in the bocage area is organized according to two distinctly different regimes of fractality. Considering short distances (less than 344 m), the buildings are distributed in space in a non-uniform manner following a fairly sparse fractal dust logic (fractal dimension close to 1); considering distances greater than 344 m, the spatial distribution of buildings approaches a uniform pattern (fractal dimension close to 2). Note that for the two box size ranges [8; 344 m] and [344; 8,868 m], the ratio between the largest and smallest box sizes is well over 10, but the difference between the minimum and maximum sizes is less than three orders of magnitude.

For the open-field zone, if the analysis is restricted to box sizes in the range of [8; 752 m], the following results are obtained: D_B ~ 1,283; R^2: 0.998383; p-value: 0.00000; 95% confidence interval: [1.251–1.316]; bootstrap confidence interval: [1.253–1.320]. The quality of estimation of the fractal dimension is slightly better than for the entire empirical curve. The estimated fractal dimension is slightly smaller (1.283 vs. 1.358). However, considering the following interval [752; 13,300 m], the quality of the estimate is not satisfactory: R^2: 0.998558; p-value: 7.80315×10^{07}; D_B~ 1.779; 95% confidence interval: [1.685–1.873]; bootstrap confidence interval: [1.736–1.884].

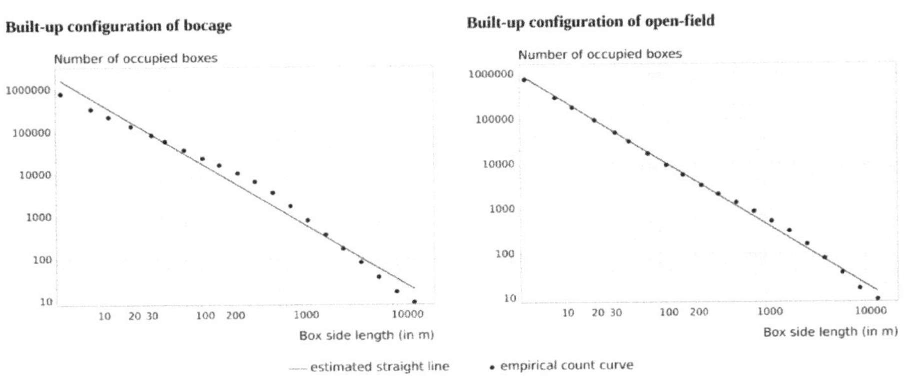

Figure 4.3. *Fractal box-counting analysis of spatial distributions of buildings in bocage and open-field areas shown in Figure 4.2: double-logarithmic graph of N(ε) as a function of ε*

4.1.3. *Correlation dimension*

Another family of fractal dimensions consists of counting the number of points jointly present in neighborhoods (generally circular in shape) of increasing size. These fractal dimensions are said to be second order, in contrast to box-counting dimensions that are first order. Formally, M_r is defined, where r corresponds to ε and M corresponds to N (equation [4.4]):

$$M_r \sim r^{D_C} \qquad [4.4]$$

The scale function M_r is the average number of neighbors of each point e of the object of study (equation [4.5]) (Grassberger and Procaccia 1983):

$$M_r = \frac{1}{n} \sum_e M_r(e) \qquad [4.5]$$

where $M_r(e)$ is the number of elements located at a distance r around e. The fractal dimension of correlation D_C is obtained by calculating the slope of the line that best fits the curve of $\log(M_r)$ as a function of $\log(r)$.

Some authors define the scale function M_r a little differently, considering, for example, the number $M_r(e)$ divided by the square of the number of points e (Schroeder 1991, p. 220) or the cumulative number of neighbors of all points e (M_r is then equal to the sum of $M_r(e)$). The choice of either of these definitions has no influence on the estimated fractal dimension.

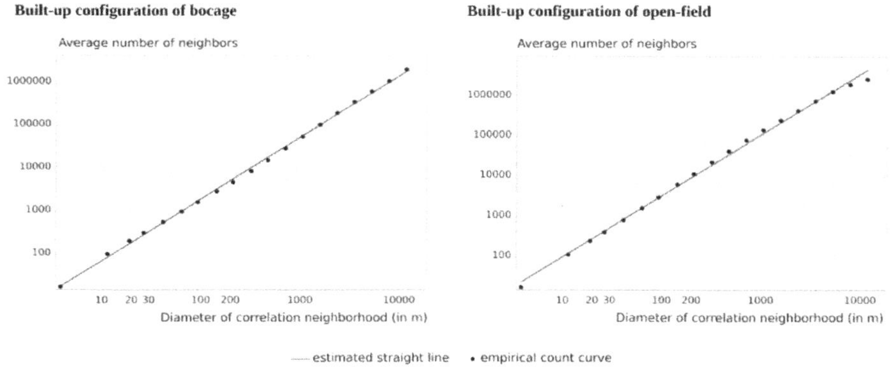

Figure 4.4. *Fractal correlation analysis of the spatial distributions of buildings in bocage and open-field zones presented in Figure 4.2: double-logarithmic plot of M_r as a function of r*

By way of example, we take the case of the two ideal-typical built configurations in bocage and open-field areas previously studied, performing the analyses again with the *Fractalyse* software. The fit of the empirical counting curve to a straight line is satisfactory for both the bocage and open-field zones (interval [4; 13,300 m]) (Figure 4.4):

– bocage: R^2: 0.998880; p-value: 0.00000; D_C ~1.426; 95% confidence interval: [1.401–1.450]; bootstrap confidence interval: [1.401–1.449];

– open-field: R^2: 0.997251; p-value: 0.00000; D_C ~1.505; 95% confidence interval: [1.464–1.545]; bootstrap confidence interval: [1.452–1.556].

Note that the curve of the open-field zone moves away from a straight line for larger neighborhood sizes. We have therefore tried to find a range of neighborhood sizes that improves the quality of the estimate for the open-field zone without degrading that for the bocage. The best compromise was found for the interval [12; 5,908 m]:

– bocage: R^2: 0.998237; p-value: 0.00000; D_C ~1.412; 95% confidence interval: [1.378–1.446]; bootstrap confidence interval: [1.373–1.451];

– open-field: R^2: 0.999181; p-value: 0.00000; D_C ~1.532; 95% confidence interval: [1.507–1.557]; bootstrap confidence interval: [1.508–1.565].

For this range of neighborhood sizes, the fractal correlation dimension D_C of the bocage is slightly reduced and that of the open-field is slightly increased compared to the range [4; 13,300 m]. In both cases, we find that the correlation dimensions are higher than the box-counting dimensions calculated previously and that they discriminate more between the two built-up fabrics.

It is possible to use a mathematical standard other than the mean for the aggregation of $M_r(e)$ (equation [4.6]).

$$M_r^q = \left[\frac{1}{n}\sum_e M_r(e)^{q-1}\right]^{\frac{1}{q-1}} \sim r^{D_q} \qquad [4.6]$$

In this case, the estimated fractal dimension is a generalized (multifractal) dimension D_q calculated using the sandbox method (Vicsek 1990) (see Chapter 6). The parameter q controls the nature of the aggregation: more importance is given to high values of $M_r(e)$, when q is high, or to low values of $M_r(e)$, when q is weak. When q is equal to 0, the values $M_r(e)$ all have the same importance. When q is equal to 2, we find the correlation dimension D_C of Grassberger and Procaccia (1983) (equations [4.4] and [4.5]).

4.1.4. *Geographically Weighted Fractal Analysis (GWFA)*

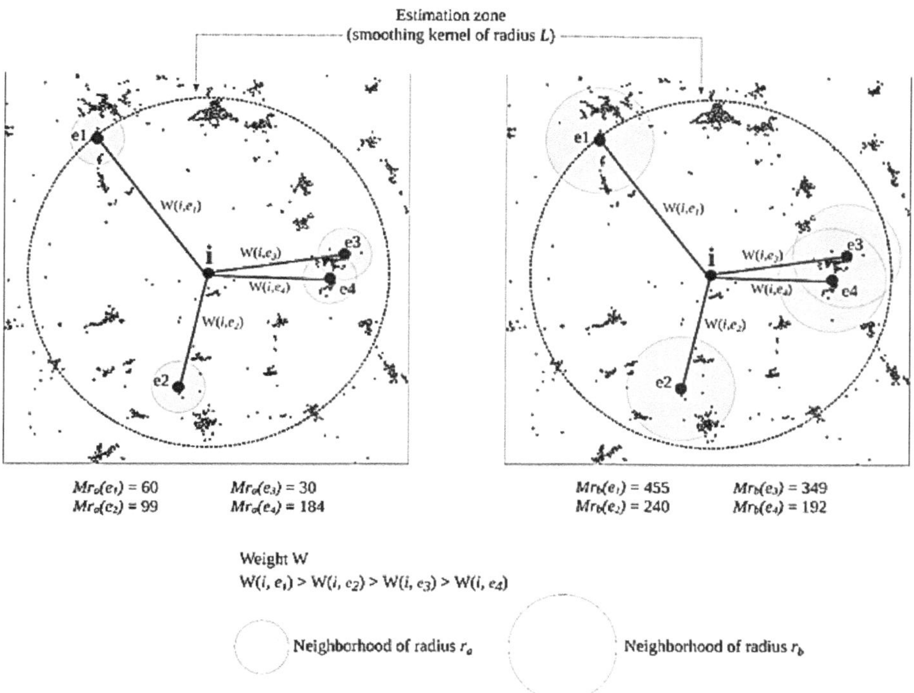

Figure 4.5. *GWFA count method. For a color version of this figure, see www.iste.co.uk/tannier/fractal.zip*

Sémécurbe et al. (2019) proposed a method called GWFA (Geographically Weighted Fractal Analysis) that expands correlation analysis by introducing a weighting W_{ie} for the contribution of points e to $M_r(i)$ based on their distance from an estimation point i (Figure 4.5). The introduction of the weighting $W_{i,e}$ is part of the logic of the geographically weighted regression introduced by Brunsdon et al. (1996) and Fotheringham et al. (1998). Subsequently, Brunsdon et al. (2002) and Harris et al. (2015) extended this method to a wide range of statistics such as medians and principal component analysis (GWPCA). Geographically weighted analyses are said to be local because of the assignment of a weight to statistical data according to their proximity to a given estimation point i. The further the statistical observations are from the estimation point considered, the less significant their contribution to the calculation of the statistic. The weights are determined using a positive decreasing function of distance, called a kernel. This function is parameterized by a smoothing radius (also called *bandwidth*) that defines the size of the neighborhood being considered in the analysis.

In GWFA, the weight $W_{i,e}$ is a function of the distance between the estimation point i and the built point e: the greater the distance, the lower the weight. This weight is calculated using a biweight kernel K: $W_{I,e} = K(d_{i,e} / L)$, where L is the smoothing radius and $d_{i,e}$ is the distance between points i and e. Different forms of kernels have been proposed in the literature, including uniform, Gaussian or quadratic forms (Diggle 2013; Baddeley et al. 2015). To implement GWFA, we chose to adopt a quadratic kernel (Figure 4.6), mainly for reasons of optimizing computation times:

$$K(x) = (1 - x^2)^2 \text{ if } x < 1;$$

$$K(x) = 0 \text{ otherwise.}$$

The quadratic kernel looks like a Gaussian kernel, but unlike the latter, it has a compact base that makes it possible to restrict calculations to points e located at a distance less than L from the observation point i and thus increase processing speed.

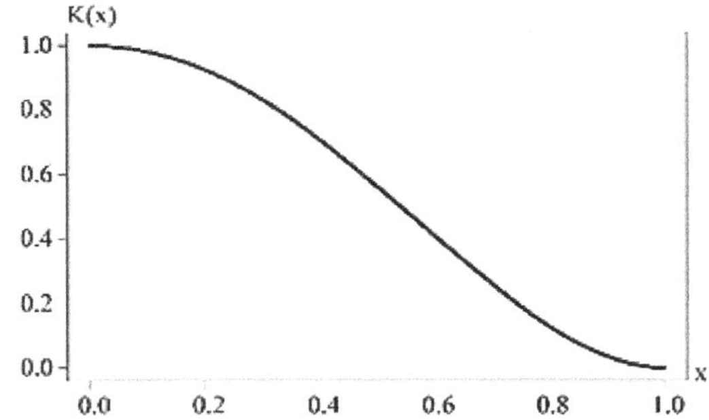

Figure 4.6. *Quadratic kernel*

Fractal correlation analysis lends itself easily to the introduction of kernel weighting, unlike the box-counting dimension that relies on the use of grids. Hence, we chose it to implement GWFA. Given an estimation point i, the local scale function $M_r(i)$ is defined as follows (Sémécurbe 2020):

$$M_r(i) = \frac{1}{n}\sum_{e} W_{i,e} M_r(e) \sim r^{D_{WC}} \qquad [4.7]$$

where i is an observation point, e is a built point, and $M_r(e)$ is the number of built-up points located at a distance less than or equal to r from the point e. The local fractal dimension D_{WC} at a point i is obtained by estimating, as for the analyses previously presented, the slope of the straight line that best fits the curve of $\log(M_r(i))$ as a function of $\log(r)$. For example, in Sémécurbe et al. (2019), the scale function used for GWFA is not that of correlation analysis, but that of sandbox analysis for q=0:

$$M_r^0(i) = \left[\sum_{e}(W_{i,e} M_r(e))^{-1}\right]^{-1} \sim r^{D_0} \qquad [4.8]$$

To implement GWFA, an R package *gwfa*[9] has been developed. A GWFA application is also available on the github repository (see: github.com/fsemecurbe/GWFA.git).

With this method, we analyzed the bocage and open-field configurations shown in Figure 4.2. The data used were the centroids of buildings from the IGN BD Topo (2012) and not the polygons of rasterized buildings, as was the case for previous analyses. The GWFA results obtained for the two ideal-typical built configurations in bocage and open-field areas are as follows:

– open-field, neighborhood sizes [25; 25,600 m]: R^2: 0.995817; p-value: 5.13389×10^{-12}; $D_{WC} \sim 1.403$; 95% confidence interval: [1.399–1.535]; bootstrap confidence interval: [1.333–1.488];

– bocage, neighborhood sizes [25; 25,600 m]: R^2: 0.994059; p-value: 2.48919×10^{-11}; $D_{WC} \sim 1.455$; 95% confidence interval: [1.370–1.540]; bootstrap confidence interval: [1.351–1.557].

D_{WC} appears slightly higher in the bocage zone than in the open-field zone, while the reverse was observed for the correlation analysis. However, the difference in dimensions is statistically insignificant. The results are therefore closer to those obtained with the box-counting analysis.

If we study the curves presented in Figure 4.7, we can see that the points are well aligned on a straight line for the range of neighborhood sizes [100; 25,600 m] for bocage and [25; 6,400 m] for open-field:

9. See: https://cran.r-project.org/web/packages/gwfa/index.html.

– open-field, neighborhood sizes [25; 6,400 m]: R^2: 0.997314; p-value: 2.92584×10^{-10}; $D_{WC} \sim$ 1,467; 95% confidence interval: [1.370–1.540]; bootstrap confidence interval: [1.419–1.549];

– bocage, neighborhood sizes [100; 12,800 m]: R^2: 0.998029; p-value: 9.89604×10^{-11}; $D_{WC} \sim$ 1,537; 95% confidence interval: [1.476–1.598]; bootstrap confidence interval: [1.462–1.616].

The spatial distribution of the centroids of buildings studied through the prism of the GWFA is therefore very different in the two cases, but it is the fractality domains of the two configurations that differ, more than their D_{WC} dimension (see Chapter 7).

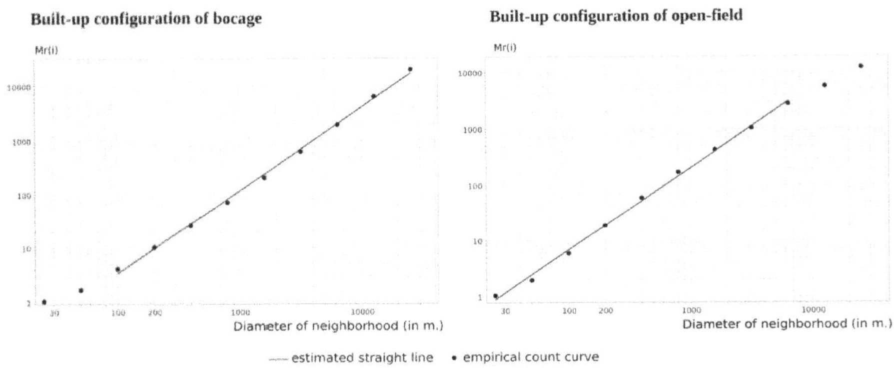

Figure 4.7. *GWFA of spatial distributions of building centroids in bocage and open-field areas shown in Figure 4.2*

4.1.5. *Clarifications and comments*

4.1.5.1. *Statistics relating to the estimation of fractal dimensions*

The statistical estimation of a fractal dimension is based on the linear regression model, most often estimated by the ordinary least squares (OLS) method. The question of the validity of the estimate is crucial and has been discussed by different authors (Gonzato et al. 1998; Lam 2009). In the examples of fractal analyses of bocage and open-field zones conducted previously, we conditioned the statistical validity of the estimate to an R^2 greater than 0.99 and a p-value less than 0.001. We also verified that the confidence interval of the estimated fractal dimension was quite small (less than 0.10). These conditions are necessary but, however, not sufficient because they do not guarantee that the empirical count curve is indeed a straight line and not, for example, a very flattened parabola. Other, more refined methods may improve the assessment of the statistical goodness-of-fit to the linear model. However, in any case, there is no substitute for visual analysis of the empirical count curve. This must, as far as possible, be carried out prior to statistical analysis of the data (Anscombe 1973; Tufte 1983). In general, in the case of fractal analyses in human geography, each empirical count curve has relatively few points (it is not a cloud of points stricto sensu but rather a curve), and it can be visually determined whether the points of the curve are aligned along a line or not. The fact is that such a visual analysis is only possible when the number of study zones considered is small. Otherwise, only statistical indicators can be used to verify the validity of the quality of the fit of empirical curves to straight lines. It is therefore very important to choose well in advance an interval of scales (sizes of boxes or diameters of neighborhoods of correlation) for which all the study areas are supposed to show scale-invariant behavior, knowing that a result of the analysis may be that some of them do not (Sémécurbe et al. 2019). Gonzato et al. (1998) stress that statistical tests are often not relevant in the case of fractal analyses because the number of points in the samples studied is too small (generally less than 15). Sémécurbe et al. (2019) defend, for their part, another position: that adding additional points (i.e. intermediate sizes of correlation boxes or neighborhoods between minimum and maximum sizes) is not necessarily useful, and that the magnitude of the standard errors of statistical estimators shows deviations from scale invariance that are interesting to study (see Chapter 7).

4.1.5.2. *Fractal analysis and spatial (statistical) analysis*

Spatial analysis of point processes is often used in economics and biology to describe point distributions. Yet it is rarely used to characterize built structures, even though its methods bear a strong resemblance to those of fractal analysis (Agterberg 2013). On the one hand, box-counting analysis is similar to the quadrat method since both rely on the use of a regular grid of squares of varying size. On the other hand, the writing of the scale function of the correlation dimension is very close to that of the Ripley function, one of the most commonly used methods of spatial analysis.

The Ripley function (1977) bears a strong resemblance to the sandbox estimation method of fractal dimensions. They are both based on the count of neighbors. The Ripley function is defined as follows:

$$K(r) = \frac{1}{n}\sum\frac{M_r^*(e)}{\lambda} \qquad [4.9]$$

where $K(r)$[10] is the Ripley function calculated for the radius r, n is the number of points in the set under consideration, λ is the density of points (reference value), and

10. No relation to the quadratic kernel *K(x)* is mentioned in the presentation of the GWFA.

$M_r^*(e)$ is the number of points neighboring the point e and located at a distance r of e. Unlike sandbox fractal analysis, the point e is not counted in $M_r^*(e) = M_r(e) - 1$.

The values of $H(r) = \sqrt{\frac{K(r)}{\pi}} - r$ are compared to those obtained for fully random point distributions of similar density. When $H(r)$ is significantly greater than 0, the distribution is more clustered than a random distribution. On the contrary, when $H(r)$ is significantly less than 0, the points are arranged regularly in space.

Applying the Ripley function to both the bocage and open-field distributions in Figure 4.2 does not give satisfactory results (Figure 4.8). The two curves are by a great margin above the values taken by the random distributions materialized by the dotted lines in the figure. Both distributions appear significantly more concentrated than a totally random distribution. The higher values taken by the open-field distribution indicate that this distribution is more concentrated than the bocage distribution. However, it is difficult to go into more detail in the analysis.

The reference model for traditional methods of spatial analysis is a totally random spatial distribution of points. A random distribution is characterized by mean local densities of points that are constant and equal to λ. On comparable surfaces, no portion of the space contains on average more points than any other. The choice of the reference density value λ of the Ripley function is crucial and determines the results of the analysis. These are also likely to vary significantly depending on the narrowness of the study zone under consideration around the objects studied, and whether or not it takes into account constraints such as rivers, elevations or flood zones. More complex spatial analysis methods have been proposed to go beyond the random distribution model. They are mainly based on the a priori estimation of a reference distribution based on the spatial distribution of the analyzed objects. But even with this refinement, the results are not satisfactory. The reference distribution is defined for a given scale and is still too regular to adequately describe the relationships linking objects with space.

With spatial analysis, a distribution of points is compared to reference distributions exogenous to it. In contrast, with fractal analyses, the information taken into account to characterize an object is not located at a given scale, but emerges from comparison of the properties of the configuration between several successive distinct scales. Spatial units containing the objects at a coarse scale determine the possible space of spatial units of comparison at finer scales. Spatial units containing no objects at a coarse scale are not further considered at finer scales. In doing so, the arbitrariness of the choice of a reference configuration is removed because the analyzed configuration is its own standard of measurement.

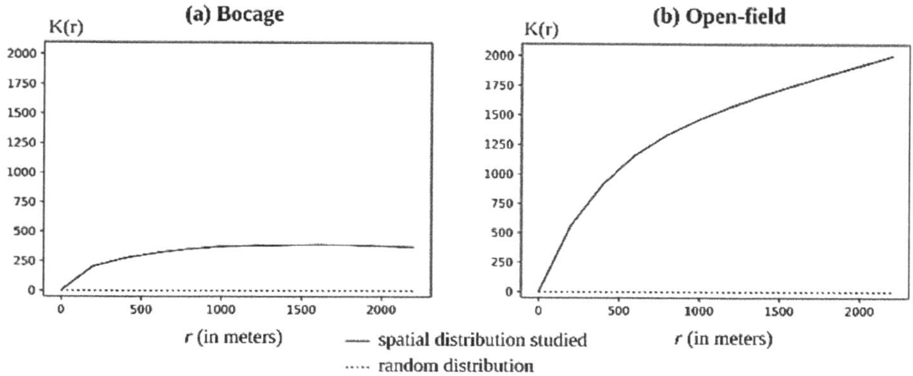

Figure 4.8. *Ripley's function for the centroids of buildings in the bocage and open-field areas shown in Figure 4.2*

The fractal dimension informs us about how spaces containing counted points become contained spaces observed at coarser scales. The higher the fractal dimension and the closer it is to 2, the more container and content are of the same nature; the lower the fractal dimension, the more the container breaks up into contents isolated from each other. Note that when a distribution of points is not strictly self-similar, the relationships between container and content vary between scales.

4.1.5.3. *Fractal dimension and built density*

Fractal dimension is strongly correlated with built density. The information provided by these two indicators is generally quite redundant. Sémécurbe (2020) highlighted this in his analysis of the built-up fabrics of mainland France. The two maps in Figure 4.9 are indeed very similar. The correlation coefficient between the two maps, greater than 0.9, confirms this visual impression. For high-density spaces, buildings occupy the entire space, and the fractal dimension is close to 2. On the contrary, in sparse spaces, the gaps dominate and the fractal dimension is small[11]. In

11. In Figure 4.9a, urban centers have fractal dimensions greater than 1.5. We find these values consistent with those obtained by geographers for the built-up fabric of urban spaces. Very sparsely populated rural areas have dimensions less than 1. Between these two types of places, fractal dimensions range between 1 and 1.5.

the case of 18 Belgian cities, Tannier and Thomas (2013) also demonstrated a strong correlation between high built-up densities and high fractal dimensions calculated over built-up areas. However, density only provides information on the average occupation of a surface by a mass, in this case a built mass, without taking into account the spatial distribution of this mass, unlike the fractal dimension. The statistical correlation between built-up density and fractal dimension results from the fact that the fractal dimension, like density, provides information on the degree to which space is filled by buildings, but obscures the fact that the fractal dimension also provides information on how space is occupied. Thomas et al. (2007) present various examples of built-up fabrics of the same density but of very different fractal dimensions (see Figure 11 of the article in question and the associated commentary).

It is possible to distinguish the information provided by the built-up density from that provided by the fractal dimension via the study of scaling indices derived from empirical count curves used as a basis for estimating fractal dimensions (see Chapter 7).

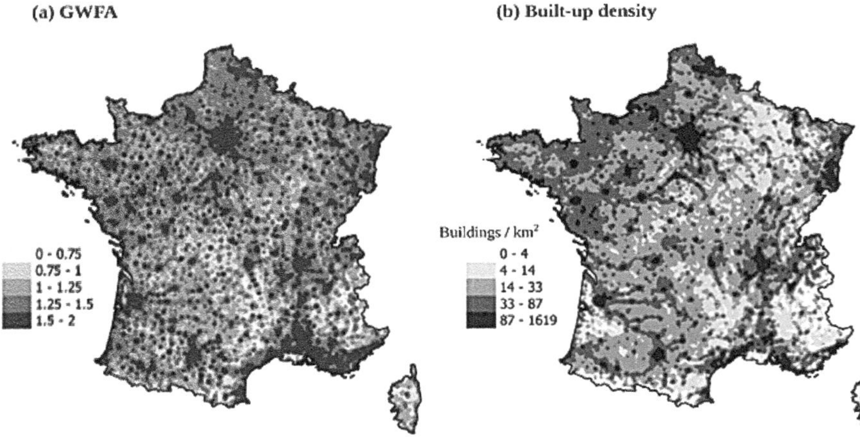

Figure 4.9. *Fractal dimension estimated using GWFA on building centroids (a) and built-up density (b) in mainland France. Fractal dimension and built-up density calculated in 2 km wide squares with a smoothing radius of 8 km. Taken from Sémécurbe (2020, p. 72). For a color version of this figure, see www.iste.co.uk/tannier/fractal.zip*

4.1.5.4. *Fractal analysis software*

The fractal box-counting and correlation analyses presented above were performed with *Fractalyse*, open-source software developed in Java and available for MS Windows, MacOS and Linux. Other fractal analysis software applications exist, of course. Among the most commonly used, we should mention the plugin *FracLac*[12] for the open-source image analysis software *ImageJ*[13], also developed in Java; *FracLab*[14] by INRIA (National Institute for Research in Digital Science and Technology, France), free software running on MS Windows, MacOS and Linux; and *Benoit*[15], proprietary software running exclusively on Windows. A comparison of some fractal analysis software applications, including *FracLac* and *Fractalyse*, was conducted by Hadzieva et al. (2015).

4.2. Fractal box-counting and correlation analyses of 50 built-up fabrics

In the previous section, we have seen that the box-counting and correlation dimensions, which are the dimensions most commonly used in geography to characterize built-up fabrics, may each take substantially different values for the same object. The objective of this section is to study these differences in more detail and to try to identify the morphological characteristics of the built-up fabrics that influence to a greater or lesser degree the value of each dimension. This study will address the need, highlighted by Lam (2009), for a thorough evaluation of the various fractal measurement algorithms so that they can be used to reliably characterize geographical objects.

Our study is based on a sample of 50 built-up fabrics, of which 25 are from the real world and have been downloaded from the *geofabrik* platform of Open Street Map (see: http://download.geofabrik.de/). The other 25 fabrics are qualified as theoretical as they were created from scratch using software dedicated to the generation of fractal figures. These theoretical built-up fabrics are strictly self-similar and their similarity dimension D_S is known exactly. They serve as references against which to compare the results of analysis obtained for real built-up fabrics.

---

12. See: https://imagej.nih.gov/ij/plugins/fraclac/fraclac.html.

13. See: https://imagej.nih.gov/ij/index.html.

14. See: https://project.inria.fr/fraclab/.

15. See: https://www.trusoft-international.com/benoit.html.

It has already been noted (see section 4.1) that some parameters of analysis, notably the range of neighborhood sizes, have an influence on the measurement and estimation of the fractal dimension. As a preliminary to our study, it is therefore necessary to fix all the parameter values likely to influence the box-counting and correlation dimensions of the 50 built-up fabrics. To this end, a variability analysis is carried out on a sub-sample of ten built-up fabrics chosen from the 50 fabrics studied. The idea is to observe the impact of the variations of each parameter taken separately on the estimated fractal dimensions in order to choose the best possible set of parameter values that will be used to study the sample as a whole. Once these parameters of analysis have been fixed and are common to all of the fabrics, it is possible to compare the fractal dimension values obtained.

This part of the chapter is divided into three sections, which: (1) present the real and theoretical build-up of the fabrics studied; (2) highlight the variability of the estimated dimensions as a function of the values of the analysis parameters and choice of these values (10 built-up fabrics); and (3) highlight the similarities and differences between the box-counting and correlation dimensions (50 built-up fabrics).

4.2.1. *The built-up fabrics studied*

4.2.1.1. *Real built-up fabrics*

The 25 real fabrics (Figure 4.10) come from the *geofabrik* platform derived from the OpenStreetMap (OSM) project, which provides free access to free-of-charge geographic databases. The selected built-up fabrics are not a reflection of the built reality of the cities under consideration, but are the result of what has been mapped by the voluntary contributors to the OSM project. However, this is not a problem here since our objective is not to characterize cities strictly speaking, but to study different forms of built organization. The 25 real built-up fabrics were chosen to provide the most diverse possible sample in terms of spatial distribution of buildings, number and surface area of buildings and built-up density[16]. Coastal and border cities were excluded from the selection so the fabrics would have built development in all directions of the plane. Cities with characteristic morphological elements such as large rivers or marked reliefs are not excluded from the selection since these morphological elements contribute to the diversity and originality of the spatial distributions of buildings in each city.

16. Built-up density is measured as the ratio of the total area covered by buildings and the area of the study zone. This index is expressed as a percentage (%).

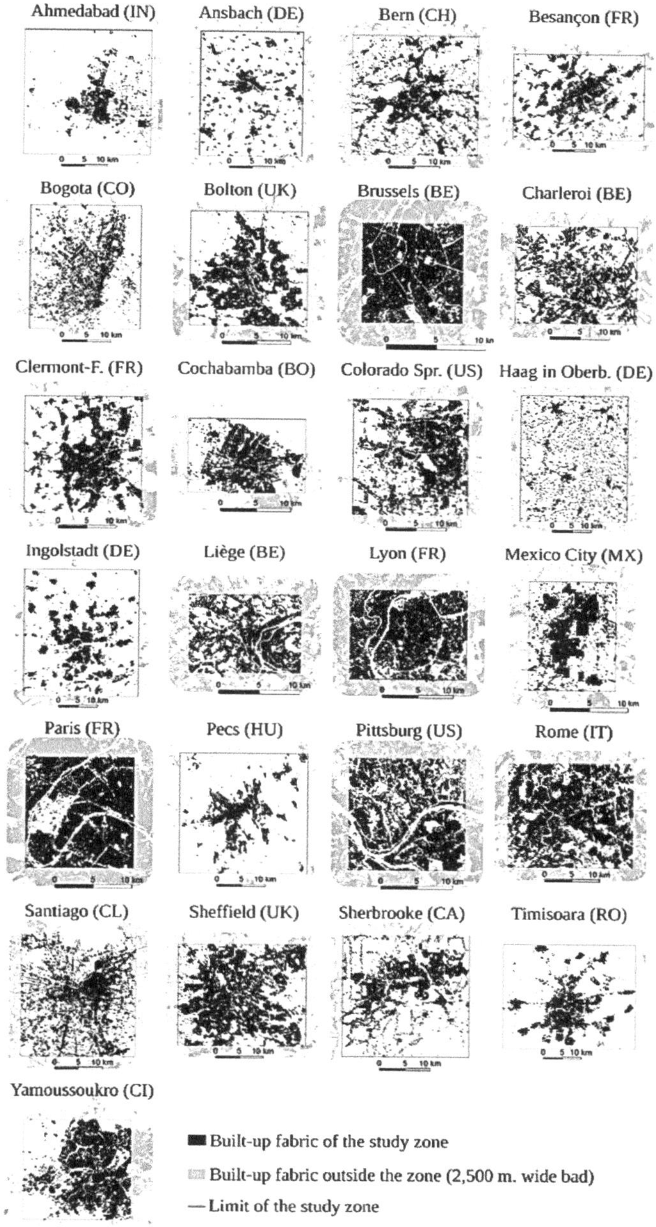

Figure 4.10. *The 25 real built-up fabrics. For a color version of this figure, see www.iste.co.uk/tannier/fractal.zip*

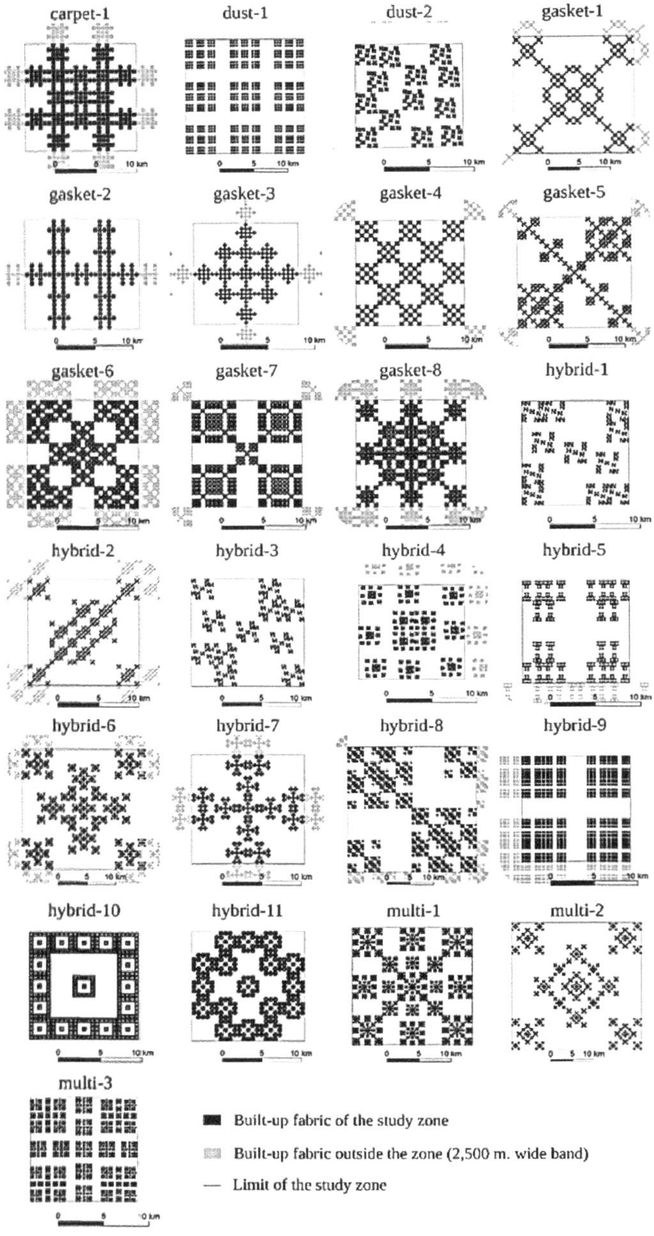

Figure 4.11. *The 25 theoretical built-up fabrics. For a color version of this figure, see www.iste.co.uk/tannier/fractal.zip*

4.2.1.2. *Theoretical built-up fabrics*

Title	Type	D_S	Iteration
hybrid-1	Hybrid fractal closer to a dust than to a carpet or a gasket	1.4898	5
hybrid-2	Hybrid fractal closer to a gasket than to a dust	1.4898	5
hybrid-3	Hybrid fractal closer to a dust than to a carpet or a gasket	1.4898	5
hybrid-4	Hybrid fractal closer to a dust than to a carpet or a gasket	1.5439	5
hybrid-5	Hybrid fractal closer to a dust than to a carpet or a gasket	1.5439	5
hybrid-6	Hybrid fractal closer to a dust than to a carpet or a gasket	1.5849	6
hybrid-7	Hybrid fractal closer to a gasket than to a dust	1.5936	5
hybrid-8	Hybrid fractal close to a dust	1.6609	6
hybrid-9	Hybrid fractal closer to a dust than to a carpet or a gasket	1.7227	5
hybrid-10	Hybrid fractal closer to a carpet than to a dust	1.7603	5
hybrid-11	Hybrid fractal closer to a gasket or a carpet than to a dust	1.7603	5
gasket-1	Fractal gasket	1.4999	6
gasket-2	Fractal gasket	1.5936	5
gasket-3	Fractal gasket	1.5936	5
gasket-4	Fractal gasket	1.5936	5
gasket-5	Fractal gasket	1.5936	5
gasket-6	Fractal gasket	1.7603	5
gasket-7	Fractal gasket	1.7603	5
gasket-8	Fractal gasket	1.7603	5
dust-1	Fractal dust	1.5849	6
dust-2	Fractal dust	1.5936	5
carpet-1	Fractal carpet	1.7603	5
multi-1	Multifractal hybrid figure	1.6547	6
multi-2	Multifractal hybrid figure	1.4566	6
multi-3	Multifractal dust	1.6747	5

Table 4.1. *Presentation of the theoretical fabrics*

The theoretical built-up fabrics (Figure 4.11) were created using iterated function systems (IFS)[17] (Barnsley 1988) via the application *GenFrac*[18] according to a number of iterations reported in Table 4.1. The figures created with *GenFrac* were then exported in shapefile (.shp) format. In order to ensure that the average surface area of the buildings corresponds to that of the real fabrics, an affine transformation was carried out a posteriori using a Geographic Information System (GIS). At the end of this transformation, the spatial extent of the theoretical fabrics and their number of buildings were often too large. The fabrics were then truncated to preserve only a section of them. The truncation was performed according to the construction logic of a fractal object, that is, by retaining a part of the object that was identical to the object as a whole.

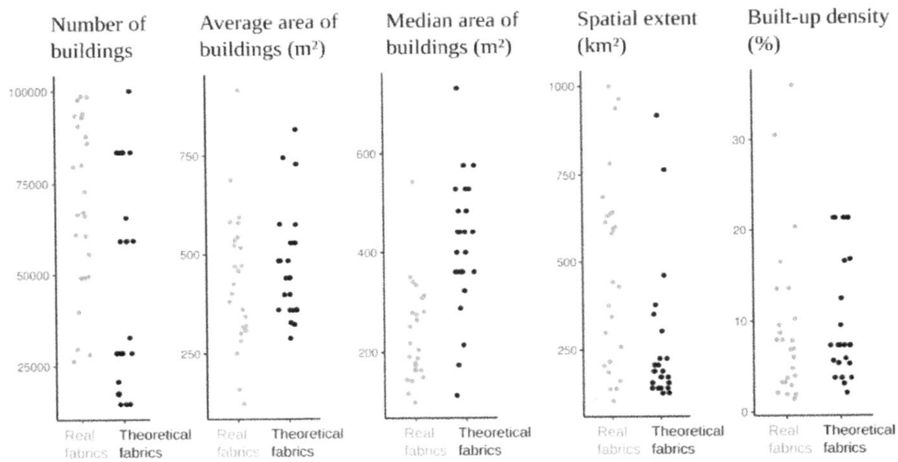

Figure 4.12. *Characteristics of the fabrics studied*

To compare the results obtained for the theoretical built-up fabrics and for the real fabrics, it is necessary that these fabrics have a certain number of common characteristics: number of buildings, average and median surface area of the buildings, built-up density and spatial extent of the fabrics. The diversity of characteristics of each fabric and also the commonalities between the real fabrics, on the one hand, and the theoretical fabrics, on the other hand, are observable in Figure 4.12. Some theoretical fabrics are fractal dusts, others carpets, yet others hybrid figures: 22 are monofractal and 3 are multifractal (Table 4.1)[19]. A good

17. See Chapter 2, section 2.1.

18. See: https://sourcesup.renater.fr/www/genfrac/.

19. For a description of the properties of different types of fractal figures, see Chapter 3.

proportion of the theoretical fabrics have a median surface area of buildings slightly larger than that of the actual fabrics. By definition, only multifractal fabrics show a diversity of size of elementary spatial entities (squares). However, the connected elementary squares of each monofractal figure have not been merged. Yet they form buildings wider than the elementary (square) spatial entity. The diversity of size of the buildings in monofractal fabrics is therefore in reality greater than that measured by the size of the elementary squares.

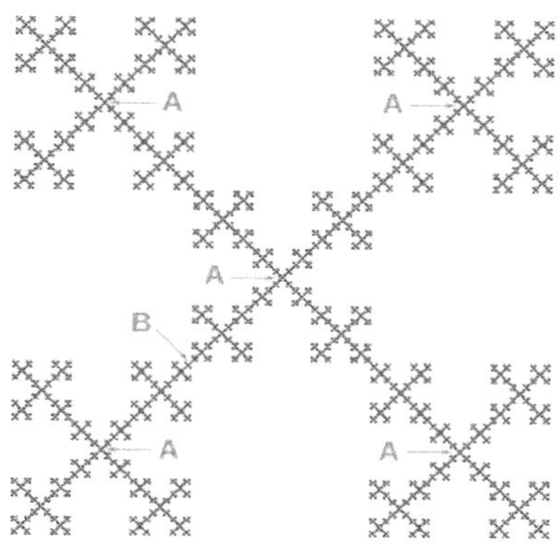

Figure 4.13. *Fractal box at iteration 5. IFS: N = 5 and r = 1/3*

In the case of theoretical fractal fabrics, we might think that it is possible to identify parameter values that allow us to estimate their fractal box-counting dimension D_B and correlation dimension D_C, knowing that the homothety (or similarity) dimension D_S must coincide with these in the case of strictly self-similar fractal objects[20]. However, this equality of the three dimensions is purely theoretical; it does in fact presuppose achieving a perfect overlay of the figures, which is rarely possible in practice. To illustrate this, consider the example of the box fractal at iteration 5 shown in Figure 4.13. Each element of the figure has a length equal to 1 m. We proceed to a fractal correlation analysis by performing the count solely on the basis of the five points denoted A in the figure. The count window is square. Given that the reduction factor r of the IFS of the fractal box is equal to 1/3, the size of the window of analysis is multiplied by 3 from one step of the count to another. The result of the analysis, considering five sizes of count window, is shown in

20. "For strictly self-similar mathematical fractals, such as the Sierpinski gasket and the Cantor dust, the mass dimension is the same as the Hausdorff dimension (and any other fractal dimension considered here). They are all given by the similarity dimension of the scaling law, defined by the initiator and the generator that generates the fractal" (Schroeder 1991, p. 216).

Figure 4.14a; the estimated fractal dimension is $D_C \sim 1.46$ equal to the similarity dimension value $D_S = \log 5 / \log 1/3$, the points being perfectly aligned in a straight line. In a second step, the count is performed starting from all points of the figure, as for fractal analyses carried out automatically by means of software such as *Fractalyse*. The result of the analysis is shown in Figure 4.14b. The measure points are not perfectly aligned in a straight line; the estimated fractal dimension is $D_C \sim 1.345$, and therefore differs from D_S. Indeed, many points do not have a number of neighbors equal to five or to a multiple of five considering the five sizes of count windows defined above. Such is the case, for example, for point B noted in Figure 4.13.

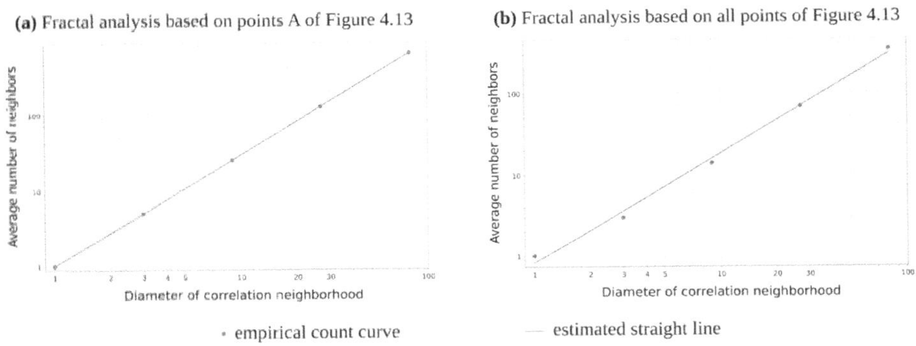

Figure 4.14. *Fractal correlation analyses based on points A in Figure 4.13 (a) and based on all points in Figure 4.13 (b)*

4.2.1.3. *Extent of study zones*

In the selection of real built-up fabrics and in the creation of theoretical built-up fabrics, particular attention was paid to ensuring that all have similar geographical extents (of the same order of magnitude). Indeed, the geographical extent of a built-up fabric, in other words the size of the study zone, can significantly influence the estimated fractal dimension, especially when different built structures are integrated as the study area grows. Conversely, the fractal dimension will vary very little if identical structures are integrated. As shown in Figure 4.15, boundaries 1, 2 and 3 of the study zone cover the same spatial structure; the estimated D_B are therefore very close or even equal for these three study zones. However, delimitation 4 of the study zone incorporates a different spatial structure, from which result different values of D_B from the dimensions obtained for the other delimitations.

It is quite easy to choose the delimitation of the study zones of the theoretical fabrics in our sample by controlling the type of spatial structures they integrate.

However, this is impossible in the case of real fabrics. Hence, given that the size of the windows of analysis directly influences the calculation of fractal dimensions, we chose to consider study zones with extents of the same order of magnitude.

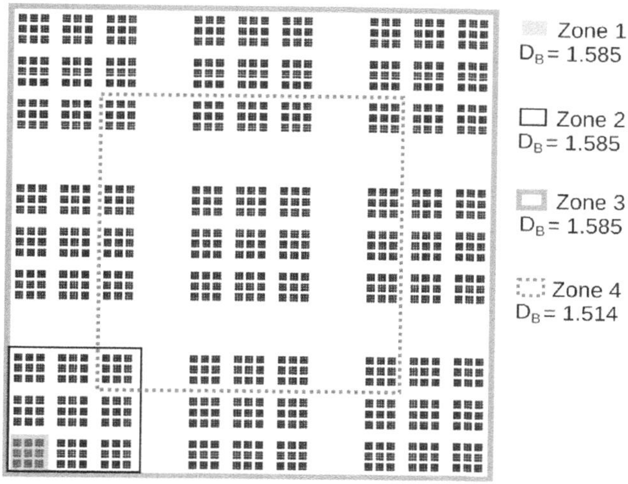

Figure 4.15. *Variation of the estimated box-counting dimension D_B as a function of the study zone under consideration*

4.2.2. Choice of parameter values for fractal analyses

In order to ensure the comparability of the results of analysis for the 50 built-up fabrics, we must choose a single set of parameter values that will be systematically applied for the analyses performed with the *Fractalyse* software.

To help in the choice of parameter values, tests were carried out on a sub-sample of 10 fabrics, five real and five theoretical, among the 50 contained in the sample. These 10 fabrics were selected so that this sub-sample would be representative of the different spatial configurations present in the sample. The 10 selected fabrics are: Besançon, Charleroi, Paris, Pecs, Santiago, carpet-1, gasket-1, dust-1, hybrid-6 and multi-1. The choice was made to retain both real and theoretical fabrics. The latter are interesting because their dimension D_S is known, but they do not reflect all of the complexity of real fabrics, for which potential sources of variability of the estimated fractal dimensions as a function of the values of the parameters chosen may be different than those of theoretical fabrics.

Three main families of parameters may influence the calculation of the fractal box-counting or correlation dimension: (1) the image format and its resolution; (2) the characteristics of the count windows; and (3) the method of fitting the theoretical curve to the empirical curve.

In this part of the chapter, we will review each of the parameters that can change the value of the box-counting and correlation dimension. We will thus show the importance of carefully choosing the values of these parameters. This will also allow us to set the parameter values for analysis of our 50 built-up fabrics. We begin the analysis with a set of parameter values chosen by default. We then vary the value of each parameter one after the other in order to assess its impact on the calculation of the fractal dimension and set the value which seems the most relevant for the rest of the analysis. In the case of theoretical fabrics, we choose the parameter values so that D_B and D_C approach D_S. We also pay attention to the significance of the dimensions obtained using the R^2 and the p-value of the estimate. As the number of fabrics and tests is large, we do not systematically analyze the confidence intervals and the associated graphs as we did previously for the study of the open-field and bocage landscapes (section 4.1).

4.2.2.1. *Image format and resolution*

Software like *Fractalyse* allows analysis of images in vector format (each element is a point, a line or a polygon) or in raster format (pixels of two colors, one for buildings and one for non-buildings). For analyzing built-up fabrics, each building is represented either by polygons (vector format) or by an association of several related pixels (raster format).

Fractalyse allows fractal box-counting analysis on both vector format and raster format. It is therefore useful to check whether the use of one format rather than another influences the value of the calculated box-counting dimension. Fractal correlation analysis can only be done on raster format. In both cases, the resolution, that is, the size of a pixel in a raster format, will also influence the estimated fractal dimension.

For the box-counting dimension, four tests were performed: one test on vector files and three on raster files with resolutions of 2 m, 4 m and 10 m. The same resolutions were tested for the correlation dimension (on raster format only).

The results obtained, shown in Figure 4.16[21], show that for a rather coarse resolution (pixels of 10 m per side), the box-counting dimension is significantly different (in general, smaller) than that of finer resolutions. It is at a fine resolution that the D_B of theoretical fabrics comes closest to their D_S. These two trends are less marked for D_C, for which a variation in resolution does not imply a particularly significant variation in the fractal dimension.

The values of D_B obtained on vector format are slightly higher than those obtained on fine-resolution raster format and tend overall to deviate a little from the values of D_S in the case of theoretical fabrics. These results are consistent with the observations of Ge and Liu (2009) who note that analyses on raster format are better than on vector format.

The remaining analyses will therefore be performed on the files in raster format at a resolution of 4 m. With this format, D_B and D_C are effectively close to D_S for the theoretical fabrics. In addition, a resolution of 4 m, less fine than that of 2 m, makes it possible to reduce calculation times, which may be significant in the case of correlation analyses.

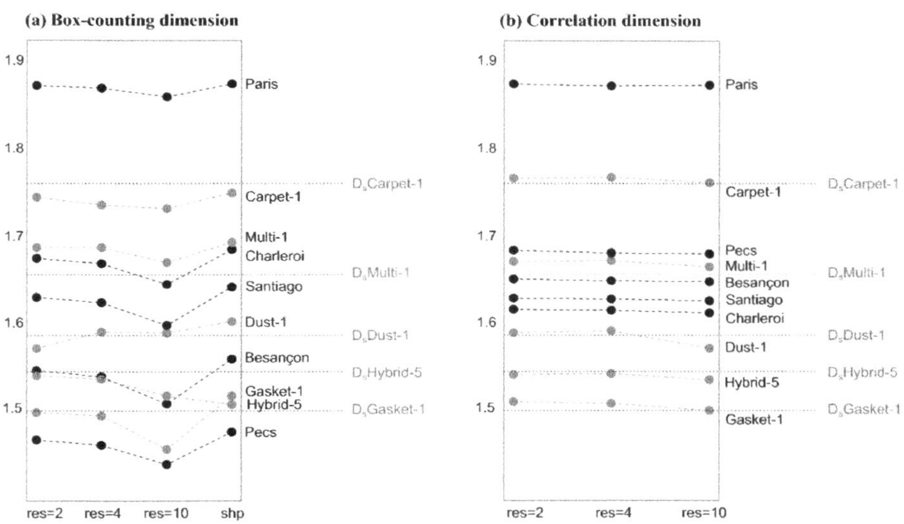

Figure 4.16. *Variation of box-counting (a) and correlation (b) dimensions depending on the format and resolution of the source image*

21. Note: The dashed lines between the points on the graphs are only intended to facilitate the reading of changes in the value of the dimensions according to the parameter values. These are in no way the values of dimensions taken for intermediate values of the parameter.

4.2.2.2. *Characteristics of the count windows*

Count windows have a decisive role in the estimation of fractal dimensions. For a given built-up fabric, it is clear that the size, position and shape of the boxes overlaying it will directly impact the calculation of the number of occupied boxes and, consequently, the estimation of the box-counting dimension. The same applies to the definition of the neighborhood of the points of the correlation analysis, which directly impacts the average number of occupied pixels. This issue is known as the MAUP (Modifiable Areal Unit Problem) (Openshaw 1983). It is therefore important to check the definition of basic spatial units (BSUs), that is, count windows, for a good estimate of fractal dimensions. For all analyses, we need to set a common shape and identical sizes for the different count windows.

For reasons of computing convenience, the square is the default shape of the count windows in the *Fractalyse* program. The fabrics are covered with a grid whose elements are square for the box-counting dimension. For the correlation dimension, the neighborhood of each point is also defined in a square.

Regarding the size of the count windows, *Fractalyse* allows us to set the minimum, maximum and intermediate sizes via a set of four parameters: the "*min*" and "*max*" parameters set the smallest and largest window sizes, respectively; the "*coef*" and "*seq*" parameters together define the number and size of the intermediate windows. The "*coef*" parameter is a coefficient that determines the size of the analysis window at each step based on the size of the window at the previous step. This coefficient is applied additively in the context of an arithmetic sequence and multiplicatively in the case of a geometric sequence. The nature of the sequence is a function of the "*seq*" parameter.

The "*min*" parameter

The "*min*" parameter sets the minimum size of ε in box-counting analyses and the minimum size of r in correlation analyses (see section 4.1). Here, we test four minimum window sizes: 4 m (equivalent to image resolution), 10 m, 20 m and 30 m.

As a function of the minimum window size, large variations of D_B for real fabrics are visible in Figure 4.17. These variations are less marked but still present for the theoretical fabrics. Variations in D_C are also visible. It is clear for the five theoretical fabrics that D_C approaches D_S as the minimum size approaches 30 m. This is also true for three out of five theoretical fabrics with the box-counting dimension, but a reverse effect is observed for the multi-1 fabric.

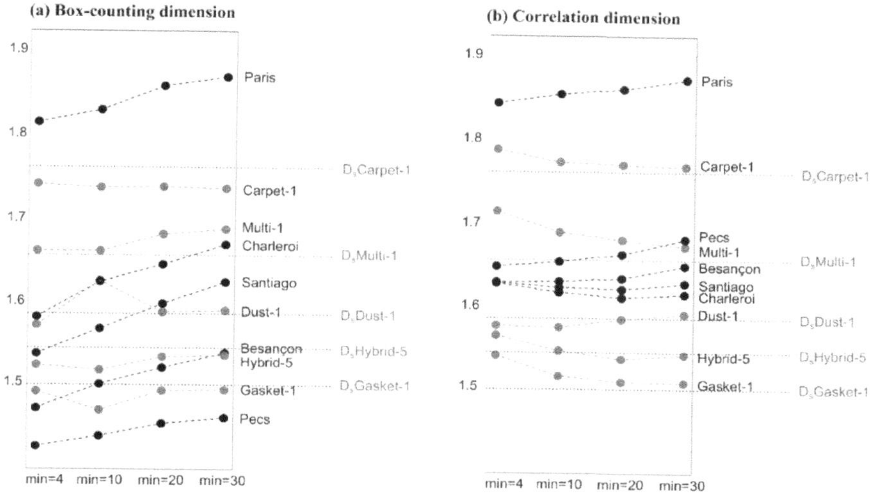

Figure 4.17. *Variation of box-counting dimensions (a) and correlation dimensions (b) according to the minimum size of the count window*

We will therefore apply, for the rest of the analyses, a minimum count window size of 30 m. This choice is consistent with the recommendations found in the literature. Indeed, Benguigui et al. (2000) note that the minimum window size should be at least larger than the smallest building. In our database of 50 built-up fabrics, the largest minimum building size is 576 m^2. A minimum analysis window of 24 m (24×24 = 576) would therefore be the minimum acceptable window size. Shen (2002) confirms that the minimum window size must be larger than the image resolution. Finally, Thomas et al. (2010) point out that a window size of less than 30 m captures variations related to the size of the buildings themselves, and that this may cause significant deviations from the estimated fractal laws.

The "*max*" parameter

The "*max*" parameter sets the maximum size of ε in box-counting analyses and the maximum size of *r* in correlation analyses. It is clear that the maximum size of the count window cannot be greater than the size of the image itself. By default, the *Fractalyse* application uses a maximum window size of half of the smallest side of the image. Since not all fabrics in the database have exactly the same spatial extent (Figure 4.12) and, moreover, do not have the same smallest side length, it follows the fact that the maximums set by default are not the same. We must therefore check that the adoption of a maximum size common to all fabrics does not modify the fractal dimension of these fabrics with regard to the dimension estimated with the default maximum size.

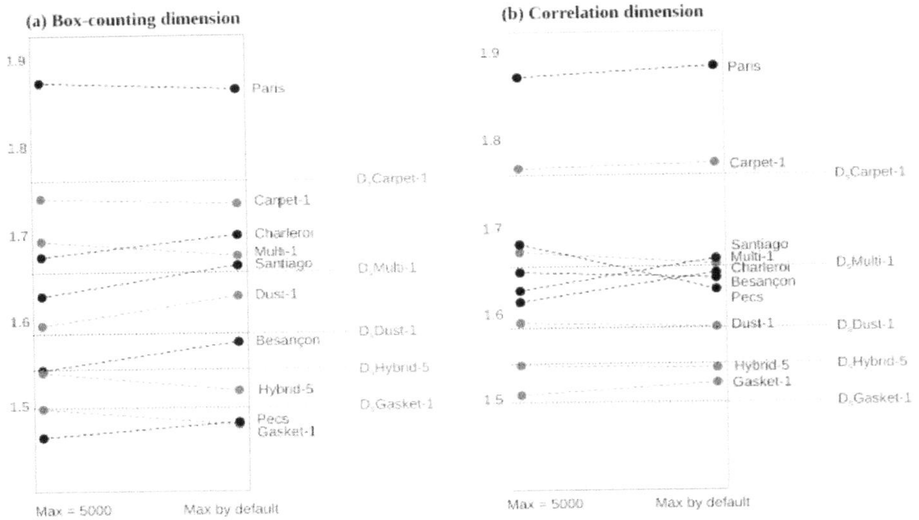

Figure 4.18. *Variation of the box-counting (a) and correlation (b) dimensions according to the maximum size of the count window*

Of the 50 built-up fabrics, the Liege fabric has the smallest side, with a length slightly greater than 10,000 m. We therefore set the common maximum at 5,000 m (i.e. 10,000 divided by 2). With this threshold, the size range of the count windows covers three orders of magnitude (ε and r range from 30 m to 5,000 m).

For the theoretical fabrics, Figure 4.18 shows that the D_C are less influenced by changes in the maximum window size than the D_B. Note that with the exception of the multi-1 fabric, the D_B of the theoretical fabrics are closer to their D_S when the size limit is set at 5,000 m. The fractal dimensions of the real fabrics seem to be more influenced by changes in the "*max*" parameter. However, the differences are quite small. Indeed, the greatest difference is observed for the Santiago fabric, for which the dimension measured with the default "*max*" parameter is greater by 0.035 than the dimension measured with the "*max*" equal to 5,000 m.

Since the influence of the maximum size of the count window is reasonable, we set this parameter to 5,000 m for all fabrics. It is important to note that the 50 built-up fabrics in the present study have different spatial extents, but all are of the same order of magnitude. The influence of the maximum size of the count window might have been quite different if a larger gap had existed between the smallest and the largest extent.

The "*coef*" parameter

The "*coef*" parameter sets the number and size of the intermediate windows (ε for box-counting dimensions and r for correlation dimensions) between the minimum value (here 30 m) and the maximum value (here 5,000 m). Assume a coefficient value equal to 2. If we decide to apply a geometric sequence (see the "*seq*" parameter) and since the first counting window has a size of 30 m, the second will have a size of 60 m (30×2), the third a size of 120 m (60×2) and so on.

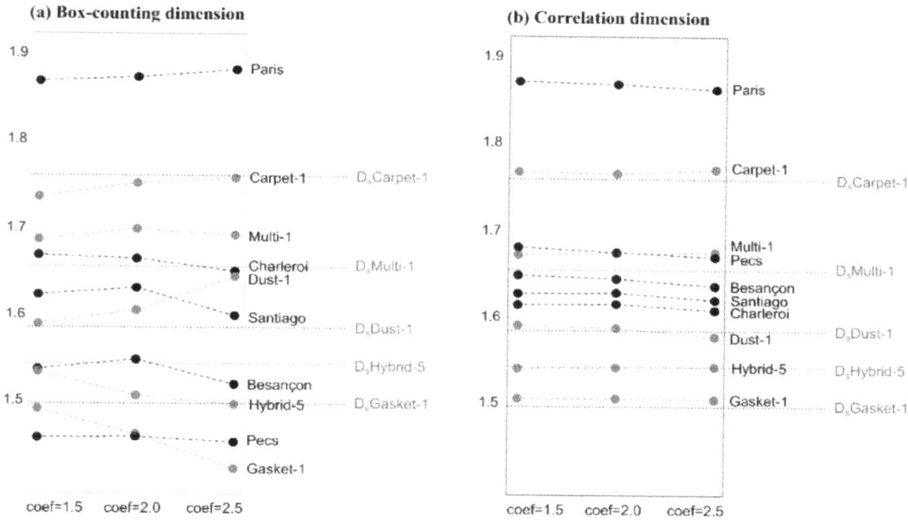

Figure 4.19. *Variation of the box-counting (a) and correlation (b) dimensions depending on the coefficient determining the size of the intermediate count windows*

It is illusory to think that a specific coefficient could make it possible to find an exact value of D_B or D_C equal to or close to D_S for all fabrics, especially theoretical fabrics. Indeed, even though the minimum size of the count window (which inevitably influences the intermediate sizes) is ignored, the coefficient would have to be equal to the reduction factor which made it possible to construct each fabric. Since these reduction factors differ from one fabric to another, it is impossible for a single coefficient to be completely satisfactory for all fabrics. We can, however, look for a coefficient that allows us to best approach D_S for a maximum number of theoretical fabrics.

Three coefficient values (1.5, 2 and 2.5) were tested. Again, the D_C are more stable than the D_B (Figure 4.19). Marked differences appear between the D_B of the theoretical tissues. It is for a coefficient value of 1.5 that the D_B are generally closer to their D_S (except for the carpet-1 fabric). It is therefore the value of 1.5 that will be fixed for the "*coef*" parameter in the remaining analyses.

Note that the quality of the adjustments, measured by the R^2 and the confidence intervals, is not fundamentally impacted by the number of intermediate windows (11 for *coef* = 1.5; 8 for *coef* = 2; 6 for *coef* = 2.5). Only the p-value is systematically lower with a smaller coefficient.

The "*seq*" parameter

The "*seq*" parameter allows us to set the sequence of sizes of the intermediate windows. The "*coef*" parameter can in fact be applied according to an arithmetic (i.e. additive) or a geometric (i.e. multiplicative) progression. Arithmetic progression is not in accord with fractal logic, which is by definition multiplicative. It is therefore more appropriate to use a geometric sequence than an arithmetic sequence.

This is confirmed by the test results (Figure 4.20). Apart from the D_C of the multi-1 fabric, all D_B and D_C are closer to their D_S with a geometric sequence. The much larger number of measurement points with the arithmetic sequence does not therefore imply a better fit.

The relevance of using a geometric sequence, which is more conducive to analyzing a fractal process than an arithmetic sequence, can be demonstrated again with the help of Figure 4.21 which presents the fitting lines of D_B for the theoretical fabric dust-1. Arithmetic progression implies that for distances ε, which are different but close to each other, the number of built pixels $N(\varepsilon)$ would be identical. This generates noise in the estimation of the line determining the D_C. This type of noise is not found when the progression is geometric. It is therefore this geometric progression that will be used for the remaining analyses.

Regarding the correlation dimension, it is important to take edge effects into account (Agterberg 2013). Indeed, a pixel at the edge of the fabric may be deprived of part of its neighborhood if this latter is located outside the image. Therefore, as the neighborhood r increases, the average number of pixels N_r is underestimated since an increasing number of neighbors located outside of the study area are not included in the count. The larger the neighborhood, the more the correlation dimension will be biased towards lower values (Carlson 1991). To counter this edge effect, it is possible to define a band around the built-up fabric being studied whose built pixels must be taken into account in counting the points in the neighborhood, but are not directly involved in the calculation of the dimension. Let us assume a built-up pixel (black) at the edge of a study area, pixel A in Figure 4.22. Pixel B, outside the study area, is counted in the vicinity of pixel A. However, the number of points in the vicinity of pixel B is not counted or taken into account, since point B is not in the study area. In the *Fractalyse* program, the *border* function allows us to define a band around the study area, of a width fixed by the user, whose pixels can be counted in the neighborhoods of the points located inside the study area.

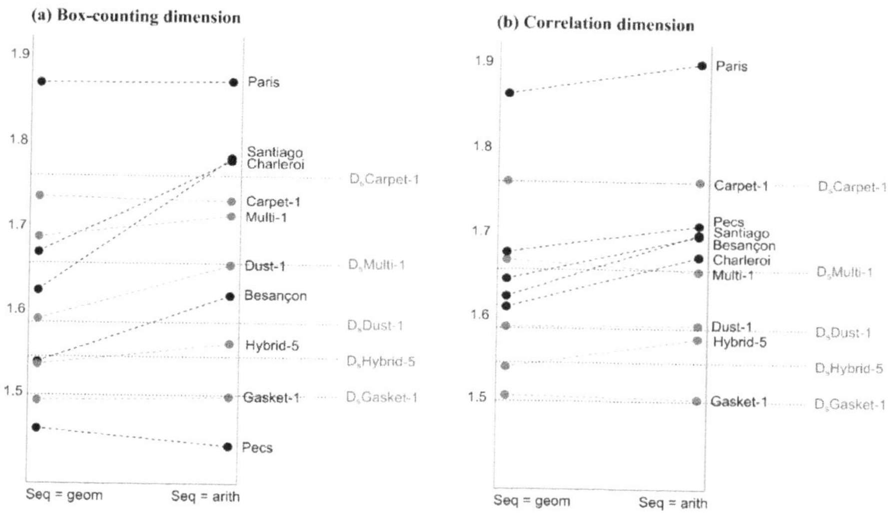

Figure 4.20. *Variation of the box-counting (a) and correlation (b) dimensions according to the type of sequence (geometric or arithmetic) determining the size of the intermediate count windows*

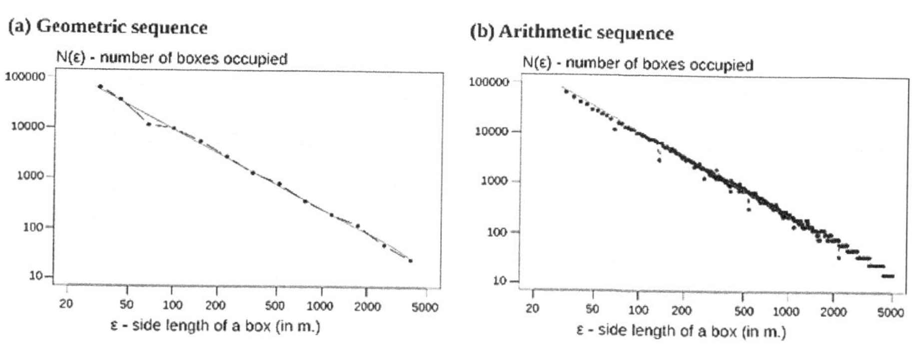

Figure 4.21. *Fractal box-counting analysis of the spatial distributions of buildings of the theoretical fabric "dust-1": geometric (a) and arithmetic (b) sequence*

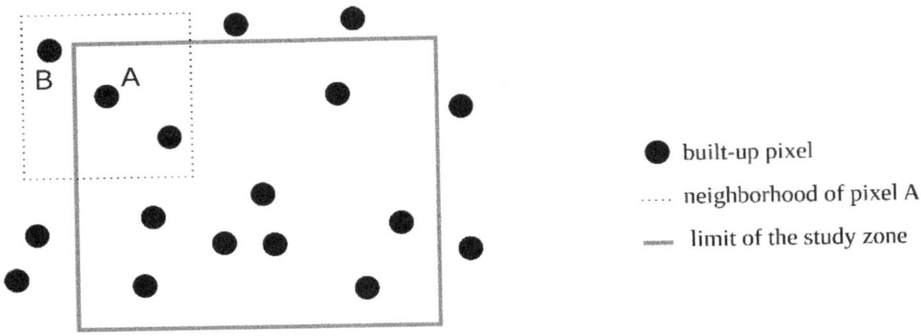

Figure 4.22. *Illustration of the function border which allows accounting for the edge effects of the correlation analysis in Fractalyse*

Since the window surrounding a pixel *i* in the calculation of a correlation dimension may have a maximum side *r* of 5,000 m, this means that any pixel *j* at a distance of 2,500 m (5,000/2) may be considered as a neighbor of *i*. For this reason, a band at least 2,500 m wide must be taken into account for each study area. The value of the *border* parameter is therefore fixed at 2,500 m[22] (Figures 4.10 and 4.11).

For the calculation of box-counting dimensions, the position of the count grid may influence the result obtained. However, it is difficult to control this parameter. With the "optimal grid position" option, *Fractalyse* allows analyses to be carried out over a predefined number of different grids. The result chosen is the one for which the position of the grid is deemed optimal, that is, the one that overlays the image with the least number of meshes. This option is only available for the analysis of files in vector format, not raster format. We performed tests on the vector formats of our sample fabrics, and these tend to show that the dimensions vary very little whether the position of the grid is optimized or not. This counterintuitive result (we expect better results for a large number of grids) is explained by the fact that the position of the grid is not identical between two steps of the count. Therefore, a grid may approach the optimum at one stage and move away from it at the next stage. Given the absence of significant differences, the box method using a single grid was retained for the analyses that follow.

22. The *border* parameter is expressed in pixels, not in meters. Its value is therefore a function of the resolution of the image being studied (*border* = (r(*max*) / resolution-1) /2). The resolution chosen for our analyses is 4 m, and the *border* parameter is set to 625.

4.2.2.3. *Fitting the theoretical curve to the empirical curve*

A count operation results in the creation of an empirical curve displaying the number of entities counted as a function of the size of the windows of analysis (number of boxes with at least one built pixel for the box-counting dimension; average number of built pixels in a certain neighborhood for the correlation dimension). To estimate the fractal dimension, it is then necessary to fit a theoretical curve to this empirical curve. In the *Fractalyse* software, it is possible to perform this adjustment directly (exponential form) or through the intermediary step of a logarithmic transformation of the empirical curve.

The type of fit used greatly influences the value of the box-counting and correlation dimensions (Figure 4.23). For these two dimensions, it is the fit after logarithmic transformation that produces the D_B and D_C closest to the D_S of the theoretical fabrics. This is explained particularly by the fact that, via a direct fit (exponential function), too much weight is given to observations made at large scales. It is therefore a logarithmic adjustment that should be preferred.

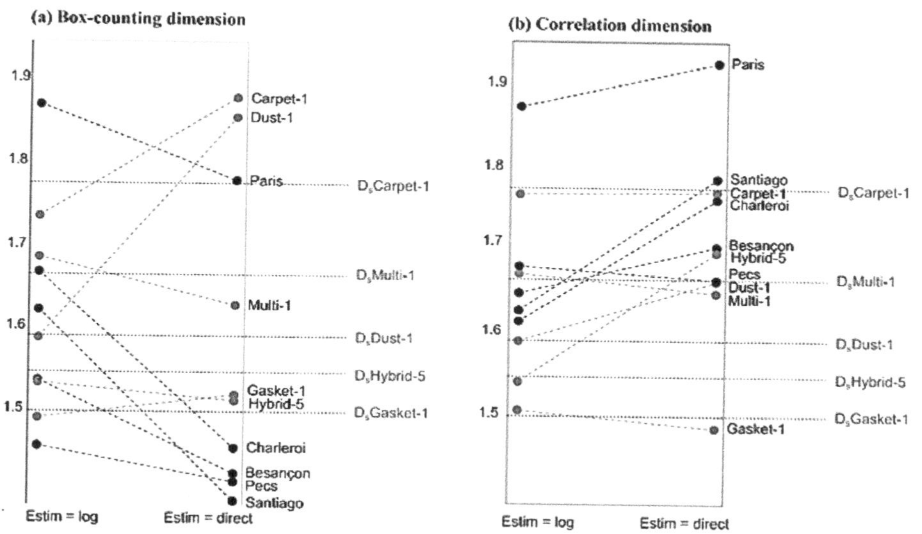

Figure 4.23. *Variation of box-counting (a) and correlation (b) dimensions depending on the type of fit (logarithmic or direct)*

4.2.2.4. *Summary of the parameter values chosen for the analyses*

Table 4.2 presents the values of parameters chosen for the calculation of the box-counting and correlation dimensions of the 50 built-up fabrics (25 theoretical fabrics and 25 real fabrics).

Box-counting dimensions are calculated on the basis of images showing only the buildings in the study area. For the correlation dimensions, in order to take edge effects into account, buildings located outside the study area over a width of 2,500 m are also taken into account as neighborhood elements. The *border* parameter, specific to the calculation of the correlation dimension in *Fractalyse*, is set at 625 pixels.

Input file (input data)	Format	Raster (.TIF)
	Resolution	4 m
Count window	Shape	Square
	Minimum size	30 m
	Maximum size	5,000 m
	Growth coefficient	1.5
	Type of growth sequence	Geometric
Fit	Estimation	Logarithmic

Table 4.2. *Summary of parameter values chosen for box-counting and correlation analyses*

4.2.3. *Comparison of the box-counting and correlation dimensions for the 50 built-up fabrics*

Table 4.3 presents the box-counting and correlation dimensions obtained for the 50 built-up fabrics with the parameter values previously set.

4.2.3.1. *Classifications for observing hierarchies*

It would be quite tedious to analyze each fabric individually. Hence, we decide to group together fabrics with identical box-counting dimensions, on the one hand, and identical correlation dimensions, on the other hand. In order to group together theoretical and real tissues that are similar in terms of their fractal dimensions, two classifications are made. The first is based on the standardized Euclidean distance between the box-counting dimensions of all the built-up fabrics. The second classification does the same for the correlation dimensions. In both cases, Ward's discretization method is chosen in order to minimize the differences between the dimensions within a single group (intra-group) and to maximize the differences between the groups (inter-group).

The dendrogram of Figure 4.24a associated with the box-counting dimension values of Table 4.3 makes it possible to distinguish seven groups of fabrics. The first group consists of three real built-up fabrics for which the box-counting dimensions are the smallest (between 1.292 and 1.414). This group merges quite quickly with groups 2 and 3 which have very similar box-counting dimensions (1.460 to 1.535 and 1.531 to 1.549, respectively). The real fabrics in these groups are Pecs, Ingolstadt (group 2) and Sherbrooke, Besançon, Timisoara and Bogota (group 3). Group 2 is mainly composed of theoretical fabrics (hybrid-1-2-3 as well as gasket-1 and multi-2). Group 3 consists of only one theoretical fabric (hybrid-5). Group 4 is the last group that can be equated with "small" box-counting dimensions (less than 1.624). It consists of nine theoretical fabrics (hybrid-4-6-7, gasket-2-3-4-5 and dust-1-2) and two real fabrics (Santiago and Bern). In total, in these four groups, there are 11 real fabrics and 15 theoretical fabrics that have the smallest box-counting dimensions in the sample. Thus, in groups 5, 6 and 7, there are 14 real fabrics and 10 theoretical fabrics with the largest box-counting dimensions. Group 5 gathers built-up fabrics whose dimension varies between 1.654 and 1.706. These are the fabrics from Bolton, Mexico City, Cochabamba, Colorado, Charleroi and Clermont-Ferrand as well as the multi-1, multi-3, hybrid-8 and hybrid-9 fabrics. Group 6 includes four real fabrics (Liège, Sheffield, Yamoussoukro, Pittsburg) and six theoretical fabrics (hybrid-10-11, gasket-1-6-8 and carpet-1). Finally, the fabrics with the highest box-counting dimensions (>1.830) are in group 7; these are Rome, Lyon, Paris and Brussels. In the end, we find that the sample of real and theoretical fabrics covers the whole spectrum of box-counting dimensions (homogeneous distribution between low- and high-dimension values).

A similar analysis can be made using the dendrogram in Figure 4.24b and the values of the correlation dimension in Table 4.3. The dendrogram suggests the presence of four groups with smaller correlation dimensions (groups 1–4) and three groups with larger correlation dimensions (groups 5, 6 and 7). Group 1 includes Haag in Oberbayern alongside the hybrid-1 and multi-2 fabrics, with correlation dimensions of less than 1.499. Group 2 has correlation dimensions varying between 1.5 and 1.548 (Ansbach, Bogota, hybrid-2-3-5 and gasket-1). With a D_C between 1.571 and 1.614, group 3 is almost exclusively composed of theoretical fabrics (gasket-2-3-4-5, hybrid-4-6-7, dust-1-2) and only one real fabric (Charleroi). Group 4 consists of nine real fabrics (Santiago, Sherbrooke, Besançon, Colorado Springs, Ahmedabad, Bern, Timisoara, Pecs and Ingolstadt) and four theoretical fabrics (multi-1-3, hybrid-8-10). Thus, 13 real tissues are in the category of smaller correlation dimensions, compared to 19 theoretical tissues. There is therefore a slight imbalance in the sample of theoretical fabrics, the vast majority of which have relatively small correlation dimensions. Group 5 (Pittsburg, Liège, Clermont-Ferrand, Bolton, Sheffield, gasket-7 and hybrid-9) has correlation dimensions between 1.713 and 1.736. In group 6, Rome, Mexico City, Yamoussoukro, Lyon, Cochabamba, gasket-6 and 8 as well as carpet-1 and hybrid-11 have correlation dimensions varying between 1.751 and 1.802. Finally, group 7 only contains the Paris and Brussels fabrics with correlation dimensions of 1.871 and 1.885, respectively.

Fabrics	D_B	D_C	Fabrics	D_B	D_C
Ahmedabad	1.414	1.668	carpet-1	1.735	1.767
Ansbach	1.385	1.548	dust-1	1.589	1.590
Bern	1.566	1.666	dust-2	1.622	1.597
Besançon	1.538	1.648	gasket-1	1.493	1.510
Bogota	1.531	1.512	gasket-2	1.581	1.606
Bolton	1.696	1.736	gasket-3	1.577	1.593
Brussels	1.922	1.885	gasket-4	1.566	1.594
Charleroi	1.668	1.614	gasket-5	1.594	1.581
Clermont-Ferrand	1.654	1.729	gasket-6	1.750	1.777
Cochabamba	1.702	1.764	gasket-7	1.721	1.734
Colorado Springs	1.706	1.654	gasket-8	1.730	1.762
Haag in Oberbayern	1.292	1.369	hybrid-1	1.485	1.471
Ingolstadt	1.512	1.673	hybrid-2	1.465	1.509
Liège	1.759	1.726	hybrid-3	1.505	1.486
Lyon	1.839	1.799	hybrid-4	1.568	1.571
Mexico City	1.704	1.788	hybrid-5	1.535	1.540
Paris	1.869	1.871	hybrid-6	1.588	1.592
Pecs	1.460	1.680	hybrid-7	1.612	1.607
Pittsburg	1.776	1.727	hybrid-8	1.704	1.667
Rome	1.830	1.779	hybrid-9	1.675	1.714
Santiago	1.623	1.627	hybrid-10	1.725	1.715
Sheffield	1.756	1.713	hybrid-11	1.758	1.749
Sherbrooke	1.549	1.651	multi-1	1.686	1.656
Timisoara	1.538	1.661	multi-2	1.464	1.444
Yamoussoukro	1.761	1.802	multi-3	1.697	1.646

Table 4.3. *Box-counting dimension D_B and correlation dimension D_C of the 50 built-up fabrics*

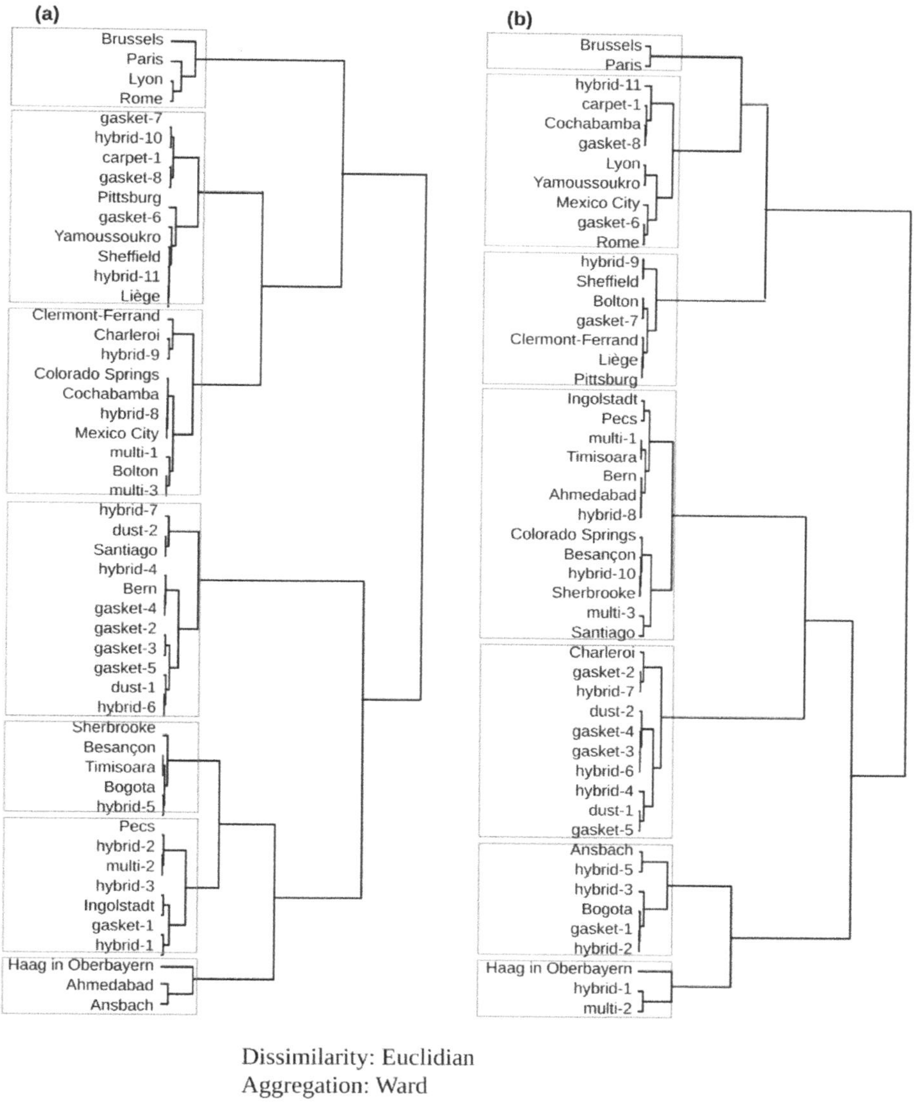

Figure 4.24. *Classification of 25 real and 25 theoretical fabrics: dendrogram based on Euclidean distance between box-counting dimensions (left) and correlation dimensions (right)*

Quite marked differences appear between the two classifications. First of all, at the level of the dimension values, the correlation dimension presents class limits that are always slightly higher than those of the box-counting dimension. Then, at the level of the hierarchy between the built-up fabrics, fabrics marked as having a fairly high box-counting dimension (respectively low) can be identified as having a fairly low correlation dimension (respectively high). This is the case, for example, for the Charleroi fabric that is in group 5 of the box-counting dimensions but in group 3 of the correlation dimensions. Let us assume that the fabrics are classified from 1 (the lowest dimension value) to 50 (the highest dimension value). The Charleroi fabric is in 28th place for D_B but only at 19th place for D_C. Conversely, a fabric like that of Ahmedabad goes from 3rd place (group 1) for D_B to 30th place (group 4) with D_C. These reversals of hierarchy are also observable for theoretical fabrics, but to a lesser extent. For example, dust-1 is in 22nd place for D_B (group 4) and 12th place for D_C (group 3). In the other direction (the ranking of the fabric in D_C is higher than that in D_B), there is almost no change in hierarchy. The most marked difference is for the hybrid-9 fabric, which "loses" five places between the hierarchy of D_C and that of D_B.

In summary, quite different hierarchies appear depending on whether the fractal dimension is calculated by box-counting or by correlation. These differences are especially marked for real fabrics. This observation should prompt researchers to be cautious when classifying geographic entities (municipalities, district, region, etc.) based on their fractal dimension, or whether they compare the value of a fractal dimension to other indicators (see, for example, Shen (2002)). Indeed, the impact of the fractal analysis method chosen on the results obtained is significant.

4.2.3.2. *Dimensions and morphological characteristics*

The box-counting and correlation dimensions theoretically have identical values for strictly self-similar objects. Figure 4.25 shows that there is a clear linear correlation between the two dimensions. However, the equality between the two dimensions is far from perfect[23] especially for real fabrics. Some built-up fabrics have a correlation dimension substantially larger than the box-counting dimension. For eight real fabrics, these dimensional differences reach values greater than or equal to 0.1. The largest difference, with a value of 0.254, is calculated for the built-up fabric of Ahmedabad. Conversely, some fabrics have correlation dimensions slightly smaller than their box dimension. However, this trend is weaker. The two largest differences are calculated here for the built-up fabrics of Charleroi (difference = 0.054) and Colorado Springs (0.052). These differences between the

23. Note here that it is difficult to assess the difference between two dimensions. Does a dimension difference of 0.05 represent a real difference, or is it solely related to inaccuracies or estimation bias? We therefore emphasize the relatively larger differences.

box dimension and the correlation dimension are at the origin of the different hierarchies noted earlier.

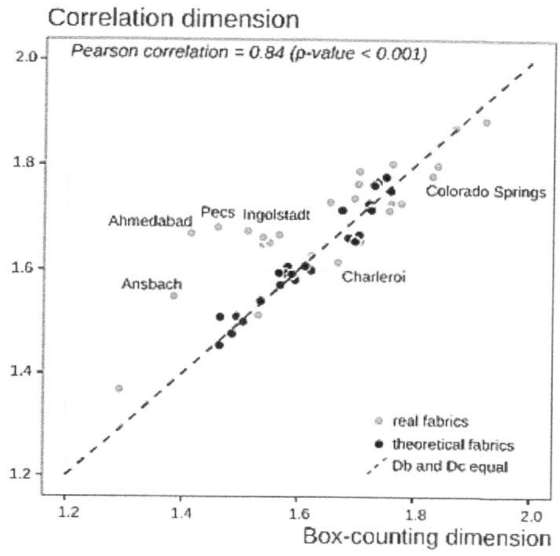

Figure 4.25. *Relationship between the box-counting and correlation dimensions for the 50 built-up fabrics*

We will try to understand what leads to the differences between the two dimensions. To do this, we will compare the values of the dimensions with those of other morphological indicators that characterize the 50 built-up fabrics. Since fractal dimensions partly describe the amount of space occupied, we compare them to the built-up density index. Figure 4.26 shows that there is a positive relationship between the dimensions and the logarithm of built-up density. This relationship is more pronounced between the built-up density and the box-counting dimension (R^2 of a simple regression = 0.84) than between the density and the correlation dimension (R^2 of a simple regression = 0.52). The box-counting dimension would therefore be comparatively more sensitive to the built-up density than the correlation dimension is.

Figure 4.26 also shows that for the same built-up density, the real fabrics have box-counting and correlation dimensions systematically larger than those of the theoretical fabrics. This can be explained using another morphological index: the Landscape Shape Index (LSI). This index is widely used in landscape ecology because it describes the complexity of a landscape by comparing it to the perfect figure of a circle (Gyenizse et al. 2014). This index is based on the ratio between the perimeter and the area of the object studied[24]. The higher the index, the longer the perimeter is in relation to the surface, and therefore, the more tortuous and complex the landscape. Although we have endeavored to design theoretical fabrics that match the characteristics of real fabrics, the latter are more complex, as shown in Figure 4.27. The box-counting and correlation dimensions are therefore also sensitive to the complexity of the built-up fabrics.

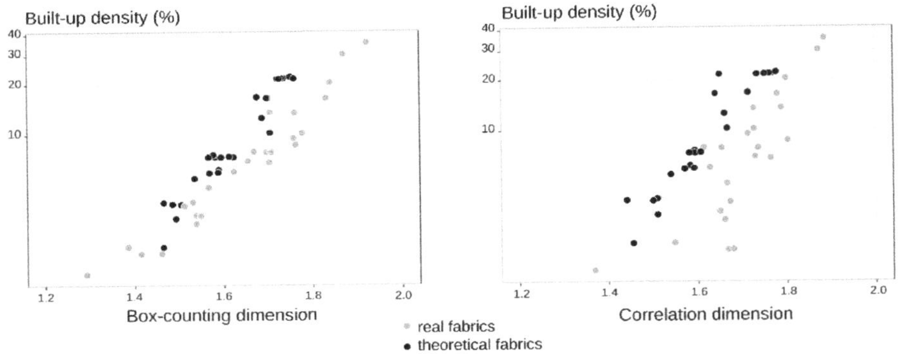

Figure 4.26. *Relationship between the logarithm of the built-up density and the box-counting (left) and correlation (right) dimensions of the 50 built-up fabrics*

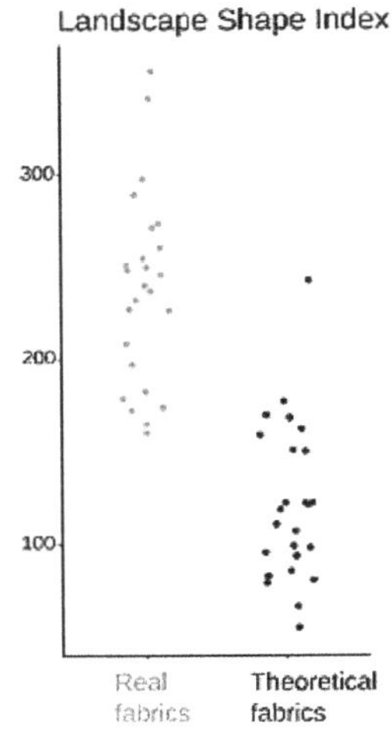

Figure 4.27. *Landscape Shape Index of real and theoretical fabrics*

24. LSI = perimeter ÷ (2 (area × π)$^{1/2}$).

In section 4.2.1.3, we saw that the delineation of the study zones in our sample is likely to influence the estimated values of the fractal dimension. In order to reduce this possible influence, we have chosen to consider study zones whose extents are of the same order of magnitude. Regarding the theoretical built-up fabrics, we further controlled the position of each study zone to prevent the relevant fractal spatial structures being truncated. With these precautions taken, Figure 4.28 shows that, in the case of real built-up fabrics, the difference in value between D_B and D_C tends to increase as the study zones increase in extent.

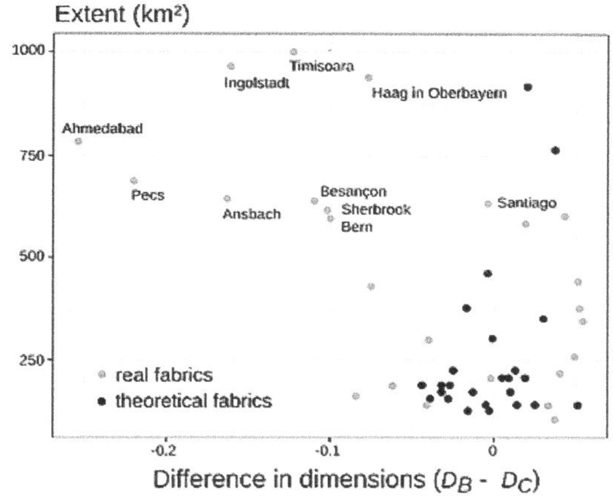

Figure 4.28. *Relationship between the difference in box-counting and correlation dimensions and the extent of the 50 built-up fabrics*

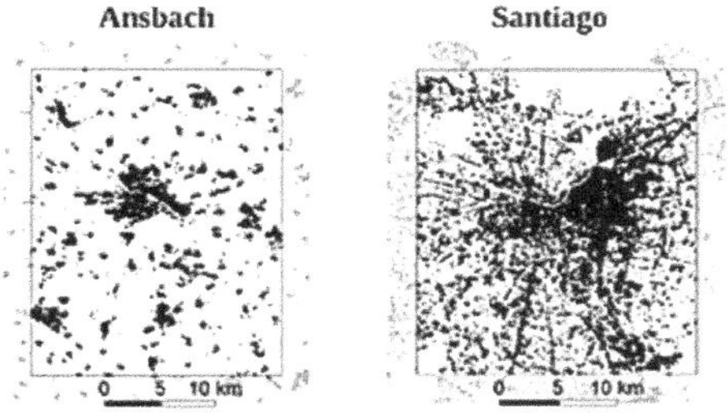

	Ansbach	Santiago
Spatial extent (km²)	643	631
Built-up density (%)	2.12	5.9
Landscape Shape Index	161	271
D_B	1.385	1.548
D_C	1.623	1.627

Figure 4.29. *Comparison of the Ansbach and Santiago fabrics. For a color version of this figure, see www.iste.co.uk/tannier/fractal.zip*

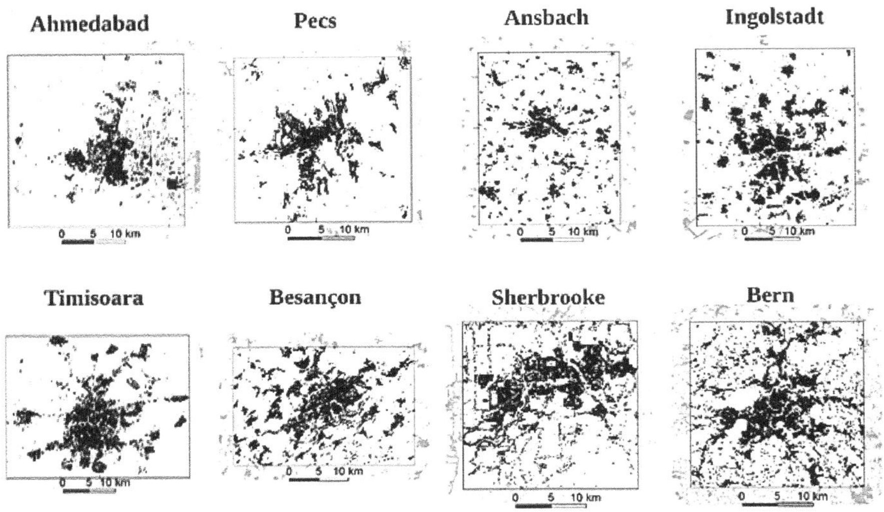

Figure 4.30. *The eight built-up fabrics for which the correlation dimension is greater than the box-counting dimension (difference greater than or equal to 0.1). For a color version of this figure, see www.iste.co.uk/tannier/fractal.zip*

Take, for example, the Ansbach and Santiago fabrics, whose spatial extents and built-up densities are quite comparable but whose LSI is different, that of Santiago being higher (Figure 4.29). Their correlation dimensions are quite close although the Santiago fabric, being more complex (larger LSI), has a slightly higher correlation dimension than that of Ansbach. However, their box-counting dimensions are quite different, the Ansbach box-counting dimension being much smaller than its correlation dimension. The built-up fabric of Santiago appears visually more homogeneous, while that of Ansbach includes a large number of unbuilt spaces of varying sizes whose spatial distribution is irregular. It might therefore be the presence of unbuilt spaces of irregular sizes and spatial distribution that causes differences between the box-counting dimension and the correlation dimension. Such a configuration would appear all the more frequently as the study areas considered grew larger. Figure 4.30, which shows the eight real built-up fabrics whose correlation dimension is greater than their box-counting dimension and whose study zone extent is relatively wide, tends to visually confirm this hypothesis.

4.3. Conclusion

In this chapter, we have presented the founding principles of estimating the fractal dimension of sets of human settlements represented in the form of points (intersections of roads, building centroids, built-up pixels, etc.), lines (linear networks, contours of buildings or built clusters) or polygons (buildings mapped in 2D). We have described in detail the methods of calculating different fractal dimensions.

The characterization and differentiation of built-up fabrics or road and public transport networks according to their fractal dimensions is relevant, subject to choosing indicators that are able to capture the types of spatial irregularities that we assume exist and perform analyses according to a rigorous protocol, using data of the same nature at a sufficiently fine spatial resolution. It is possible to use the ordinal classification of fractal dimensions to identify different types of built-up fabrics or networks, but it is not known what credence should be given to a small difference in dimension values (e.g. D =1.56 and D =1.58)[25]. Nor is it possible to thematically interpret differences in dimensions that we know, by experience, to be significant (e.g. D = 1.50 and D = 1.65).

Fractal dimensions calculated by box-counting and correlation methods considering the footprint of buildings are frequently used for the characterization of built-up fabrics. Both reflect a high built-up density and complex spatial organization of buildings. However, depending on the case, these two dimensions may or may not take different values for the same object. They are therefore not susceptible to the same types of irregularities and not directly comparable. Estimation of the fractal box-counting and correlation dimensions is dependent on a series of parameters. The analyses performed in the second part of this chapter show that for built-up fabrics (both theoretically constructed and real) with a spatial extent of the same order of magnitude, it is possible to set many of these parameters in order to allow comparisons between built-up fabrics, considering one or the other dimension, without biasing the estimates of fractal dimensions. It is particularly interesting to perform the analyses based on fine-resolution raster files (e.g. 4 m). A minimum count window size of 30 m allows robust results to be obtained. Starting from this minimum size, provided that all fabrics have a geographical extent of the same order of magnitude, the number and size of the intermediate count windows can be fixed by adopting an identical multiplier for all fabrics, applied in a geometric sequence up to the same maximum window size. Finally, a logarithmic fit of the empirical count curve to a theoretical curve must be performed in order to obtain reliable results close to the similarity dimension.

25. "If two surfaces are computed with D values of 2.21 and 2.29, would they be considered to be derived the same, thus indicating the same degree of complexity?" (Lam 2009).

4.4. References

Agterberg, F.P. (2013). Fractals and spatial statistics of point patterns. *Journal of Earth Science*, 24(1), 1–11.

Anscombe, F.J. (1973). Graphs in statistical analysis. *American Statistician*, 27, 17–21.

Arlinghaus, S.L. (1985). Fractals take a central place. *Geografiska Annaler, Series B, Human Geography*, 67(2), 83–88.

Arlinghaus, S.L. (1987). *Essays on Mathematical Geography II*. Institute of Mathematical Geography, Ann Arbor.

Baddeley, A., Rubak, E., Turner, R. (2015). *Spatial Point Patterns: Methodology and Applications with R*. Chapman and Hall/CRC Press, London.

Barnsley, M. (1988). *Fractals Everywhere*. Academic Press, New York.

Batty, M. and Longley, P.A. (1986). The fractal simulation of urban structure. *Environment and Planning A*, 18, 1143–1179.

Batty, M. and Longley, P.A. (1987). Urban shapes as fractals. *Area*, 19(3), 215–221.

Batty, M. and Longley, P.A. (1994). *Fractal Cities: A Geometry of Form and Function*. Academic Press, San Diego and London.

Batty, M. and Xie, Y. (1996). Preliminary evidence for a theory of the fractal city. *Environment and Planning A*, 28, 1745–1762.

Benguigui, L., Czamanski, D., Marinov, M., Portugali, Y. (2000). When and where is a city fractal? *Environment and Planning B: Planning and Design*, 27, 507–519.

Bouligand, G. (1929). Sur la notion d'ordre de mesure d'un ensemble plan. *Bulletin des sciences mathématiques*, II(53), 185–192.

Brunsdon, C., Fotheringham, A.S., Charlton, M. (1996). Geographically weighted regression: A method for exploring spatial nonstationarity. *Geographical Analysis*, 28(4), 281–298.

Brunsdon, C., Fotheringham, A.S., Charlton, M. (2002). Geographically weighted summary statistics – A framework for localised exploratory data analysis. *Computers, Environment and Urban Systems*, 26(6), 501–524.

Carlson, C.A. (1991). Spatial distribution of ore deposits. *Geology*, 19(2), 111–114.

Chen, Y. (2011). Derivation of the functional relations between fractal dimension of and shape indices of urban form. *Computers, Environment and Urban Systems*, 35, 442–451.

De Keersmaecker, M.-L., Frankhauser, P., Thomas, I. (2003). Using fractal dimensions for characterizing intra-urban diversity. The example of Brussels. *Geographical Analysis*, 35(4), 310–328.

Diggle, P.J. (2013). *Statistical Analysis of Spatial and Spatio-Temporal Point Patterns*. Chapman and Hall/CRC Press, London.

Falconer, K.J. (1990). *Fractal Geometry: Mathematical Foundations and Applications*. John Wiley & Sons, Chichester.

Falconer, K.J. (2013). *Fractals. A Very Short Introduction*. Oxford University Press.

Feng, J. and Chen, Y. (2010). Spatiotemporal evolution of urban form and land-use structure in Hangzhou, China: Evidence from fractals. *Environment and Planning B: Planning and Design*, 37(5), 838–856.

Fotheringham, A.S., Batty, M., Longley, P.A. (1989). Diffusion-limited aggregation and the fractal nature of urban growth. *Papers of the Regional Science Association*, 67, 55–69.

Fotheringham, A.S., Charlton, M.E., Brunsdon, C. (1998). Geographically weighted regression: A natural evolution of the expansion method for spatial data analysis. *Environment and Planning A*, 30(11), 1905–1927.

Frankhauser, P. (1988). Fractal aspects of urban systems. *Sonderforschungsbereich 230 "Natürliche Konstruktionen"*, 1, 67–76.

Frankhauser, P. (1990). Aspects fractals des structures urbaines. *L'Espace géographique*, 19–20(1), 45–69.

Frankhauser, P. (1994). *La fractalité des structures urbaines*. Anthropos, Paris.

Frankhauser, P. (2003). Morphologie des "villes émergentes" en Europe à travers les analyses fractales. Rapport de recherche pour le Ministère de l'équipement, des transports et du logement. Report, lettre de commande n°F0001 PUCA, 242.

Ge, M. and Liu, Q. (2009). Realizing the box-counting method for calculating fractal dimension of urban form based on remote sensing image. *Geo-spatial Information Science*, 12(4), 265–270.

Gonzato, G. (1998). A practical implementation of the box counting algorithm. *Computational Geosciences*, 24, 95–100.

Gonzato, G., Mulargia, F., Marzocchi, W. (1998). Practical application of fractal analysis: Problems and solutions. *Geophysical Journal International*, 132(2), 275–282.

Goodchild, M. (1980). Fractals and the accuracy of geographical measures. *Mathematical Geology*, 12(2), 85–98.

Goodchild, M. and Mark, D. (1987). The fractal nature of geographic phenomena. *Annals of the Association of American Geographers*, 77(2), 265–278.

Grassberger, P. and Procaccia, I. (1983). Measuring the strangeness of strange attractors. *Physica D*, 9, 189–208.

Gyenizse, P., Bognár, Z., Czigány, S., Elekes, T. (2014). Landscape shape index, as a potential indicator of urban development in Hungary. *Landscape & Environment*, 8(2), 78–88.

Hadzieva, E., Capeska Bogatinoska, D., Gjergjeska, L., Shuminoska, M., Petroski, R. (2015). Review of the software packages for estimation of the fractal dimension. *ICT Innovations 2015 Web Proceedings*, 22, 201–211.

Harris, P., Clarke, A., Juggins, S., Brunsdon, C., Charlton, M. (2015). Enhancements to a geographically weighted principal component analysis in the context of an application to an environmental data set. *Geographical Analysis*, 47(2), 146–172.

Hausdorff, F. (1919). Dimension und äusseres Mass. *Mathematische Annalen*, 79, 157–179.

Lam, N.S. (2009). Fractal analysis. In *International Encyclopedia of Human Geography*, Kitchin, R. and Thrift, N. (eds). Elsevier, Amsterdam.

Longley, P.A. and Mesev, V. (1997). Beyond the analogue models: Space filling and density measurement of an urban settlement. *Papers in Regional Science*, 76(4), 409–427.

Longley, P.A. and Mesev, V. (2000). On the measurement and generalisation of urban form. *Environment and Planning A*, 32, 473–488.

Makse, H.A., Andrade, J.S., Batty, M., Havlin, S., Stanley, H.E. (1998). Modeling urban growth patterns with correlated percolation. *Physical Review E*, 58(6), 7054–7062.

Mandelbrot, B. (1975). *Les objets fractals. Forme, hasard et dimension*. Flammarion, Paris.

Minkowski, H. (1903). Volumen und Oberfläche. *Mathematische Annalen*, 57, 447–495.

Openshaw, S. (1983). *The Modifiable Areal Unit Problem. Concepts and Techniques in Modern Geography*. Geobooks, Norwich.

Pumain, D. (2010). Commentaire : transition fractal-non fractal et objets géographiques. *L'Espace géographique*, 39(2), 113–116.

Ripley, B.D. (1977). Modelling spatial patterns. *Journal of the Royal Statistical Society: Series B (Methodological)*, 39(2), 172–192.

Schroeder, M.R. (1991). *Fractals, Chaos, Power Laws: Minutes from an Infinite Paradise*. W.H. Freeman & Co, New York.

Sémécurbe, F. (2020). Analyse de la distribution spatiale des implantations humaines – apports et limites d'indicateurs multi-échelles et trans-échelles. PhD Thesis, Université de Franche-Comté, Besançon.

Sémécurbe, F., Tannier, C., Roux, S.G. (2019). Applying two fractal methods to characterise the local and global deviations from scale-invariance of built patterns throughout mainland France. *Journal of Geographical Systems*, 21(2), 271–293.

Shen, G. (2002). Fractal dimension and fractal growth of urbanized areas. *International Journal of Geographical Information Science*, 16(5), 437–519.

Stewart, I. (1989). *Does God Play Dice? The New Mathematics of Chaos*. Basil Blackwell, Oxford and New York.

Tannier, C. and Pumain, D. (2005). Fractals in urban geography: A theoretical outline and an empirical example. *Cybergeo: European Journal of Geography*, 307. doi: 10.4000/cybergeo.3275.

Tannier, C. and Thomas, I. (2013). Defining and characterizing urban boundaries: A fractal analysis of theoretical cities and Belgian cities. *Computers, Environment and Urban Systems*, 41, 234–248.

Theiler, J. (1990). Estimating fractal dimension. *Journal of the Optical Society of America Part A*, 7(6), 1055–1073.

Thomas, I., Frankhauser, P., De Keersmaecker, M. (2007). Fractal dimension versus density of the built-up surfaces in the periphery of Brussels. *Papers in Regional Science*, 86(2), 287–307.

Thomas, I., Frankhauser, P., Biernacki, C. (2008). The morphology of built-up landscapes in Wallonia (Belgium): A classification using fractal indices. *Landscape and Urban Planning*, 84, 99–115.

Thomas, I., Frankhauser, P., Badariotti, D. (2010). Comparing the fractality of European urban neighbourhoods: Do national contexts matter? *Journal of Geographical Systems*, 14, 189–208.

Tufte, E. (1983). *The Visual Display of Quantitative Information*. Graphics Press, Cheshire, Connecticut.

Vicsek, T. (1990). Mass multifractals. *Physica A: Statistical Mechanics and Its Applications*, 168(1), 490–497.

5

Morphological Delineation of Urban Agglomerations: Comparison of Fractal and Non-Fractal Methods

Gaëtan MONTERO[1], Cécile TANNIER[2] and Isabelle THOMAS[1]

[1] *Centre for Operational Research and Econometrics (CORE), Université catholique de Louvain-la Neuve, Belgium*

[2] *Laboratoire ThéMA, CNRS-Université de Franche-Comté, Besançon, France*

The delimitation of cities is a challenge for planners and scientists. The former need to know the limits of urban agglomerations in order to govern them in the best possible way (organizing transport systems, collecting taxes, etc.). The latter need it to study the growth of cities retrospectively and prospectively (see, for example, Tayyebi et al. 2011; Chakraborti et al. 2018; Wang et al. 2021; de Olde and Oosterlynck 2021) and measure the impacts that different boundaries may have on the values of social, economic and ecological indicators (see, for example, Ding et al. 1999; Han et al. 2009; Haisch and Müller 2015; Uchiyama and Mori 2017; Thomas et al. 2018). The challenge is significant, and many researchers are working to provide solutions. However, there is no consensus on the most appropriate method for delineating cities. Different methods have been applied, which tend to yield different results (for examples and literature reviews, see: Ferreira et al. 2010; Rozenfeld et al. 2011; Thomas et al. 2018; Cottineau et al. 2019).

Fractal Geometry in Human Geography and Planning,
coordinated by Cécile TANNIER.

Two main strategies are generally adopted. The first is called the "morphological" strategy, because it focuses on purely physical elements such as buildings or roads (see, for example, Medda et al. (1998), Thomas et al. (2018) and many other articles cited later in this chapter). The second is called the "functional" strategy as it involves characteristics related to the functions of places (e.g. the number of sewers and garbage cans or the number of parks and public gardens) or to the individuals who live and move around in those places (see, for example, Blondel et al. 2010; Kauffmann 2012; Guérois et al. 2014; Fujishima et al. 2020; Sotomayor-Gómez and Samaniego 2020; Chen 2021; Rybnikova et al. 2021; Chen and Gar-On Yeh 2022; Dong et al. 2022). The two strategies complement each other in that they each highlight the specific realities of the form and function of cities (Thomas et al. 2012; Adam et al. 2019).

From a morphological point of view, it is possible to use the fractality of the built-up fabrics of cities as a basis for identifying their limits. Indeed, analysis of geographical phenomena on several scales makes it possible to highlight crucial spatial discontinuities (Lam and Quattrochi 1992). The limits of a city then separate spaces that are morphologically different from a fractal point of view, namely the city and its periphery. Based on this principle, Tannier et al. (2011) and Tannier and Thomas (2013) proposed a method called *MorphoLim*, which uses the study of the fractality of built-up fabrics as a basis for identifying the morphological limits of cities. In this chapter, this method is compared to two other non-fractal methods of morphological delineation of cities (section 5.1). It is then applied to 82 French urban areas (section 5.2), in order to study the morphological characteristics of cities related to the delimitations obtained and propose a typology of cities based on these characteristics.

5.1. Comparison of three fractal and non-fractal methods of morphological delimitation of cities

Methods of morphological delimitation of cities consider their elementary physical structures: buildings, roads, plots (Moudon 1997) or types of land use. Based on the scientific literature in the field, Montero et al. (2021) identified four broad categories of methods:

– *Density measures*: a space is considered "urban" if its density is greater than a fixed value. Density is the mathematical relationship between a mass (e.g. a number of buildings) and the surface area of a basic spatial unit (e.g. an administrative entity or a cell of a regular grid) (Schoning et al. 1999; Borruso 2003; Hu et al. 2015; Oliveira et al. 2015; Zhou 2015; Zhou and Guo 2018; Harig et al. 2021; Tan and Huang 2022).

– *Classifications of satellite images or aerial photographs*: contiguous pixels belonging to the same class are grouped together. Among these groups of pixels, some are considered urban, while others are considered suburban or rural (Weber 2001; Abel and Kaysi 2003; Henderson et al. 2003; Zhang et al. 2008; Jiang et al. 2016; Nicolau and Cavaco 2018; Zhou et al. 2018; Ban et al. 2021).

– *Clustering methods*: depending on their proximity and possibly other properties (e.g. density, spatial auto-correlation), physical elements (e.g. buildings, road intersections) are grouped together. The clusters thus formed represent one or more morphological agglomerations (i.e. cities) (Chaudhry and Mackaness 2008; Rozenfeld et al. 2008, 2011; Deng et al. 2011; Liu et al. 2012, 2013; Zhong and Duckham 2016; Caruso et al. 2017; Ma et al. 2017; Maisonobe et al. 2018; Wu et al. 2018; Arribas-Bel et al. 2019; Usui 2019; Xu et al. 2021; Montero et al. 2022).

– Analyses of the statistical distribution, across scales, of a particular morphological variable, such as the distance between buildings. Identification of one or more thresholds in the distribution and application of the threshold(s) to delimit the city or cities (Jia and Jiang 2010; Tannier et al. 2011; Tannier and Thomas 2013; Jiang and Miao 2015; Massuci et al. 2015; Wang et al. 2015; Arcaute et al. 2016; Long 2016; Chen et al. 2017; Chen and Jiang 2018; Behnisch et al. 2019; Liu and Huang 2019; Liu et al. 2019; Jayasinghe et al. 2021).

The methods in the first two categories rely on clustered data, that is, the elements under consideration (pixels, buildings, etc.) are clustered into basic spatial units (BSUs). These BSUs are generally administrative entities (neighborhoods, municipalities etc.) or cells of a regular grid. By using BSUs, the boundaries obtained with these methods are inevitably dependent on the level of clustering adopted, as well as on the shape and position of BSUs. These dependencies are well known to geographers as the *modifiable areal unit problem* (MAUP) (Openshaw 1983). To limit the impact of the MAUP, researchers sometimes perform smoothing on the characteristics of BSUs via, for example, kernel-smoothed density estimates (Zhou 2015; de Bellefon et al. 2019; Ban et al. 2021). However, smoothing functions are themselves dependent on parameters, namely the bandwidth and the kernel density. Zhou and Guo (2018) compared the delimitations of New Zealand cities obtained for different spatial resolutions of the input data (pixels of varying sizes) and different settings of the smoothing function. They found that the threshold values of the most appropriate parameters for the delimitation of different cities are mostly identical or very close to each other. In some cases, however, these threshold values differ significantly.

Morphological delimitation methods based on clustering (third category of methods) or analysis of a statistical distribution across scales (fourth category of methods) have the advantage of avoiding the MAUP problems related to the clustering of elements (buildings, pixels, etc.) in BSUs from the start. Hence, we chose to compare the *MorphoLim* fractal delimitation method of Tannier et al. (2011), which falls into the fourth category, to two methods that also involve disaggregated data. The first method, *A-DBSCAN*, was proposed by Arribas-Bel et al. (2019). This is a clustering method (third category) based on a criterion of density and minimum distance between the centroids of buildings. The second method belongs to the fourth category and delimits *Natural Cities*, based on the study of a statistical distribution of distances between points of interest, which will, in our case, be the intersections of roads. The *Natural Cities* method consists of a combination of procedures proposed in different articles (Jiang and Jia 2011; Jiang 2013, 2015, 2019; Jiang and Miao 2015).

There are few studies comparing different methods of morphological delimitation of cities, and a fortiori even fewer comparing methods whose basic principles are different (belonging to one of the four categories listed above). For instance, Cetinkaya et al. (2015) compare four delimitation algorithms, which are all applied to graphs (third category of methods). Boruso (2003) compares the results of a delimitation obtained using road density via a grid and the application of a kernel smoothing function (first category). Montero et al. (2021) compare three methods, all of which fall into the fourth category: *MorphoLim*, *Natural Cities* and *Hierarchical Percolation* (Massuci et al. 2015; Arcaute et al. 2016).

As the *MorphoLim* and *Natural Cities* methods have already been compared in the article by Montero et al. (2021), in this chapter we will focus on comparing them with the *A-DBSCAN* method. The city studied is the same as in the Montero et al. (2021) study, namely Brussels, the capital of Belgium, but the cadastral data we use are more recent. Comparing the results of this study with that conducted in 2021 should make it possible to highlight certain changes in the morphological agglomeration of Brussels.

5.1.1. *Presentation and critique of the three methods*

5.1.1.1. *A-DBSCAN*

A-DBSCAN is an algorithm derived from the *DBSCAN* method (Density-Based Spatial Clustering of Application with Noise), which is a method of delimiting clusters proposed by Ester et al. (1996). *DBSCAN* clusters together those points whose neighborhood is sufficiently dense. The density is defined using two parameters: the neighborhood distance around a point *i* (*eps*) and the number of points *pts* contained in a circle of radius *eps*. A point *i* belongs to a cluster if, in a circle of radius *eps* centered at *i*, there are a number of neighbors greater than or equal to a minimum number of neighbors (*minpts*). The method can be applied to centroids of buildings to delineate clusters within which a minimum density of buildings is measured.

To delimit the clusters, *DBSCAN* proceeds as shown in Figure 5.1. A radius equal to the maximum distance at which two building centroids are considered to be neighbors (*eps*) is traced around each point. If the number of points within a particular circle is greater than or equal to the minimum number of points to form a cluster (*minpts*), the center point of that circle is called the "Core Point". If a point at the center of a particular circle does not meet the minimum number of points criterion but is at a lesser distance from a "Core Point" than the distance *eps* fixed at the start, this point is called a "Border Point". Points that do not meet any of these criteria are called "Noise Points". The delimitation of clusters is performed around "Core Points" and adjacent "Border Points", excluding "Noise Points".

DBSCAN has the advantage of being based on a simple criterion, often used in spatial planning: density. This is indeed a metric widely used for describing a city and discussing its sustainability (Boyko and Cooper 2011; Ewing et al. 2018; Berghauser-Pont et al. 2021; etc.). The method also has the advantage of using disaggregated spatial data (building centroids), which avoids the problems encountered in conventional density measurements that aggregate data according to BSUs such as administrative entities or grid cells. However, the method is not free from a priori parameters. Indeed, researchers and urban planners who use it must choose a minimum number of buildings in advance that will form a cluster, as well as the maximum distance between these buildings. These choices, whether reasoned and motivated or not, inevitably influence the final result (Sémécurbe 2020). Let us return to the built-up fabric of Figure 5.1 but, rather than applying a maximum distance of 120 m, let us apply a threshold of 90 m. The surface area of the agglomeration is then reduced (Figure 5.2). This dependence of the results on threshold values that are fixed a priori makes comparisons between different studies and/or different cities difficult. How do we compare delimitations when the thresholds set are different? For a fair comparison, a single threshold should probably be applied to all study areas. But is it relevant to use identical threshold values between a medieval European city and a sprawling North American city?

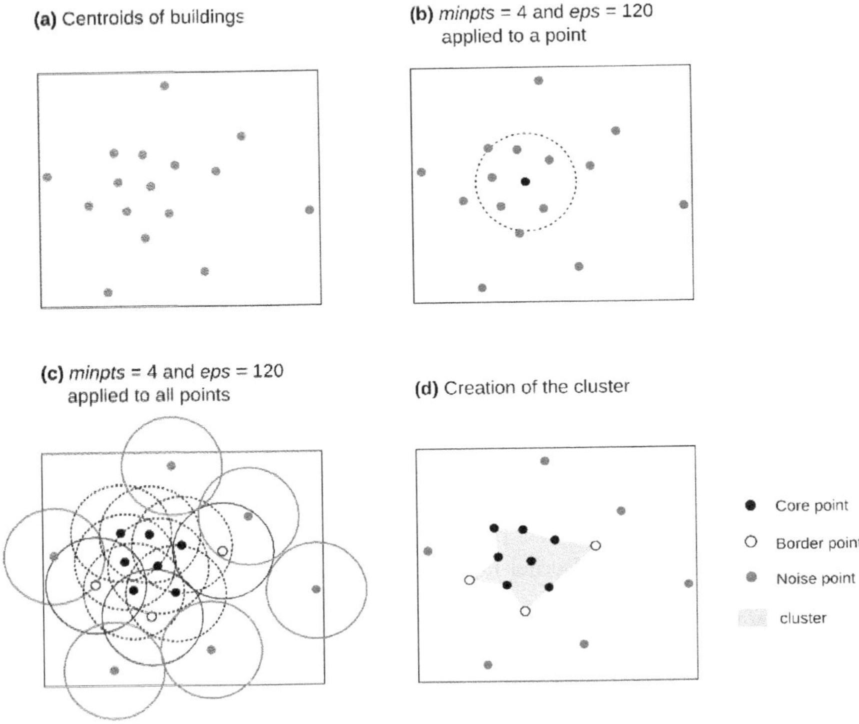

Figure 5.1. *DBSCAN method for the delimitation of clusters*

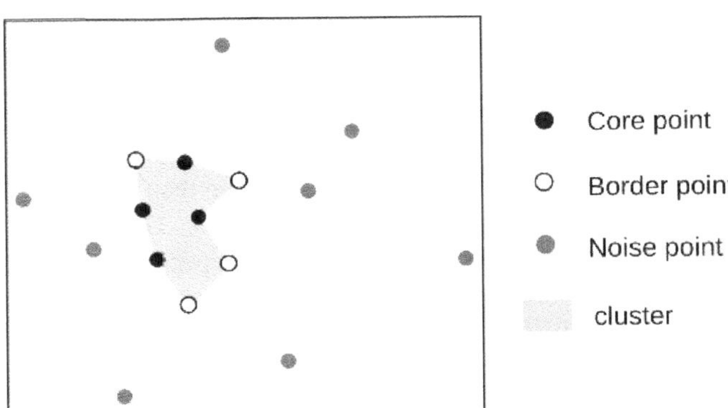

Figure 5.2. *Application of the DBSCAN method to the centroids of the built-up fabric of Figure 5.1 with minpts = 6 and eps = 10*

On the contrary, the *DBSCAN* method depends very little on the extent of the study zone. Indeed, whether a point belongs to an agglomeration depends on its number of neighbors within a certain distance, which is a priori much smaller than the size of the study zone. The extent of the latter will therefore have little impact on how this point is categorized. This is true for all points in a cluster localized far enough away from the boundaries of the study zone. For example, the extension of the study zone between Figure 5.3(b) and (c) does not imply a change in the delineation of the agglomeration. However, if the morphological agglomeration extends to the edge of the study zone (Figure 5.3(a)), it is very likely that its surface area will increase as the study zone grows (Figure 5.3(b)). Points on the edge of the study zone may change their status from "Noise Point" to "Border Point" or "Core Point" if a number of neighboring points are added to the database as the study zone is expanded. It is therefore appropriate to ensure that the extent of the study zone is sufficient to identify the boundaries of an agglomeration with *DBSCAN*.

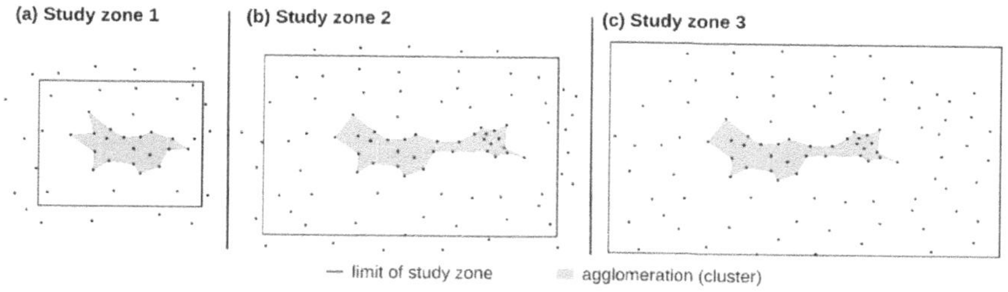

Figure 5.3. *Influence of the spatial extent of the study zone on the delimitation of an agglomeration with the DBSCAN method*

Arribas-Bel et al. (2019) found that the *DBSCAN* method leads to a less robust delimitation of clusters since whether a border point belongs to a cluster depends on the order in which the method processes each point. These authors have therefore developed a new algorithm called *A-DBSCAN* for Approximate – Density-Based Spatial Clustering of Application with Noise. The algorithm works in three steps:

– applying the *DBSCAN* method to a random sub-sample of points;

– assigning a label ("Core Point", "Border Point" or "Noise Point") to each point outside the sub-sample, corresponding to that of its nearest neighbor in the sub-sample;

– replicating steps 1 and 2 a large number of times.

A point is then part of a cluster if it is identified as belonging to that cluster for a large number of repetitions (90%).

5.1.1.2. *Natural Cities*

The *Natural Cities* method aims to delimit so-called “natural” cities from point data such as road intersections or centroids of buildings. The idea is to separate the many points that are close to each other from the less numerous points that are distant from each other. The method is based on the analysis of two rank-size distributions: that of the distances between the points and that of the area of the point clusters created from applying a maximum distance threshold between them. These distributions must be “heavy-tailed”; that is, for the method to be applicable, the distributions analyzed must have few observations with large values (head of the distribution) and many observations with low values (tail of the distribution).

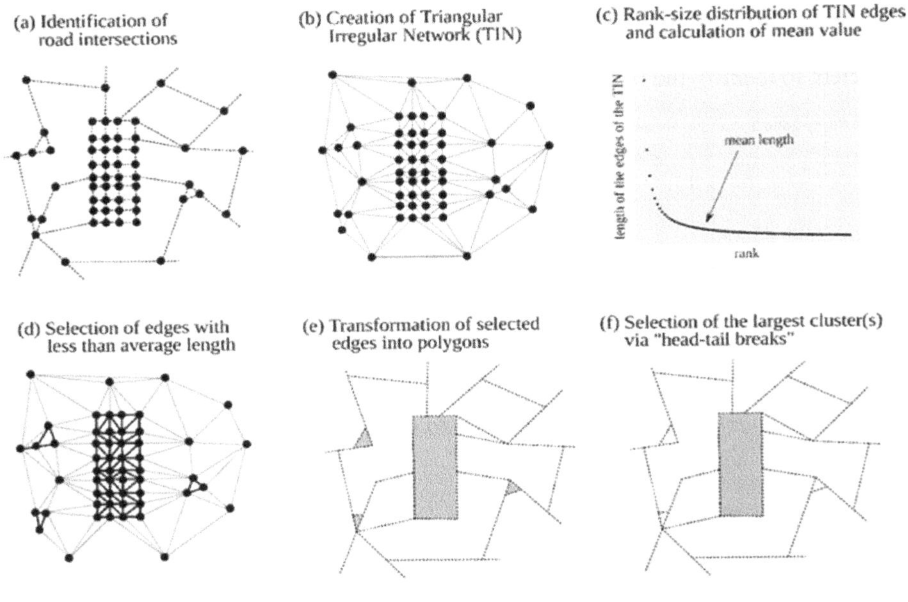

Figure 5.4. *Natural Cities process of delimitation*

The different steps of the method are described in Figure 5.4. The first step is to identify the points of interest (road intersections in the case of Figure 5.4) and trace, from these points, a triangular irregular network (TIN). The rank-size distribution of the length of the edges of the TIN is studied. If this distribution is indeed heavy-tailed, then the mean of the values of the distribution (in meters) splits the distribution into two parts: a majority of the edges of the TIN have short lengths (less than the average length) and a minority have long lengths (greater than the average) (Jiang and Liu 2012). Edges lower than the average length are selected and transformed into polygons to identify different clusters (sets of points) representing places where many road intersections are close to each other. The rank-size distribution of the surface of these clusters is then analyzed. If it is heavy-tailed, then a method of discretization by head–tail breaks is performed (Jiang 2013, 2015, 2019). This method again relies on the mean of the distribution to discriminate between small clusters, which are very numerous, and large clusters, which are much less numerous. This time it is a question of selecting the largest clusters. The *head–tail breaks* method is a method of discretization by the mean. It iteratively selects observations from the head of the rank-size distribution (larger than the average size) until these observations represent less than 40% of the total observations. These, which are therefore the largest clusters among the large clusters, are considered to be *Natural Cities*.

5.1.1.3. *MorphoLim*

The *MorphoLim* method considers the footprint of buildings mapped in 2D. Starting from the hypothesis of the fractality of built-up fabrics, the principle is to identify, across the scales, clusters of buildings whose logic of spatial distribution differs the most from that of the surrounding buildings.

To understand this principle, let us take the example of a highly schematized built-up fabric of 144 buildings, all identical in shape, surface and orientation (Figure 5.5(a)). Note that all buildings have their nearest neighbor located at a distance "*d*" equal to the length of their side. The first step of overlay consists of overlaying each building with a square of length $l_1 = 2\ d$, so that the overlaying squares of nearby buildings touch each other without being superimposed. These squares merge into clusters whose side length is $L_1 = 2\ L_1$ (Figure 5.5(b)). At this stage, the number of clusters (L_1) is 36, that is, one-fourth of the initial number of buildings.

In the second step of overlay (Figure 5.5(c)), clusters L_1 are overlaid by squares whose length $l_2 = 2\ l_1 + 2\ d$. In the same way as in step 1, since clusters L_1 have their closest neighbor at a distance of l_1, the nearest overlaying squares l_2 touch and merge to give new clusters L_2 whose size $L_2 = 2\ l_2$. At this step, the number of clusters is equal to 9, that is, one-fourth of the number of clusters in the preceding step (L_1).

The same principle is applied at the third step of overlay. The clusters L_2 are overlaid by squares of size $l_3 = 2\ l_2 + 2\ d$. The nearest overlaying squares (distance equal to l_2) merge. However, it can be seen in Figure 5.5(d) that the clusters L_3 are not all located at a distance l_3 from their nearest neighbor. Therefore, the number of clusters L_3 (7) is not equal to one-fourth of the number of clusters L_2 (9). The logic of nested scales observed during steps 1 and 2 of overlay ceases during stage 3. Step 3 thus marks a break in the self-similarity of the built-up fabric. It is at this step that the morphological limit of the clusters is traced. The largest cluster can be assimilated to a city, while the other clusters may be assimilated, depending on their size, to villages or towns. Note that if the built-up fabric shown in Figure 5.5(a) is a perfectly self-similar fractal figure, no change would have been detected in the overlaying process and the ratio of the number of clusters at one step to the number of clusters at the previous step would have been the same at all steps.

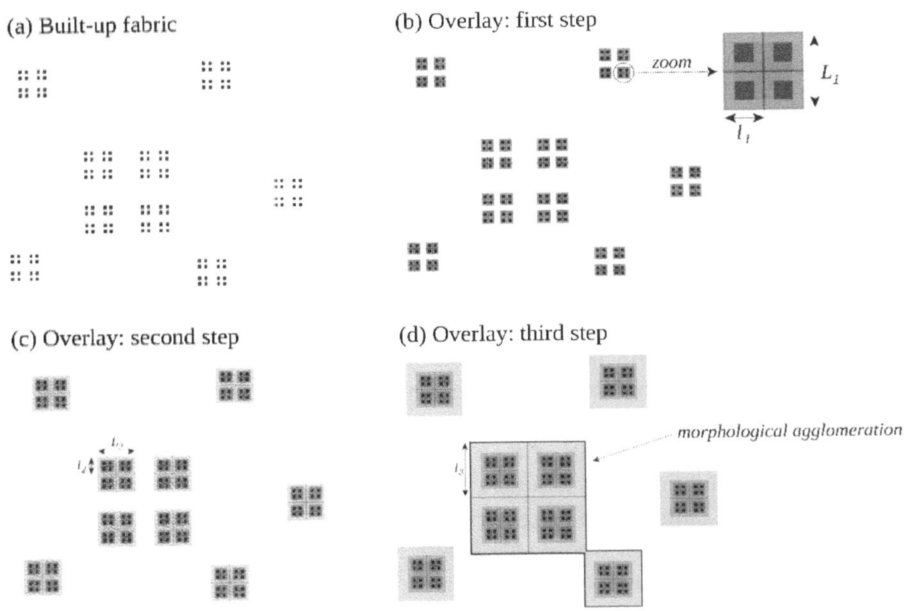

Figure 5.5. *Identification of a major change in the concentration of buildings across scales by means of an overlaying process. Source: Tannier (2017)*

For real fabrics, it is clearly impossible to use such an overlaying process since the spatial distribution of buildings is not strictly self-similar. On the contrary, it is possible to use the dilation process imagined by Minkowski (1903) and developed by Bouligand (1929). Buildings in real fabrics (or pixels in the case of non-vector images) are surrounded by an increasingly large buffer zone (dilation) in an iterative process. At each step of dilation, buildings or clusters of buildings merge if they touch each other and form a new, larger cluster. It is possible to count, at each step of dilation, the number of clusters as a function of the size of the buffer zone applied. If the buildings of the studied zone are spatially distributed according to a fractal logic, then the number of clusters at each step of dilation is related to the width (thickness) of the buffer zone by a hyperbolic law. This law corresponds to a straight line in a double logarithmic graph. Therefore, the point where the dilation curve (in double logarithmic coordinates) (Figure 5.6(a)) deviates most from a straight line marks a crucial change in the spatial distribution of buildings (Tannier et al. 2007, 2011). To locate this point on the dilation curve, we must identify the main value among the maxima of a curvature function, that is, the maximum having the greatest absolute value (Figure 5.6(b)). The dilation distance (width of the buffer zone) corresponding to this break point represents the threshold value (distance threshold) that marks the limit of the morphological agglomerations of the fabric studied. The *MorphoLim* computer application is used to implement the method (see: https://sourcesup.renater.fr/*MorphoLim*).

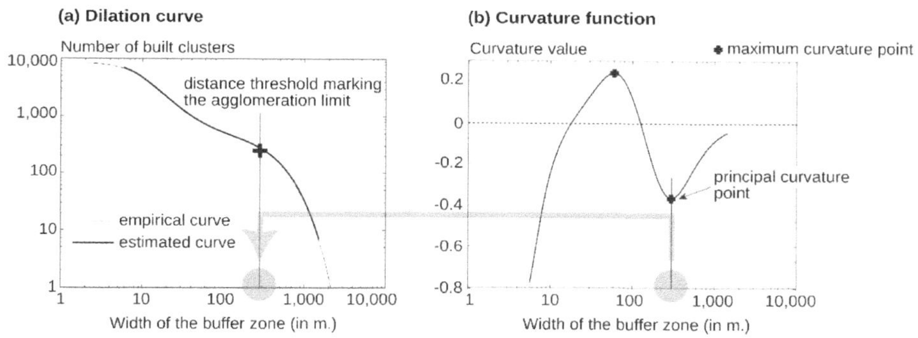

Figure 5.6. *Identification of the dilation threshold (gray circle) defining the limit of morphological clusters according to MorphoLim*

Once the dilation distance threshold has been identified, it is necessary to select the cluster(s) that will form the morphological agglomeration. This is done via a double logarithmic graph representation of the rank-size distribution of the different clusters. The largest cluster or clusters that detach from the rest of the distribution form the morphological agglomeration.

5.1.2. *Comparison of methods*

The *A-DBSCAN* method is a spatial clustering method. By definition, it implies the use of thresholds fixed a priori (maximum distance between points and minimum number of points in a cluster). This is not the case for the *Natural Cities* and *MorphoLim* methods, which both rely on the study of a statistical distribution to set a posteriori thresholds. The *Natural Cities* method first considers the rank-size distribution of the lengths of the edges of a TIN and uses the mean value of these lengths as a threshold, then, in a second step, it considers the rank-size distribution of road intersection clusters and identifies a threshold in this distribution via the *head–tail breaks* method. The *MorphoLim* method identifies the principal curvature point on the curve of the number of built clusters at each step of dilation. The delimitations of different cities made with either the *Natural Cities* or the *MorphoLim* methods are comparable, provided that the threshold identification procedure has been correctly followed. In the case of *A-DBSCAN*, on the contrary, comparing delimitations of different cities is more delicate or irrelevant if the thresholds chosen are not suitable for all the cities considered.

The three methods have the advantage of being applied to fully disaggregated data. *Natural Cities* applies to all types of points of interest, including road intersections, *A-DBSCAN* applies to building centroids and *MorphoLim* applies to buildings directly. This makes it possible to avoid measurement biases arising from the MAUP and due to the collection of disaggregated data in BSUs whose shape, size and position are likely to strongly influence the results of the clustering. Considering the footprint of buildings, *MorphoLim* differs from *A-DBSCAN* in that building centroids are abstractions that do not take into account the size (surface area) of buildings or their orientation.

According to Goodchild (2016), measurement biases related to the extent of the study zone are an integral part of the MAUP. Montero et al. (2021) showed that, in all cases, the extent of the study zone influences the results obtained with the *Natural Cities* method and, in some cases, those obtained with the *MorphoLim* method. Indeed, for the *MorphoLim* and *A-DBSCAN* methods, a change in the extent of the study zone does not influence the results as long as the extension of the study zone does not lead to the integration of new structures of point distributions.

In the case of the existence of a built-up corridor between two morphological agglomerations, the *Natural Cities* and *MorphoLim* methods will not separate the two agglomerations, but will integrate them into the same entity, as long as the intersections of roads are at a distance less than the average length (*Natural Cities*), and as long as the distance between the buildings is less than the threshold detected by *MorphoLim* (Figure 5.7(a) and (b)). With the *A-DBSCAN* method, the presence of a built-up corridor in which the distance between centroids of buildings is less than the maximum neighborhood radius of the method does not automatically imply the fusion of the two neighboring agglomerations in the same cluster. Indeed, the number of neighbors in the maximum neighborhood radius must still be greater than a predefined value. Therefore, two agglomerations may be grouped together by their built-up corridor (Figure 5.7(c)) or they may be delimited separately (Figure 5.7(d)).

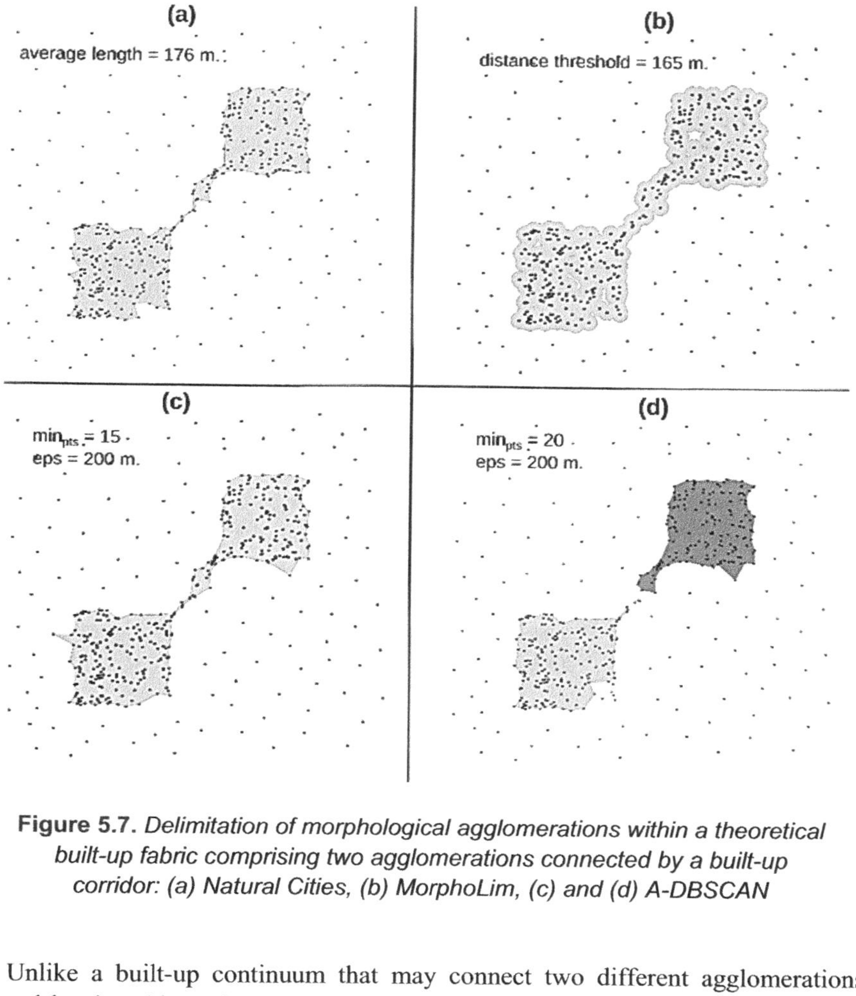

Figure 5.7. *Delimitation of morphological agglomerations within a theoretical built-up fabric comprising two agglomerations connected by a built-up corridor: (a) Natural Cities, (b) MorphoLim, (c) and (d) A-DBSCAN*

Unlike a built-up continuum that may connect two different agglomerations, natural barriers (river, forest, etc.) or human constructions (railway, highway, etc.) may separate an a priori single agglomeration into two. This will be the case with all the three methods if the distance between two road intersections is greater than the average length of the edges of the TIN (*Natural Cities*), if the distance between two centroids of buildings is greater than the maximum distance (*A-DBSCAN*) or if the distance between two buildings is greater than the dilation distance threshold (*MorphoLim*) (Figure 5.8).

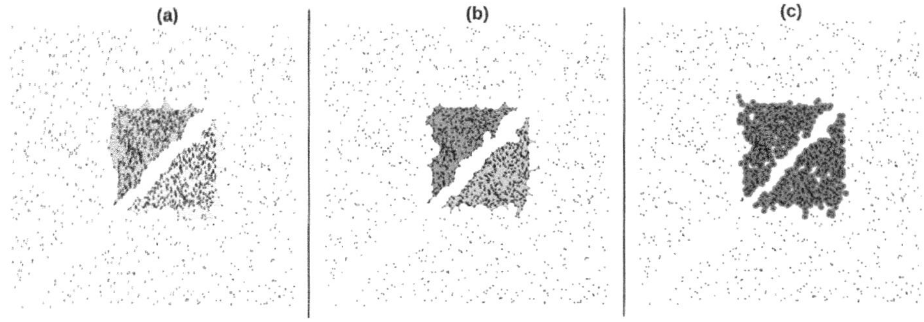

Figure 5.8. *Delineation of morphological agglomerations within a theoretical built-up fabric traversed by a large unbuilt zone by means of (a) A-DBSCAN, (b) Natural Cities and (c) MorphoLim*

5.1.3. *Application of the three methods to identify the morphological limits of the city of Brussels (Belgium)*

Belgium has three regions (Brussels-Capital Region, Flemish Region and Walloon Region) (Figure 5.9). The urban agglomeration of Brussels is located in the Brussels-Capital Region, but its spatial extent is not scientifically fixed. Depending on the studies and databases used, it takes different forms and may extend well beyond the borders of the Brussels-Capital Region (see, for example, Dujardin et al. 2007; Van Hecke et al. 2009; Boussauw et al. 2012; Thomas et al. 2012; Vandermotten 2016).

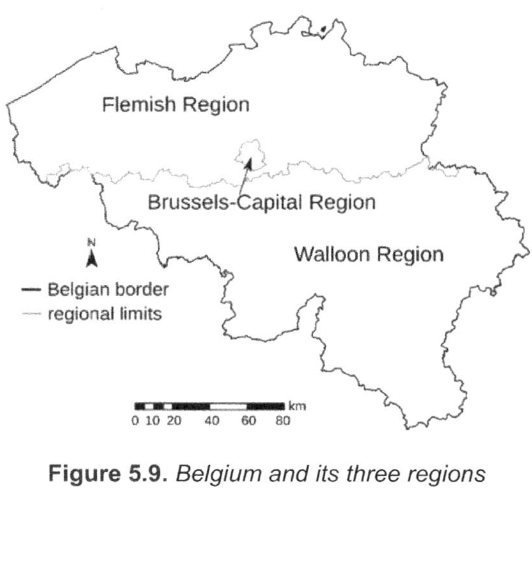

Figure 5.9. *Belgium and its three regions*

5.1.3.1. *Study zone and data*

We have seen previously that the extent of the study zone may influence the results obtained with each of the methods. The *A-DBSCAN* and *MorphoLim* methods also require the chosen study zone to fully include the targeted morphological agglomeration and its hinterland. It turns out that Belgian urbanization is characterized by a strong suburbanization that connects urban centers (Vanneste et al. 2008; Vanderstraeten and Van Hecke 2019) to such an extent that Vandermotten et al. (2008) consider Belgium to be an urbanized continuum. For this reason, we have chosen to take the entire country as a study zone, within which we will seek to identify the morphological boundaries of the city of Brussels.

To implement the *A-DBSCAN, Natural Cities* and *MorphoLim* methods, two different databases are needed: a database containing the footprint of buildings and a database containing road intersections. Building data are provided by the General Administration of Heritage Documentation. These include the cadastral survey (2D) of buildings over 4 m^2 located in Belgium in 2018. In the study by Montero et al. (2021), cadastral data from 2009 were used. Note that we have removed from this database all buildings smaller than 12 m^2 in order to limit the noise induced by the presence of small buildings such as garden sheds. This database can be used directly in the case of *MorphoLim*. For the application of *A-DBSCAN*, the centroids of buildings are extracted from the database. The road network, which allows the extraction of road intersections for the application of the *Natural Cities* method, is taken from the Geofabrik platform of the OpenStreetMap project (data downloaded on 21/08/2018 from: http://download.geofabrik.de). Road intersections, including terminal nodes and junction nodes between roads, were extracted according to Ren's (2018) tutorial. This same dataset was used in the study by Montero et al. (2021).

5.1.3.2. *Parameterization of A-DBSCAN*

As a reminder, the maximum distance at which two centroids of buildings can be considered to belong to the same cluster and the minimum number of centroids that a cluster must contain to form part of an agglomeration must be fixed a priori. In their article, Arribas-Bel et al. (2019), who seek to delineate Spanish urban agglomerations, note that commuter journeys in large Spanish cities are 4 km on average. Hence, they set the maximum distance between two building centroids at 2 km (2 km one way and 2 km back). These authors also believe that to be considered an agglomeration, a cluster must contain at least 5,000 people. The average number of people per household is 2.5 in Spain and by making the simplifying assumption that a building is occupied by a single household, they estimate that 2,000 building centroids are required as a minimum for each cluster. If we apply this reasoning to Belgium, we could keep the distance of 2,000 m. Indeed,

Duprez and Nautet (2019) show that 80% of residents of the Brussels-Capital Region travel less than 4 km per day for their home-to-work journeys. However, applying this distance threshold would amount to delimiting the region's employment pool rather than the built-up morphological agglomeration. In their study for Belgium, Vanderstraeten and Van Hecke (2019) adopt the definition of agglomeration recommended since 1978 by the United Nations, which we also choose to adopt: "an agglomeration [...] is a group of buildings, none of which is more than 200 meters from the nearest building." These authors also consider that a minimum threshold of 500 inhabitants/km² within a morphological agglomeration in Belgium is appropriate. As the average number of individuals per household in Belgium is 2.3, we can estimate that a minimum of 63 building centroids within a maximum radius of 200 m are needed to form a cluster. We set the share of centroids selected by the algorithm at each repetition to 10% and set the number of repetitions to 1,000.

5.1.3.3. *Identifying thresholds for the delimitation of Natural Cities*

The distribution of the lengths of TIN edges connecting road intersections is indeed heavy-tailed (Figure 5.10). The mean of this distribution is equal to 250 m. The *head–tail breaks* method is applied until only five clusters remain, as shown in Table 5.1. Of these five clusters, only one delimits the morphological agglomeration of Brussels (Figure 5.12).

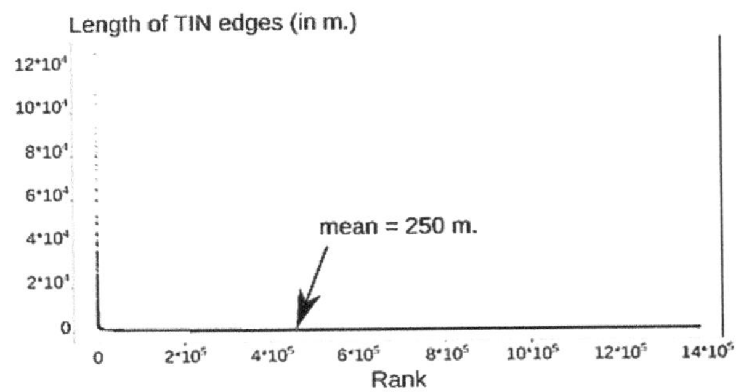

Figure 5.10. *Rank-size distribution of the length of the TIN edges connecting road intersections in Belgium*

Step of head–tail breaks	Number of clusters	Average surface area of clusters (in m²)	Number of clusters in the head of the distribution	Percentage of clusters in the head of the distribution
1	23,770	0.1082	2,435	10
2	2,435	0.8980	402	17
3	402	4.0189	75	19
4	75	13.8951	17	23
5	17	39.2771	5	29
6	5	81.3112	2	40
7	2	118.8025	1	50

Table 5.1. *Different steps of the head–tail breaks process (the process stops at the penultimate line since the last line no longer meets the 40% criterion)*

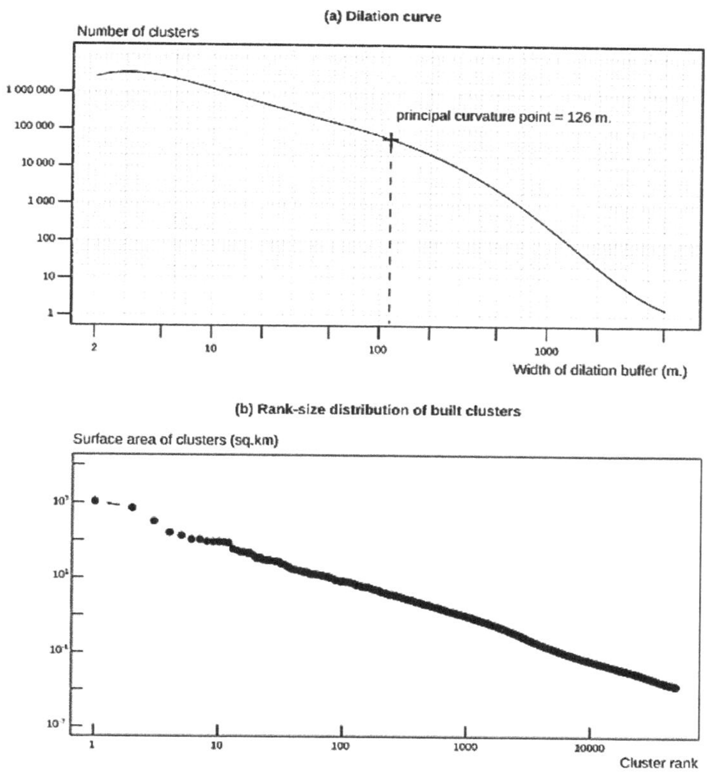

Figure 5.11. *(a) Dilation curve and (b) rank-size distribution of the built-up clusters obtained with MorphoLim for buildings located in Belgium*

5.1.3.4. *Identification of a distance threshold in the dilation curve obtained from MorphoLim*

The dilation curve is very close to a straight line, but a principal curvature point may be identified (-0.109), corresponding to a distance threshold of 126 m (Figure 5.11(a)). The largest cluster in the head–tail distribution of clusters is that of Brussels (Figure 5.11(b)). It therefore forms, in itself, the morphological agglomeration of Brussels.

5.1.3.5. *Analysis of the results*

According to the *A-DBSCAN* method applied to the centroids of buildings and the threshold values adopted, the morphological agglomeration of Brussels extends slightly beyond the borders of the Brussels-Capital Region (Figure 5.12). The *Natural Cities* method applied to road intersections delimits a slightly larger agglomeration that is almost limited to the borders of the region. Compared to the delimitation obtained with *A-DBSCAN*, the result of *Natural Cities* presents an agglomeration extending a little further away to the west and east of the Brussels-Capital region. The area of the morphological agglomeration of Brussels is 148 km^2 with the *Natural Cities* method, whereas it is 136 km^2 with *A-DBSCAN* (Table 5.2). As the limits of the morphological agglomeration of Brussels are relatively identical with the *A-DBSCAN* method applied to building centroids and the *Natural Cities* method applied to road intersections, this tends to confirm (in any case, for the *A-DBSCAN* criteria that we have retained) that road intersection data are a good proxy to describe urbanization phenomena (Masucci et al. 2015).

The morphological agglomeration of Brussels delimited by means of *MorphoLim* extends well beyond the borders of the Brussels-Capital Region, over an area of 1,063 km^2. The significant conurbation between Brussels and the city of Ghent (in the north-west) means that the two cities are grouped together in the same morphological agglomeration. This is a novelty compared to the results presented by Montero et al. (2021) based on the 2009 data. Then, the Brussels agglomeration stopped at the gates of the city of Ghent. It is likely that new constructions of buildings between the two agglomerations have now merged them. Moreover, as was already the case from the 2009 data, a development of the Brussels agglomeration is observed to the south of the Brussels-Capital region, where it crosses the regional border between Flanders and Wallonia to extend as far as the towns of Tubize, Braine-l'Alleud and Ottignies-Louvain-la-Neuve.

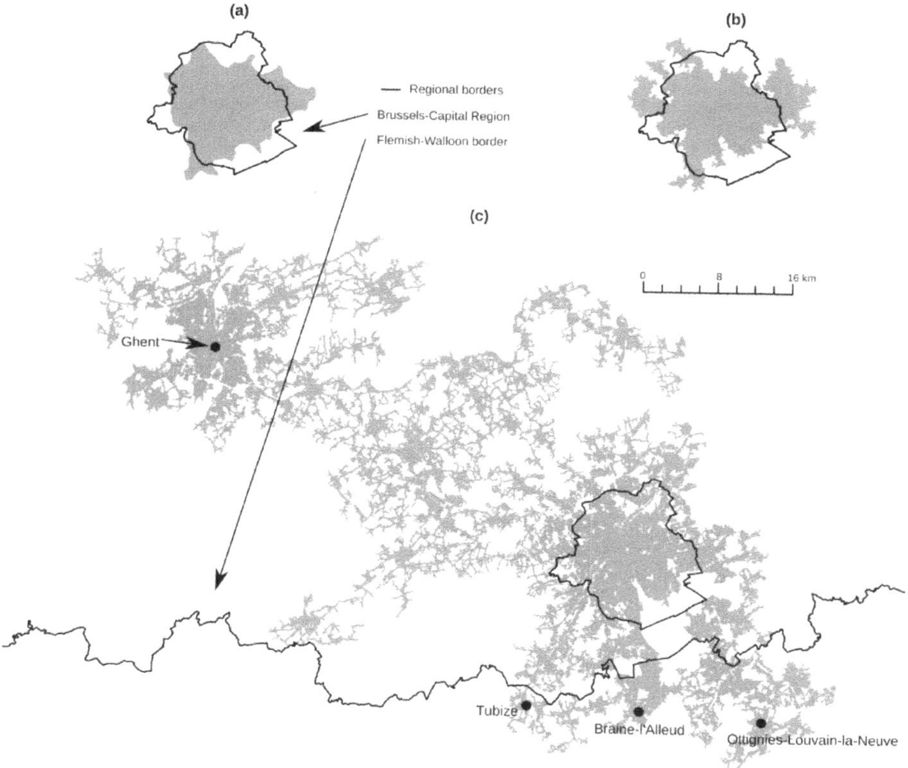

Figure 5.12. *Delimitation of the morphological agglomeration of Brussels according to the three methods: (a) A-DBSCAN (eps = 200 m; minpts = 63), (b) Natural Cities and (c) MorphoLim*

Method	Data	Threshold(s)	Area of the agglomeration
A-DBSCAN	Building centroids	Max distance = 200 m Min number = 63	136 km²
Natural Cities	Road intersections	Average length of TIN edges = 250 m	148 km²
MorphoLim	Footprint of buildings (2D)	Dilation distance threshold = 126 m	1,063 km²

Table 5.2. *Parameters and quantitative results of each method applied to the morphological delimitation of the city of Brussels, taking the whole of Belgium as the study zone*

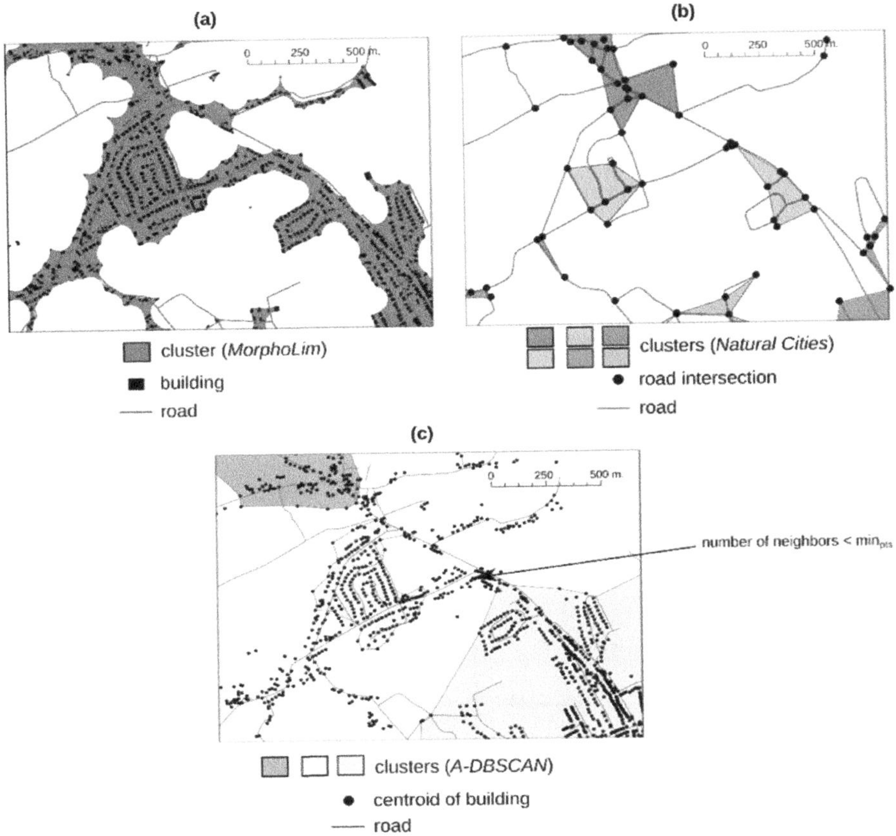

Figure 5.13. *Delimitation of clusters along a built-up corridor by MorphoLim (a) and non-delimitation of these corridors by Natural Cities (b) and A-DBSCAN (c)*

A-DBSCAN and *Natural Cities* present much smaller agglomerations than *MorphoLim* because they do not detect the phenomenon of conurbation (built-up continuity along the main roads). When there is urban development along a main road, the density constraint imposed by *A-DBSCAN* cannot be satisfied: although there is sufficient proximity between the buildings, there are too few of them. The same is true for *Natural Cities*: the distance between road intersections becomes too great when urban development becomes a ribbon. This is illustrated in Figure 5.13, which shows a zoom in on the results of the three methods north-west of Brussels.

The result of the *A-DBSCAN* method is a function of the choices of thresholds fixed a priori. The maximum distance of 200 m, although in line with United

Nations recommendations, may not be entirely suitable for Belgium. Indeed, *MorphoLim* shows a break at a distance of 126 m between the buildings. The minimum threshold of 63 centroids per cluster is also quite questionable. The choice of different thresholds would have led to other morphological delimitations of the city of Brussels. For example, we may imagine that with a lower density threshold (minimum number of points less than 63), *A-DBSCAN* would delimit a large Brussels agglomeration that would encompass built-up corridors and tend to resemble the delimitation obtained with *MorphoLim*.

5.2. Morphological characterization of cities delimited by means of MorphoLim within 82 French urban areas

As a reminder, *MorphoLim* morphologically delimits cities by performing a step-by-step dilation of the built-up fabric and by identifying the dilation distance at which a major break appears with regard to the spatial distribution of the buildings (major change in the distances between buildings). The delimitation made corresponds to the boundary between two morphological subsets that are distinct from a fractal point of view, one of which is the city.

In order to analyze the morphological characteristics of cities identified by means of *MorphoLim*, we will measure different indices relating to the spatial distribution of buildings in morphological subsets, on the one hand, and to the shape of the boundary drawn between these subsets (i.e. the morphological envelope of cities), on the other hand. We will apply the *MorphoLim* method to a large sample of French urban areas in order to highlight possible covariations between the different indices calculated, and then try to establish a morphological typology of the cities studied. Tannier and Thomas (2013) already carried out these analyses for six theoretical cities and 18 Belgian cities. They show in particular that the relationship between the shape of the perimeter of an agglomeration and the built morphology within this agglomeration cannot be easily and simply characterized. Indeed, for the 18 Belgian cities, no correlation appears, for example, between the built density of cities and the complexity of their borders measured using a fractal box-counting dimension. Tannier and Thomas (2013) observe, however, among other elements to which we will return in more detail later in the chapter, that the larger the surface area of an agglomeration, the more indentations its perimeter has. In addition, Sémécurbe et al. (2019) showed, for 20 French cities, that the distance threshold identified by means of *MorphoLim* and characterizing each city is not statistically correlated with the extent of each urban area. From a larger sample of 82 French cities, we will seek to confirm or deny these different results.

5.2.1. *The 82 urban areas studied*

According to the definition of INSEE (National Institute of Statistics and Economic Studies, France), an urban area is "a collection of municipalities, in one piece and without enclaves, constituted by an urban pole of more than 10,000 jobs, and by rural municipalities or units where at least 40% of the resident population with a job works in the pole or in associated municipalities" (INSEE 2016). In 2020, France had 161 urban areas in mainland France and the overseas territories. We chose to study only urban areas of more than 100,000 inhabitants located in mainland France. We also removed urban areas extending to other countries from our sample, for example Lille, which extends to Belgium, or Strasbourg, which extends to Germany. In the end, we retained a sample of 82 urban areas (Figure 5.14). The data, that is, all buildings with a footprint greater than 1 m^2, come from the BD TOPO published in March 2021 by the National Institute of Geographic and Forest Information, France (see: geoservices.ign.fr).

The urban area of Paris is by far the largest in terms of area (17,278 km^2) and number of buildings (3,654,901). The urban areas of Lyon, Bordeaux and Toulouse have an area between 5,500 km^2 and 6,000 km^2 for a number of buildings between 890,000 and 985,000. Aside from these four particularly large urban areas, the others have areas between 157 km^2 and 3,755 km^2 and a number of buildings between 30,976 and 838,781. There is therefore significant variability in area and the number of built entities between the different study zones.

5.2.2. *Methodology*

5.2.2.1. *Delimitation of morphological agglomerations*

We use the *MorphoLim* program in accordance with the recommendations of Tannier et al. (2011) and Tannier and Thomas (2013). A dilation of the buildings of each urban area is carried out according to the same geometric progression (coefficient of 1.2). The empirical curves of the numbers of clusters at each step of dilation are fitted using polynomial functions. The degree of the polynomial is specific to each urban area and is determined by visually minimizing, using a graph, a Bayesian information criterion (BIC). The degree of the polynomial should be large enough to best fit the expansion curve, but not too large in order to limit overfitting bias. The dilation distance threshold for each fabric is obtained by maximizing a curvature function (main point of curvature in absolute value) applied to the estimated dilation curve. If the principal curvature point indicates a threshold at a distance less than 30 m or greater than 1,000 m, the second maximum (in absolute value) of the curvature feature is chosen. A distance threshold of less than

30 m would in fact describe a break in the distances between buildings due to the width of the roads, while a threshold greater than 1,000 m is beyond the orders of magnitude of intra-urban distances between buildings.

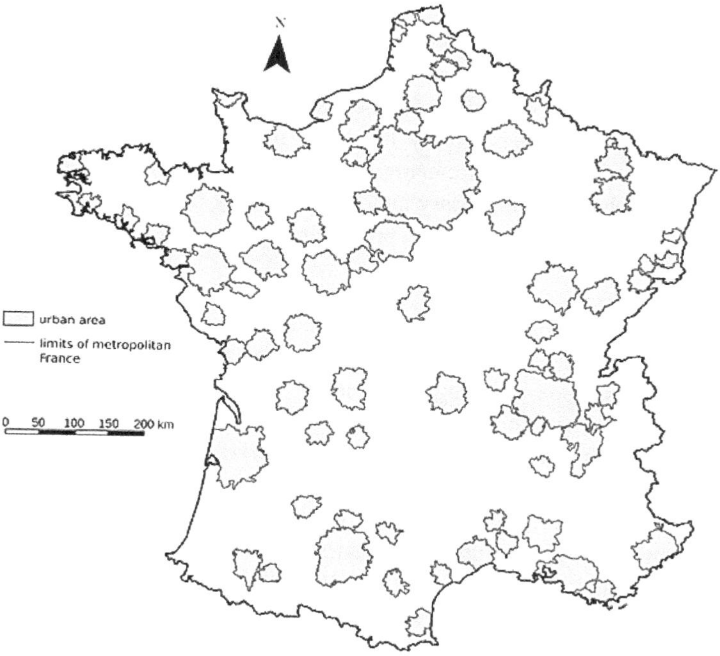

Figure 5.14. *The 82 urban areas of the sample*

Once the distance threshold has been determined, *MorphoLim* produces two shapefiles. The first is called "buffer", which corresponds to the built-up fabric dilated according to a buffer zone with a thickness equal to half of the distance threshold. The second shapefile is named "envelope", which corresponds to the "buffer" shapefile eroded by applying a buffer zone of thickness equal to half of the distance threshold. The surface areas of the clusters of the "buffer" shapefile are calculated to produce the rank-size distributions of the clusters of each urban area. The visual analysis of these distributions makes it possible to select the main cluster(s) forming the morphological agglomeration. Next, for the analysis of the characteristics of the morphological agglomeration, we consider the eroded building clusters of the "envelope" shapefile contained within the boundaries of the morphological agglomeration identified from the "buffer" shapefile.

5.2.2.2. *Characteristics of the built-up fabrics of the 82 cities studied*

Given that the morphological limit of cities identified with *MorphoLim* corresponds to a break with respect to a fractal spatial distribution of buildings, it would be relevant to measure the fractal dimension of the built-up fabrics on either side of this limit. However, although it is possible to calculate the fractal dimension of the built-up fabric of the morphological agglomeration, it is not possible to measure the fractal dimension of the built-up area located outside the morphological agglomeration, since the gap left by the latter in the middle of the fabric being studied would distort the measurements. The difference in fractality of the built-up fabric on either side of the boundary can therefore only be evaluated by comparing the fractal dimension of the entire urban area with that of its morphological agglomeration.

Like Tannier and Thomas (2013), we chose to calculate the fractal *box-counting dimension*. In our analyses, we adopt the same measurement parameters for the estimation of the box-counting dimension of urban areas (D_s(UA)) and morphological agglomerations (D_s(MA)) in order to allow the comparison between them (see Chapter 4). The built-up fabrics are rasterized at a resolution of 4 m. The count windows extend from 30 m to 2,600 m according to a geometric progression with a coefficient of 1.5. The adjustment of the empirical curve takes place via a logarithmic transformation. Some of these parameters are slightly different from those used by Tannier and Thomas (2013). We will therefore not compare the values of the dimensions obtained with those of Belgian cities, but we will observe whether the relationship between the dimensions of urban areas and morphological agglomerations are identical in Belgium and France.

In addition to D_s(UA) and D_s(MA), more traditional morphological indices are calculated: the built density of the morphological agglomeration (Dens), the surface area of the agglomeration (SMA) and the proportion of buildings in the urban area that are part of the morphological agglomeration (%B).

5.2.2.3. *Characteristics of the perimeter of the 82 cities studied*

In order to characterize the shape of the perimeter (morphological limits) of cities, we use three indices: the fractal box-counting dimension (D_b(MA)), dendricity (δ) and compactness (CO).

The morphological limits (perimeters) of the cities identified using *MorphoLim* are presented as a set of closed lines (Figure 5.22). The more that a perimeter is made up of many closed lines of varying lengths and with many indentations, the higher its fractal dimension. The fractal box-counting dimension of these perimeters is estimated from the same parameter values as the box-counting dimensions of the built-up fabrics. The sole difference concerns the minimum size of the count windows. According to the recommendations of Tannier and Thomas (2013), for a given city, the smallest count window size must be greater than the dilation distance threshold applied to delimit the city. Indeed, below this distance, the envelope of the city does not show any fractal characteristics. The largest distance threshold in the sample of 82 urban areas is that of Marseille-Aix-en-Provence, with a value of 457 m. The minimum size of the count windows is therefore fixed at 450 m for all fractal analyses of the perimeters of the cities in the sample.

The dendricity index (δ) expresses the relationship between the fractality of the built-up fabric of the morphological agglomeration and the fractality of its perimeter. According to equation 5.1, an index δ close to zero indicates that the fractal dimension of the built-up fabric is much greater than that of the perimeter. Conversely, an index δ greater than 1 indicates that the fractal dimension of the built tissue is smaller than that of the perimeter; the fabric is relatively uniform, while the morphological envelope is divided into several closed lines and has many indentations of different sizes. If δ is close to 1, this means that the two dimensions are identical.

$$\delta = 2 - [D_s(MA) / D_b(MA)] \qquad [5.1]$$

The compactness index (CO) expresses the ratio between the area of a morphological agglomeration and the length of its perimeter (equation [5.2]). CO tends towards 1 when the perimeter is smooth, as in the case of simple geometric figures such as circles or squares. CO tends towards 0 when the perimeter is tortuous (many closed lines and indentations of various sizes). Indeed, in this case, the perimeter is very long relative to the surface area of the agglomeration.

$$CO = [2(\pi \times A)^{1/2}] / P \qquad [5.2]$$

with A being the area of the morphological agglomeration and P being the length of its perimeter.

Correlations between the different morphological indices are calculated with the Pearson correlation coefficient. As this index measures only linear correlations between two variables, we ensure the absence of other nonlinear relationships through visual observation of graphical representations of relationships between variables (not presented in this chapter).

5.2.3. *Results*

5.2.3.1. *Delimitation of morphological agglomerations*

It was possible to delimit the morphological agglomeration of each of the 82 urban areas. The principal curvature values vary from -0.07 (Toulon) to -0.89 (La Rochelle) (see Figure 5.15 for a sample of six dilation curves with very different principal curvature points). The principal curvature values closest to 0 testify to a very hierarchical structure in the built-up fabric in which the built forms of the morphological agglomeration differ relatively little from those of its periphery. Conversely, a principal curvature value far from 0 shows that the built-up fabric of the morphological agglomeration is clearly distinguishable from that of its periphery from a fractal point of view.

In the vast majority of cases, the empirical dilation curve was correctly estimated with polynomials of degree 4–6. For 10 urban areas, however, it was necessary to increase the degree of the polynomial to 7, 8 or even 9 for Annecy and Rennes. Over the sample of 82 urban areas, the degree of the polynomial is not correlated with the principal curvature values nor with the dilation distance threshold.

The broad spectrum of dilation distance thresholds, from 68 m to 457 m, shows the morphological diversity of the selected French urban areas. Distance thresholds are not correlated with the size of urban areas. These results are consistent with those obtained by Sémécurbe et al. (2019) for 20 urban areas in mainland France. We have already indicated that methods using a distance threshold fixed a priori to delimit morphological sets do not always allow the morphological specificities of different urban spaces to be taken into account. We add here that a threshold of 100–200 m, generally adopted in studies (Zahn 1971; Yu et al. 2014; Caruso et al. 2017), may only be relevant for 48% of the urban areas in our sample. Indeed, only 40 urban areas out of the 82 have a distance threshold between 100 and 200 m.

For 69 urban areas, one morphological cluster appears clearly larger than the others. The morphological agglomerations of these urban areas are therefore monocentric. The morphological agglomeration of 10 urban areas is made up of two clusters. In these cases, the agglomeration is monocentric if the two aggregates are, in a certain way, contiguous (Bordeaux, Perpignan, Nantes, Agen, Blois, Angers and Toulouse). On the contrary, if the two clusters are distant from each other, the agglomeration is polycentric (Bethune, Quimper and Vienna). The morphological agglomeration of Saint-Etienne consists of three spatially distinct clusters. This agglomeration is therefore also polycentric. The agglomeration of Tours is composed of four contiguous clusters, and the agglomeration of Avignon has eight non-contiguous clusters. Figure 5.16 shows examples of rank-size distributions corresponding to these different scenarios.

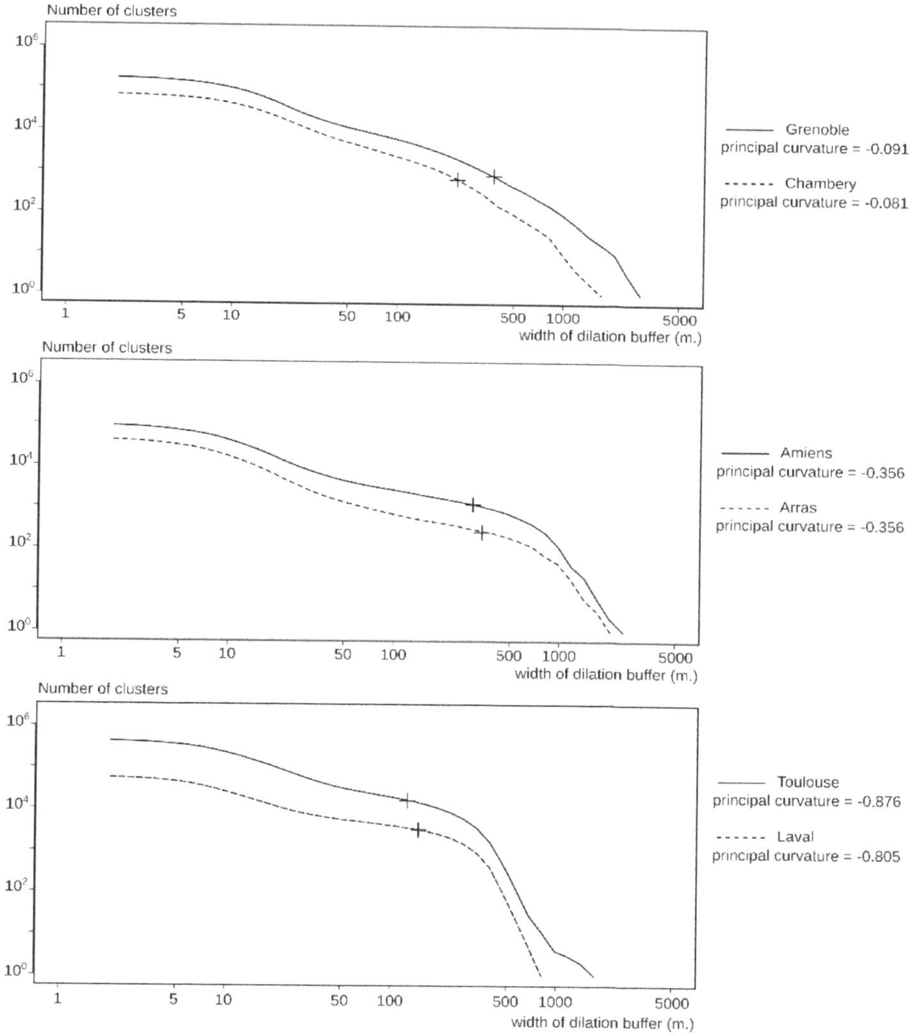

Figure 5.15. *Dilation curves and principal curvature points of six French urban areas*

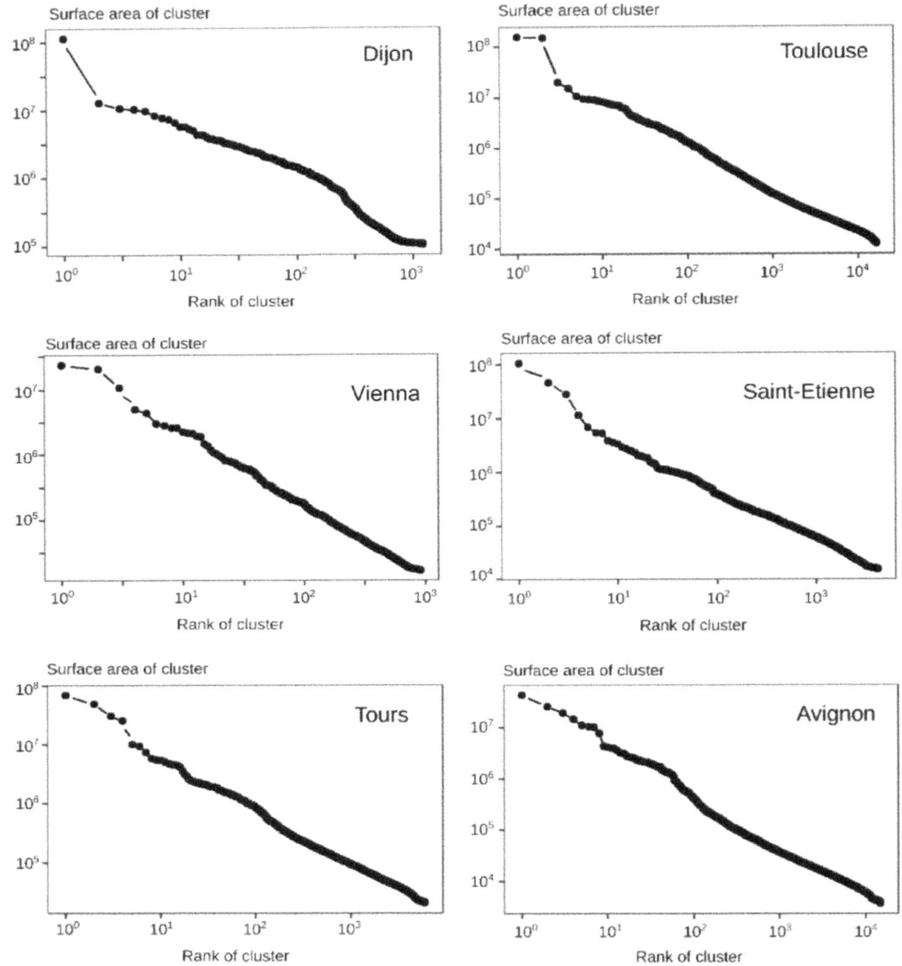

Figure 5.16. *Rank-size distribution of built clusters from six morphologically very different urban areas*

Morphological agglomerations have areas ranging from 1,654 km^2 for the largest (Paris) to 12 km^2 for the smallest (Lorient). The great diversity of sizes of the morphological agglomerations testifies to the great morphological variety of the built-up fabrics of the urban areas of the sample.

Groups	Urban areas	D_s(MA)–D_s(UA) (interval)	Description
1	Creil, Marseille-Aix-en-Provence, Béthune, Thionville, Douai–Lens, Le Havre, Saint-Nazaire, Avignon, Saint-Brieuc, Toulon, Mulhouse, Nice, Lyon	0.045–0.22	Small difference between D_s(UA) and D_s(MA) D_s(UA) high compared to those of all the cities
2	Grenoble, Belfort, Calais, Bordeaux, Vienna, Montpellier, Ales, Chambéry, Quimper, Chalon-sur-Saône, Macon	0.123–0.218	Small difference between D_s(UA) and D_s(MA) D_s(UA) and D_s(MA) average compared to those of all the cities
3	Montauban, Saint-Etienne, Tarbes, Rouen, Boulogne-sur-Mer, Valence, Annecy, Caen, Cherbourg-en-Cotentin, Colmar, Brive-la-Gaillarde, Pau, Albi, Perpignan, Vannes, Nîmes	0.225–0.316	Average difference between D_s(UA) and D_s(MA) D_s(MA) rather high and D_s(UA) average compared to those of all the cities
4	Metz, Montbéliard, Périgueux, Tours, Roanne, La Roche-sur-Yon, Arras, Agen, Besançon, Poitiers, Niort, Angoulême, Bourg-en-Bresse, Evreux, La Rochelle, Béziers, Limoges, Cholet, Nancy, Laval, Orléans	0.255–0.391	Large difference between D_s(UA) and D_s(MA) Low D_s(UA) compared to those of all the cities
5	Lorient, Nantes, Paris, Brest, Clermont-Ferrand, Dunkerque, Le Mans, Toulouse, Angers, Rennes	0.281–0.390	Large difference between D_s(UA) and D_s(MA) D_s(MA) very high compared to those of all the cities
6	Charleville-Mézières, Saint-Quentin, Troyes, Blois, Beauvais, Bourges, Carcassonne, Amiens, Chartres, Dijon, Reims	0.292–0.467	Large difference between D_s(UA) and D_s(MA) D_s(UA) very low compared to those of all the cities

Table 5.3. *Urban areas classified according to their fractal box-counting dimension D_s(UA) and that of their morphological agglomerations D_s(MA)*

5.2.3.2. *Morphological difference between cities (i.e. morphological agglomerations) and their periphery*

The fractal box-counting dimensions of the built-up fabric of urban areas (D_s(UA)) and morphological agglomerations (D_s(MA)) were calculated considering 12 sizes of count windows spread over three orders of magnitude (30–2,600 m).

This responds to the established recommendations for estimating fractal dimensions (see Chapter 4). The estimation of fractal dimensions is generally quite robust with respect to p-values and R^2. It should be noted, however, that, for the built-up fabrics of certain urban areas, the R^2 of the estimate is slightly below the threshold of 0.99. These are the urban areas of Toulouse, Rennes, Angers, Le Mans, Limoges, Laval, Agen, Roanne, Brive-la-Gaillarde, Montauban, Périgueux and Albi. The values of D_s(UA) are between 1.232 and 1.617. The values of D_s(MA) are, for their part, between 1.524 and 1.784.

As expected (Thomas et al. 2007; Tannier and Thomas 2013; Sémécurbe 2020), the built density of morphological agglomerations is proportional to their fractal box-counting dimension (D_s(MA)) (Pearson's correlation index = 0.63, Table 5.4). The built density is also inversely proportional (correlation = -0.61) to the proportion of buildings in the urban area located in the morphological agglomeration (%B). However, contrary to what Tannier and Thomas (2013) observe, D_s(MA) and %B also have an inversely proportional relationship (correlation = -0.44). Thus, when a morphological agglomeration is relatively small compared to its urban area, its built-up fabric tends to be dense and rather uniform. Conversely, if a morphological agglomeration contains a large proportion of the buildings in its urban area, the spatial distribution of its buildings will be more heterogeneous and less dense. Note that here we are discussing the difference in relative size of a morphological agglomeration compared to its urban area. Indeed, as shown by Tannier and Thomas (2013), the surface area (in absolute value) of an agglomeration is not correlated with its fractal dimension. The same applies to the total extent of the urban area.

Tannier and Thomas (2013) also observe for 18 Belgian agglomerations that the difference in fractal dimension between the morphological agglomeration and the urban area is not significantly associated with the value of the principal curvature index or with %B. Conversely, our results show that these measures are statistically interlinked in the case of the 82 French urban areas. First, the principal curvature index is correlated with the difference in fractal dimension with a Pearson correlation coefficient of 0.64. Thus, in general, for French urban areas, the more marked the difference in dimensions between a morphological agglomeration and its urban area, the higher the absolute value of the principal curvature index. The principal curvature index measures the most significant break in the fractal-scale nesting logic of a built-up fabric. If the break is very marked (high curvature index), the difference in dimensions on either side of the break point is therefore also significant. Note that D_s(UA) does not measure the fractal dimension of the built-up fabric outside the morphological agglomeration, but that of the entire urban area. The difference between the fractal dimensions of the built-up fabrics of the morphological agglomeration and outside the agglomeration may therefore be even more marked than the difference from the fractal dimension of the urban area as a whole. Moreover, within the same urban area, the logics of the spatial distributions of buildings are sometimes very different between the morphological agglomeration and outside the agglomeration, which may explain why estimates of dimensions $D_s(UA)$ are generally less robust (R^2 lower) than those for $D_s(MA)$. In addition, since the proportion of buildings in the urban area located in the agglomeration is inversely proportional to the difference in fractal dimension (-0.70), the difference between $D_s(MA)$ and $D_s(UA)$ is potentially even greater, as an agglomeration occupies a limited part of the urban area.

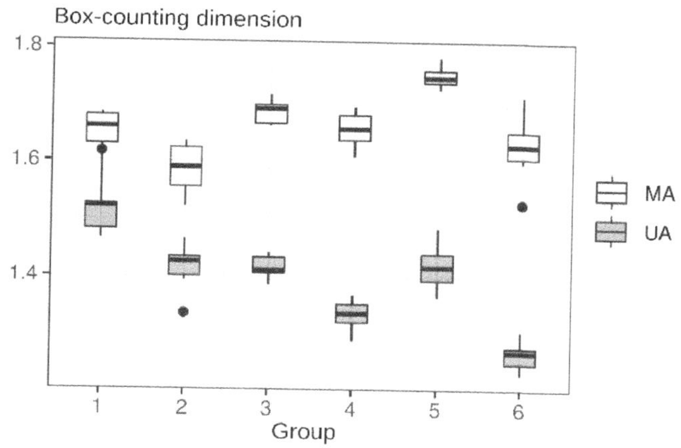

Figure 5.17. *Fractal box dimensions of urban areas (UA) and morphological agglomerations (MA) for each group, derived from the classification performed on these dimensions*

We established an initial typology of the 82 urban areas studied based solely on the morphological characteristics of the built-up fabrics. This typology results from a classification (Euclidean dissimilarity and Ward aggregation) based on fractal box-counting dimensions of urban areas and morphological agglomerations. Six groups were identified (Table 5.3, Figures 5.17 and 5.23(a)). Groups 1, 3 and 6 include morphological agglomerations whose box-counting dimensions are relatively identical. Conversely, the dimensions of their urban areas are quite different (Figure 5.17). Built-up fabrics belonging to group 1 have a $D_s(UA)$ slightly weaker than their $D_s(MA)$. The difference between $D_s(MA)$ and $D_s(UA)$ is greater in group 3, with a reduction in the UA dimensions. Group 6 is the one for which the difference is most marked between $D_s(MA)$ and $D_s(UA)$, with the smallest $D_s(UA)$. Overall, the built-up fabrics of groups 1, 3 and 6 are relatively concentrated in clusters distant from each other, this trend increasing from group 1 to group 6 (group 1 – D_s(UA) around 1.51; group 3 – D_s(UA) around 1.41; group 6 – D_s(UA) around 1.26). The built-up fabrics of the morphological agglomerations are more connected while displaying internal gaps of different sizes (D_s(MA) around 1.64). Groups 2 and 5 have very similar D_s(UA) but different D_s(MA). The difference between D_s(UA) and D_s(MA) was less marked for fabrics in group 2 than for those in group 5. In the urban area, the built-up fabrics of groups 2 and 5 are quite similar to those of groups 1, 3 and 6 (D_s(UA) around 1.41). The same is true for the morphological agglomerations in group 2 (D_s(MA) around 1.58), while those in group 5 fill the space more and their buildings are spatially distributed more evenly (D_s(MA) around 1.75). Finally, in group 4 urban areas, buildings tend to be concentrated in clusters that are distant from each other (D_s(UA) around 1.33), while built-up fabrics in morphological agglomerations tend to be more connected while exhibiting internal gaps (D_s(MA) around 1.65). Figure 5.21 shows an example of a fabric characteristic of each group.

5.2.3.3. *Morphological characteristics of the perimeters of cities*

The values of the dendricity index are all close to 1, included in an interval of 0.77–1.07, thus suggesting that the perimeters of agglomerations have a fractality comparable to that of the built-up fabrics. This is confirmed by the compactness index, which is between 0.022 and 0.32 with an average value of 0.1 for all the morphological agglomerations. The box-counting dimensions calculated on the perimeters are also well above 1, in an interval between 1.371 for the least complex perimeter (Troyes) and 1.709 for the most complex perimeter (Limoges). Unlike some of the results obtained by Tannier and Thomas (2013) for 18 Belgian cities, the three indices are correlated with each other (Table 5.4). As a general rule, the more complex the perimeter (high D_b(MA)), the more dendritic the morphological agglomeration is (high δ) and the less compact it is (low CO).

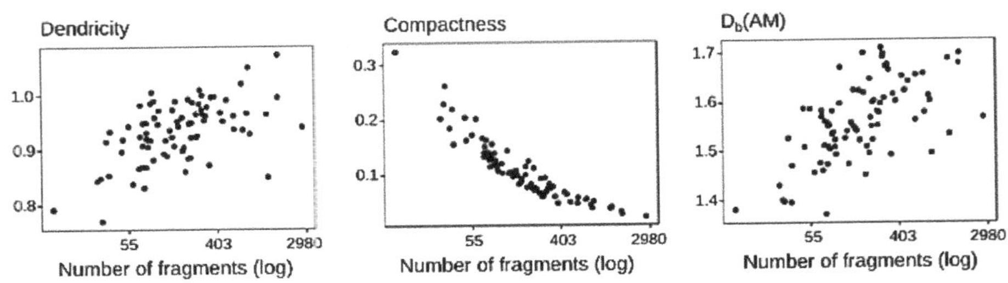

Figure 5.18. *Relations between the number of closed lines (fragments) composing the perimeter of each morphological agglomeration and the different indices characterizing it*

It is important to keep in mind that the perimeter of a morphological agglomeration is generally not composed of a single line, but a set of closed lines that are more or less numerous and indented, whose number and length determine the fragmentation and complexity of this perimeter. The perimeter of the 82 morphological agglomerations of our sample has a number of closed lines (fragments) varying between 10 and 2,364 (median value = 152). The statistical relationship between the number of fragments and the values of the three morphological indicators considered is quite clear, as shown in Figure 5.18.

	D_s(UA)	D_s(MA)	Dens	%B	S	L	D_b(MA)	δ
D_s(MA)	n.s							
Dens	n.s	0.63						
%B	0.50	-0.44	-0.61					
S	0.28	n.s	-0.30	0.44				
L	0.35	n.s	-0.25	0.42	0.89			
D_b(MA)	0.31	0.36	0.27	n.s	n.s	0.25		
δ	0.24	-0.28	n.s	n.s	n.s	0.26	0.79	
CO	-0.46	n.s	n.s	-0.27	-0.34	-0.53	-0.66	-0.69

Table 5.4. *Pearson's correlation coefficients between the different morphological indices characterizing the 82 urban areas studied. n.s: not significant at the significance level 0.05; S = surface area of MA; Dens = built density of MA; %B = proportion of UA buildings that are in the MA; L = perimeter length; D_b(MA) = fractal box-counting dimension; δ = dendricity; CO = compactness*

Contrary to the results obtained by Tannier and Thomas (2013), the fractal dimension of the perimeter (D_b(MA)) is not correlated with the surface area of the agglomerations. In this sense, our results are consistent with the hypothesis put forward by Chen (2011) for whom an increase in the area of an agglomeration does not imply an increase in the complexity of its perimeter. Tannier and Thomas (2013) also observe a negative correlation between dendricity and built density in Belgian agglomerations. Such a result is not found in the case of the 82 French agglomerations studied.

The classification of urban areas on the basis of standardized indices of dendricity, compactness and fractal dimension of their perimeter makes it possible to identify six groups (Figure 5.23(b)). From group 1 to group 6, the values of the dendricity index δ and the fractal dimension of the perimeters D_b(MA) increase. These two values are highly correlated (Table 5.4), that is, the more that a perimeter is fragmented with many indentations (high δ), the more it is characterized by a high fractal dimension value. Conversely, the values of the compactness index decrease from group 1 to group 6. This index is in fact inversely proportional to the other two (Table 5.4): the more indentations and fragments of varying magnitude that a perimeter has, the lower the value of the compactness index. An exception should, however, be noted for group 4, for which the compactness index appears abnormally high with regard to the dendricity and $D_b(MA)$ values. Indeed, although the perimeters of the morphological agglomerations of group 4 have values of $D_b(MA)$ and δ more or less equal to those of group 3, their compactness index is much higher. This difference is explained by the lesser fragmentation of the perimeters of the cities in group 4 compared to those in group 3. The envelopes of group 4 are therefore more compact than those of group 3. Figure 5.22 shows an example of the characteristic perimeter of each group of cities.

Groups	Urban areas	Indices (interval)
1	Reims, Troyes, Chartres, Colmar, Beauvais, Saint-Quentin, Evreux	δ: 0.77–0.85 CO: 0.14–0.32 D_b(MA): 1.371–1.456
2	Paris, Nice, Rouen, Nancy, Dijon, Amiens, Mulhouse, Nîmes, Vienna, Roanne	δ: 0.85–0.90 CO: 0.03–0.17 D_b(MA): 1.451–1.534
3	Marseille-Aix-en-Provence, Rennes, Clermont-Ferrand, Orléans, Caen, Metz, Le Havre, Poitiers, Valence, Saint-Brieuc, Chalon-sur-Saône, Cherbourg-en-Cotentin	δ: 0.89–0.96 CO: 0.05–0.11 D_b(MA): 1.493–1.619
4	Besançon, Lorient, La Rochelle, Montbéliard, Béziers, Niort, Bourges, Boulogne-sur-Mer, Thionville, Arras, Creil, Bourg-en-Bresse, La Roche-sur-Yon, Cholet, Carcassonne	δ: 0.89–0.95 CO: 0.11–0.20 D_b(MA): 1.471–1.587
5	Toulon, Douai–Lens, Avignon, Saint-Etienne, Tours, Béthune, Pau, Annecy, Chambéry, Saint-Nazaire, Calais, Blois, Quimper, Laval, Tarbes, Belfort, Ales, Charleville-Mézières, Montauban, Perigueux, Macon	δ: 0.94–1.00 CO: 0.02–0.13 D_b(MA): 1.523–1.624
6	Lyon, Toulouse, Bordeaux, Nantes, Grenoble, Montpellier, Angers, Le Mans, Brest, Perpignan, Limoges, Dunkerque, Angoulème, Vannes, Agen, Brive-la-Gaillarde, Albi	δ: 0.93–1.07 CO: 0.03–0.10 D_b(MA): 1.603–1.709

Table 5.5. *Urban areas classified according to dendricity, compactness and fractal dimension of perimeters of their morphological agglomeration. $D_b(MA)$ = fractal box-counting dimension; δ = dendricity; CO = compactness*

The typology obtained based on the three indices characterizing the perimeters (Table 5.5) is quite different from that obtained based on the indices characterizing the built-up fabrics. This shows that the morphological characteristics of a built-up fabric are not identical to those of its perimeter, which was already noted by Tannier and Thomas (2013). Indeed, we observe a very weak correlation between the dimensions D_s(MA) and D_b(MA) or between the built density of an agglomeration and the D_b(MA) of its perimeter. Like Tannier and Thomas (2013), we observe no statistical relationship between D_s(MA) and CO (compactness), nor between δ (dendricity) and S (surface area of agglomeration) (Table 5.4).

5.2.3.4. *Typology of the 82 cities studied based on eight morphological indices characterizing their built-up fabric and perimeter*

A principal component analysis (PCA) followed by a classification were performed on a set of variables characterizing both the built-up fabrics and the perimeters of cities. The following variables (previously standardized) were used: fractal box-counting dimension of the built-up fabric (D_s(MA)), surface area (S), built density of the morphological agglomeration (Dens), perimeter length (L), fractal box-counting dimension of the perimeter (D_b(MA)), percentage of the buildings in the urban area present in the morphological agglomeration (%B), dendricity (δ) and compactness (CO).

Three main PCA factors account for 86% of the variance in the data (Table 5.6). Factor 1 is strongly correlated with the "Surface Area" and "Perimeter Length" variables. This initial factor therefore describes the size of cities (i.e. morphological agglomerations). Factor 2 is correlated with the variables Dens, D_s(MA) and %B. This factor 2 thus synthesizes information relating to the characteristics of the built-up fabrics. Finally, factor 3 synthesizes variables relating to the characteristics of the perimeters, as it is correlated with the variables D_b(MA), δ and CO. The separation of the variables describing the built characteristics from those describing the perimeters into two different factors demonstrates once again their difference and therefore their complementarity in describing the shape of cities.

A classification (Euclidean dissimilarity, Ward aggregation, number of classes defined by the minimization of an entropy index and the maximization of the Dunn, Calinski and Harabasz indices) performed on the factorial scores of the PCA results in a typology of the 82 urban areas into six groups (Figure 5.24 and Table 5.7). Each group is characterized by the size of the agglomerations (factor 1) that it brings together, as well as by the characteristics of the built-up fabrics (factor 2) and the characteristics of the perimeters (factor 3) of these agglomerations (Figure 5.20). Group 1 consists only of Paris, which is by far the largest agglomeration. Group 2 brings together medium-sized agglomerations with fairly complex built-up fabrics and built perimeters. Group 3 consists of medium-sized agglomerations whose built-up fabric appears not very complex in contrast to their perimeter. The agglomerations in groups 4, 5 and 6 are of identical size. They feature built-up fabrics of decreasing fractality from group 4 to group 6. The agglomerations in group 5 differ from those in groups 4 and 6 by having the least complex perimeters.

Starting variables	Factor 1 Size of the morphological agglomeration	Factor 2 Morphological characteristics of the built-up fabrics	Factor 3 Morphological characteristics of the perimeters
S	**0.97**	-	-
L	**0.91**	-0.11	0.27
Dens	-0.22	**0.86**	-
D_s(MA)	0.17	**0.91**	-
%B	0.45	**- 0.69**	-
D_b(MA)	-	0.34	**0.90**
δ	-	-0.24	**0.95**
CO	-0.38	-	**- 0.81**
Share of variance explained (total = 86%)	28%	28%	30%

Table 5.6. *Saturations of the principal component analysis (correlation between the principal factors and the input variables)*

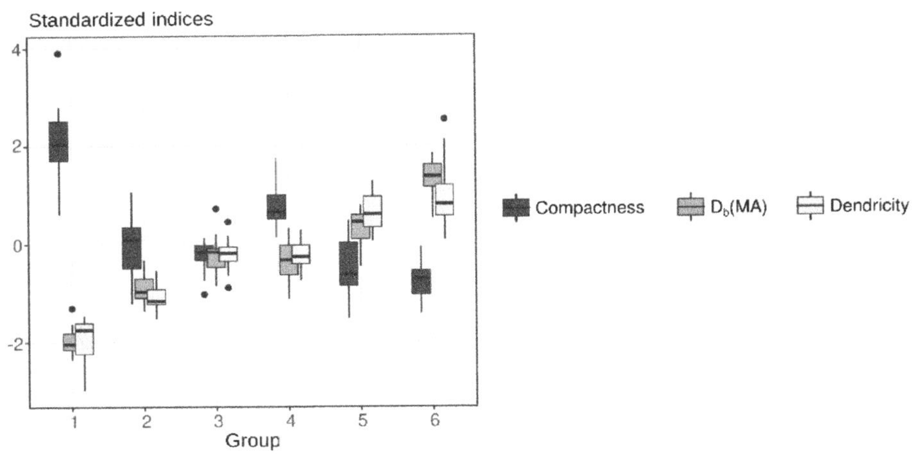

Figure 5.19. *Standardized fractal box-counting dimension D_b(MA), dendricity index and compactness index for the perimeter of the 82 morphological agglomerations for each group resulting from the classification performed on these indices*

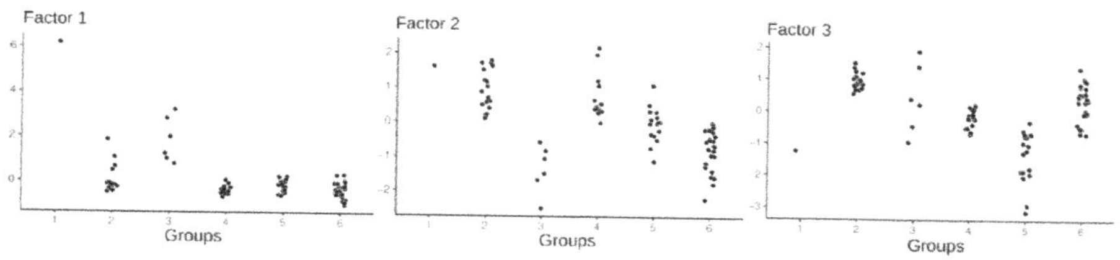

Figure 5.20. *Factorial scores of morphological agglomerations according to classification groups (third typology)*

Groups	Morphological agglomerations	Characteristics
1	Paris	Largest MA, uniform built fabric, fairly uncomplicated perimeter
2	Lyon, Toulouse, Nantes, Avignon, Angers, Le Mans, Brest, Perpignan, Limoges, Dunkerque, Pau, Annecy, Agen, Brive-la-Gaillarde, Montauban, Albi	Medium-large MA, uniform built-up fabric, complex perimeter
3	Marseille-Aix-en-Provence, Bordeaux, Nice, Grenoble, Toulon, Douai–Lens	Medium-sized MA, built-up fabric with local concentrations of buildings at different scales, complex perimeter
4	Rennes, Clermont-Ferrand, Caen, Lorient, La Rochelle, Valence, Saint-Brieuc, Béziers, Niort, Boulogne-sur-Mer, Cherbourg-en-Cotentin, Laval, Bourg-en-Bresse, La Roche-sur-Yon, Cholet	Medium-sized MA, fairly uniform built-up fabric, relatively uncomplicated perimeter
5	Rouen, Nancy, Orléans, Dijon, Reims, Amiens, Mulhouse, Nîmes, Troyes, Chartres, Arras, Colmar, Beauvais, Saint-Quentin, Evreux, Roanne, Carcassonne	Medium-sized MA, moderately uniform built-up fabric, very uncomplicated perimeter
6	Montpellier, Saint-Etienne, Tours, Metz, Béthune, Le Havre, Poitiers, Besançon, Chambéry, Saint-Nazaire, Angoulême, Montbéliard, Bourges, Thionville, Chalon-sur-Saône, Calais, Blois, Quimper, Creil, Tarbes, Belfort, Vienna, Alès, Charleville-Mézières, Périgueux, Mâcon	Medium-sized MA, built fabric with local concentrations of buildings at different scales, relatively uncomplicated perimeter

Table 5.7. *Morphological agglomerations classified according to the factorial scores from the principal component analysis*

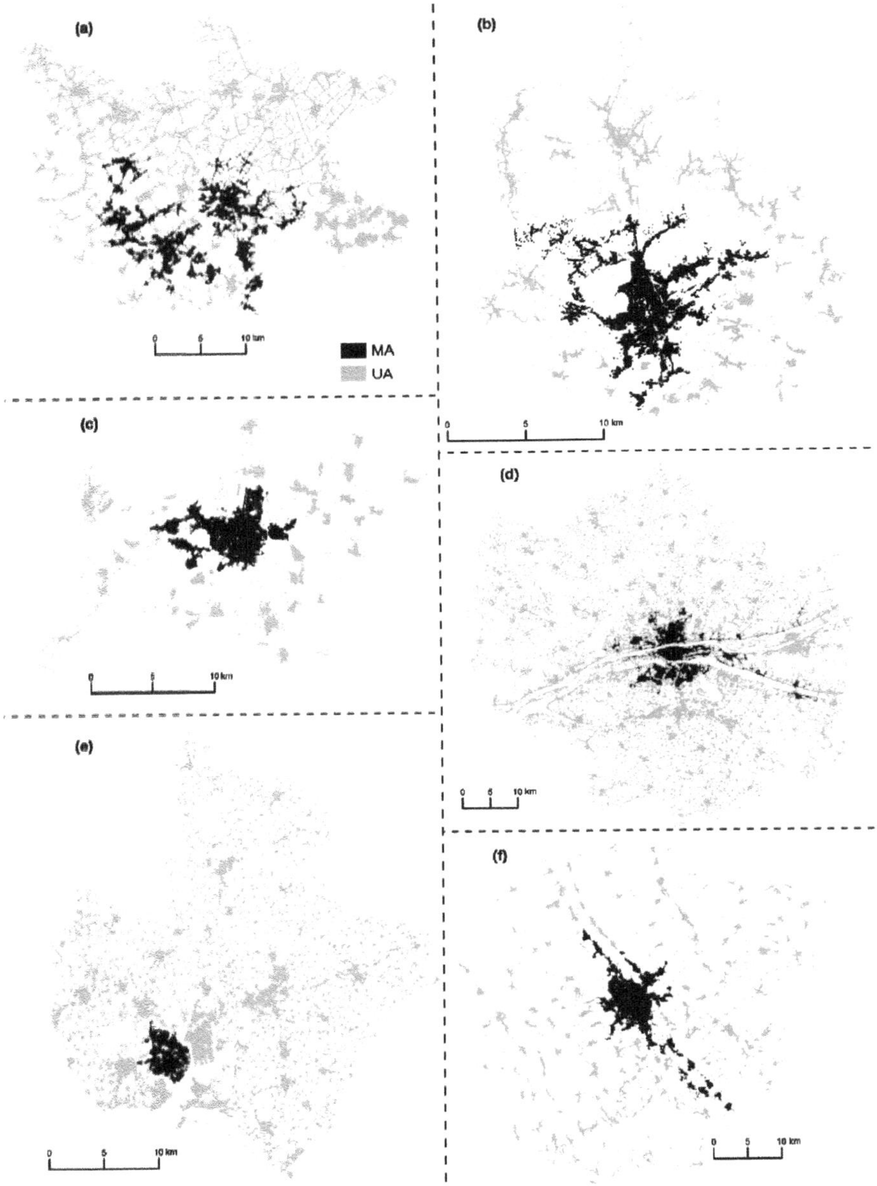

Figure 5.21. *Six built-up fabrics representative of each group of the classification according to $D_s(MA)$ and $D_s(UA)$: (a) Bethune – group 1; (b) Belfort – group 2; (c) Colmar – group 3; (d) Tours – group 4; (e) Lorient – group 5; (f) Troyes – group 6*

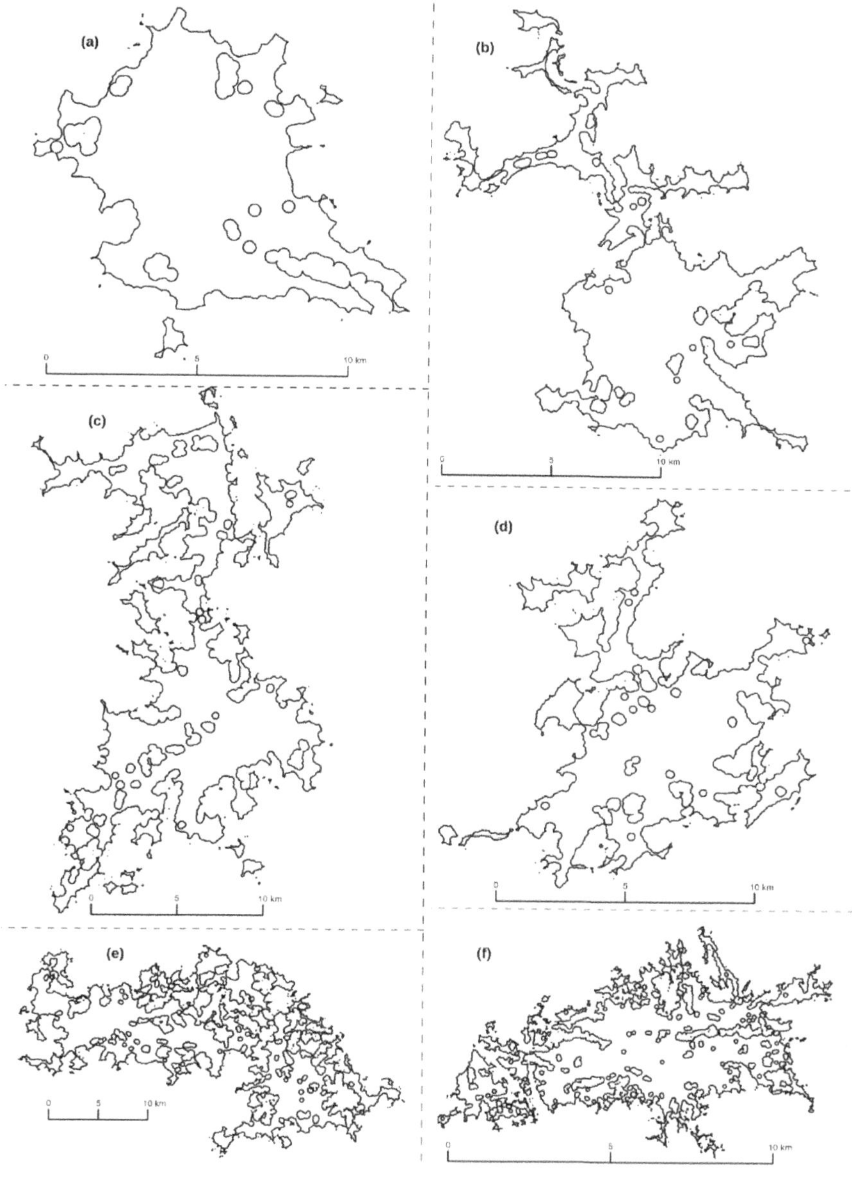

Figure 5.22. *Six perimeters of morphological agglomerations representative of each group of the classification according to δ, CO and $D_b(MA)$: (a) Reims – group 1; (b) Nancy – group 2; (c) Metz – group 3; (d) Besançon – group 4; (e) Douai–Lens – group 5; (f) Brive-la-Gaillarde – group 6*

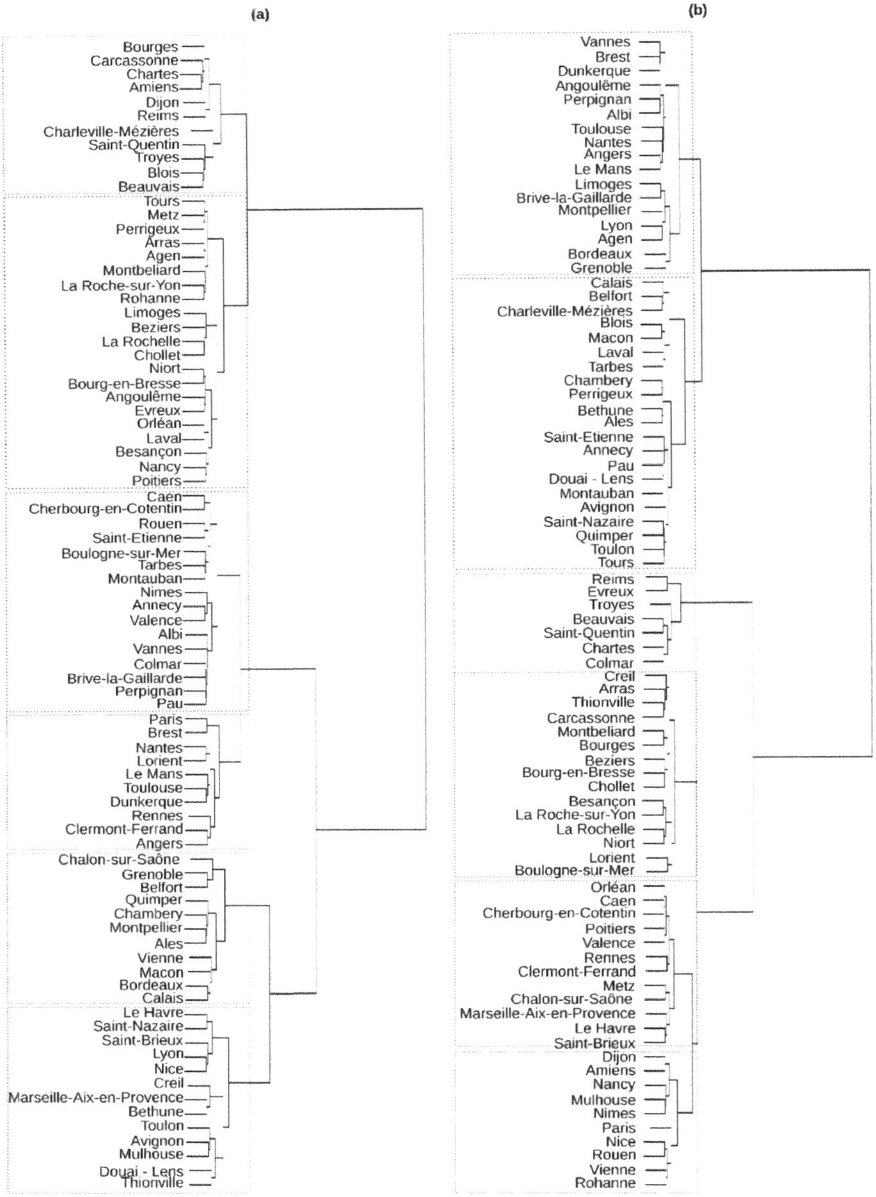

Figure 5.23. *Dendrogram of the classifications of 82 French urban areas with regard to the characteristics of their built-up fabric (a) and their perimeter (b). Dissimilarity: Euclidean (standardized data); aggregation: Ward*

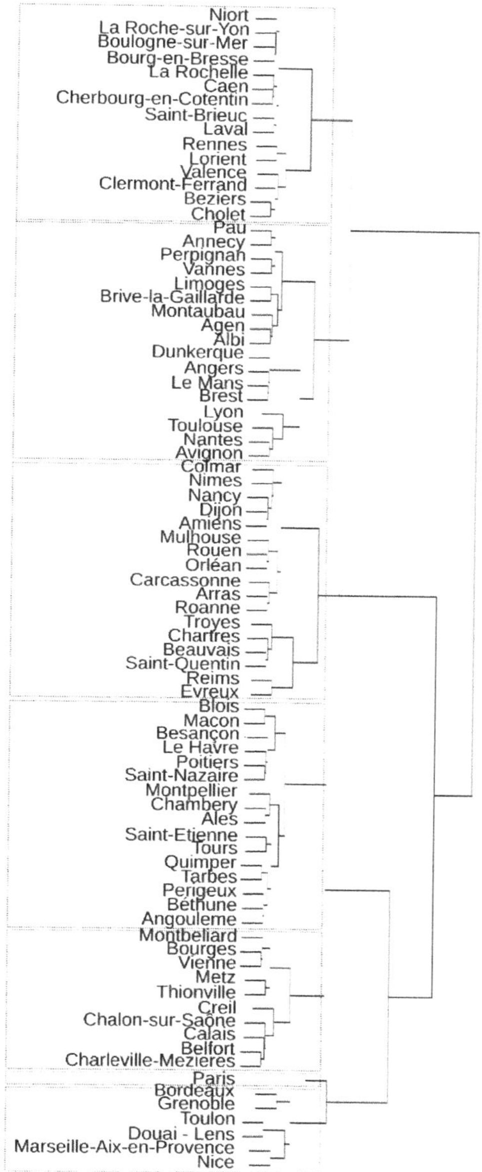

Figure 5.24. *Dendrogram of the classification of 82 French urban areas with regard to the factorial scores from a principal component analysis carried out on a set of variables characterizing both the built-up fabrics and the perimeters of cities. Dissimilarity: Euclidean; aggregation: Ward*

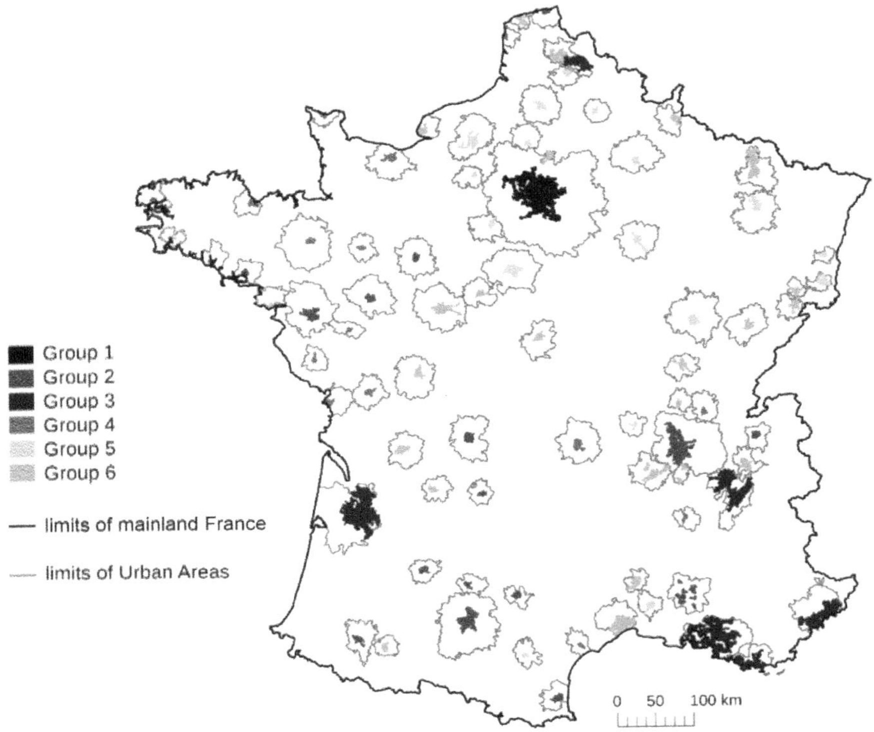

Figure 5.25. *Typology of 82 French urban areas based on their shape (principal component analysis followed by an ascending hierarchical classification based on eight variables characterizing the built-up fabrics and the perimeters of cities). For a color version of this figure, see www.iste.co.uk/tannier/fractal.zip*

Using the *MorphoLim* fractal method, we were able to identify the morphological boundaries of 82 monocentric or polycentric French cities. We have seen that the clearer the difference in spatial organization of buildings between a morphological agglomeration (i.e. a city) and its urban area, the more likely that *MorphoLim* would measure a high absolute value for the principal curvature index, marking a clear break in the fractal logic of the spatial distribution of buildings. Between our study and that of Tannier and Thomas (2013), we noted common points but also differences concerning the covariation between the morphological variables characterizing the cities. These differences mainly involve indices that include a measure of fractal dimension. One of the reasons that may explain the differences in the covariation of the indices between the two studies lies in the different ways they have measured the fractal dimensions. Indeed, in this chapter, we have strictly adopted the same parameters for estimating fractal dimensions for all study zones (see Chapter 4). In Tannier and Thomas's (2013) analyses, on the contrary, the number of count windows varied slightly from city to city. Nevertheless, a major source of difference between our results and those of Tannier and Thomas (2013) is surely that the urban areas studied, namely Belgian urban areas, on the one hand, and French urban areas, on the other hand, have very different morphological characteristics. Tannier et al. (2011) already delineated Belgian and French cities using *MorphoLim* and noted that Belgian cities generally had a significantly lower principal curvature index than French cities. A less clear delineation of Belgian cities compared to French cities, due to significant urban sprawl (Caruso 2001), may explain the differences in results between the present study and that of Tannier and Thomas (2013). It should also be noted that Tannier and Thomas's (2013) analyses were performed on a sample of 18 cities, while our study involved a sample of 82 cities.

5.3. Conclusion

As stated by Montero et al. (2021), it is essential to be aware of the importance of methodological choices (data, methods and extent of study zones) governing the morphological delimitation of cities. Each aspect must be carefully considered to propose a delimitation of the city that responds to the intended questions and objectives. The use of different methods, with different fundamental principles, makes it possible to measure and therefore understand equally different aspects of urbanization. A comparative approach to the delimitations obtained with different methods thus makes it possible to grasp different facets of the shape of cities. This may be described using three types of variables: describing their size (surface area and length of the perimeter), their built-up fabric (built density, D_s(MA), %B) and the shape of their perimeter (D_b(MA), δ, CO). These three categories of variables each provide complementary information and together make it possible to describe each city better than could be done separately.

5.4. References

Abel, J. and Kaysi, I.A. (2003). Identifying urban boundaries: Application of remote sensing and geographic information system technologies. *Canadian Journal of Civil Engineering*, 30(6), 992–999.

Adam, A., Montero, G., Finance, O., Verhetsel, A., Thomas, I. (2019). Revisiting urban models with information and communication technology data? Some examples from Brussels. In *Digital Social Networks and Travel Behaviour in Urban Environments*, Plaut, P. and Shach-Pinsly, D. (eds). Routledge, London.

Arcaute, E., Molinero, C., Hatna, E., Murcio, R., Vargas-Ruiz, C., Masucci, A.P., Batty, M. (2016). Cities and regions in Britain through hierarchical percolation. *Royal Society Open Science*, 3(4), 150691.

Arribas-Bel, D., Garcia-López, M.-A., Viladecans-Marsal, E. (2019). Buildings and cities: Delineating urban areas with a machine learning algorithm. *Journal of Urban Economics*, 125, 103217.

Ban, P., Zhan, W., Yuan, Q., Li, X. (2021). Delineating the urban areas of a cross-boundary city with open-access data: Guangzhou–Foshan, South China. *Sustainability*, 13(5), 2930.

Behnisch, M., Schorcht, M., Kriewald, S., Rybski, D. (2019). Settlement percolation: A study of building connectivity and poles of inaccessibility. *Landscape and Urban Planning*, 191, 103631.

de Bellefon, M.-P., Combes, P.-P., Duranton, G., Gobillon, L., Gorin, C. (2019). Delineating urban areas using building density. *Journal of Urban Economics*, 125, 103226.

Berghauser Pont, M., Haupt, P., Berg, P., Alstäde, V., Heyman, A. (2021). Systematic review and comparison of densification effects and planning motivations. *Buildings and Cities*, 2(1), 378–401.

Blondel, V., Krings, G., Thomas, I. (2010). Regions and borders of mobile telephony in Belgium and in the Brussels metropolitan zone. *Brussels Studies*, 42 [Online]. Available at: http://journals.openedition.org/brussels/806.

Borruso, G. (2003). Network density and the delimitation of urban areas. *Transactions in GIS*, 7(2), 177–191.

Bouligand, G. (1929). Sur la notion d'ordre de mesure d'un ensemble plan. *Bulletin des Sciences Mathématiques*, II-53, 185–192.

Boussauw, K., Neutens, T., Witlox, F. (2012). Relationship between spatial proximity and travel-to-work distance: The effect of the compact city. *Regional Studies*, 46(6), 687–706.

Boyko, C.T. and Cooper, R. (2011). Clarifying and re-conceptualising density. *Progress in Planning*, 76(1), 1–61.

Caruso, G. (2001). Peri-urbanisation: The situation in Europe. A bibliographical note and survey of studies in the Netherlands, Belgium, Great Britain, Germany, Italy and the Nordic Countries. Report prepared for the DATAR (with contributions from Cavailhès, J., Peeters, D., Rounsevell, M., Thomas, I.). Ministère de l'Aménagement du Territoire et de l'Environnement, France.

Caruso, G., Hilal, M., Thomas, I. (2017). Measuring urban forms from inter-building distances: Combining MST graphs with a local index of spatial association. *Landscape and Urban Planning*, 163, 80–89.

Cetinkaya, S., Basaraner, M., Burghardt, D. (2015). Proximity-based grouping of buildings in urban blocks: A comparison of four algorithms. *Geocarto International*, 30(6), 618–632.

Chakraborti, S., Das, D.N., Mondal, B., Shafizadeh-Moghadam, H., Feng, Y. (2018). A neural network and landscape metrics to propose a flexible urban growth boundary: A case study. *Ecological Indicators*, 93, 952–965.

Chaudhry, O. and Mackaness, W.A. (2008). Automatic identification of urban settlement boundaries for multiple representation databases. *Computers, Environment and Urban Systems*, 32(2), 95–109.

Chen, Y. (2011). Derivation of the functional relations between fractal dimension of and shape indices of urban form. *Computers, Environment and Urban Systems*, 35(6), 442–451.

Chen, W. (2021). Delineating the spatial boundaries of megaregions in China: A city network perspective. *Complexity*, 2021, 1–10.

Chen, Z. and Gar-On Yeh, A. (2022). Delineating functional urban areas in Chinese mega city regions using fine-grained population data and cellphone location data: A case of pearl river delta. *Computers, Environment and Urban Systems*, 93, 101771.

Chen, Y. and Jiang, B. (2018). Hierarchical scaling in systems of natural cities. *Entropy*, 20(6), 432.

Chen, Y., Wang, J., Long, Y., Zhang, X., Li, X. (2017). Defining urban boundaries by characteristic scales. *Computers, Environment and Urban Systems*, 94, 101799.

Cottineau, C., Finance, O., Hatna, E., Arcaute, E., Batty, M. (2019). Defining urban agglomerations to detect agglomeration economies. *Environment and Planning B: Urban Analytics and City Science*, 46(9), 1611–1626.

Deng, M., Liu, Q., Cheng, T., Shi, Y. (2011). An adaptive spatial clustering algorithm based on Delaunay triangulation. *Computers, Environment and Urban Systems*, 35(4), 320–332.

Ding, C., Knaap, G.J., Hopkins, L.D. (1999). Managing urban growth with urban growth boundaries: A theoretical analysis. *Journal of Urban Economics,* 46, 53–68.

Dong, Q., Qu, S., Qin, J., Yi, D., Liu, Y., Zhang, J. (2022). A method to identify urban fringe area based on the industry density of POI. *ISPRS International Journal of Geo-Information*, 11(2), 128.

Dujardin, C., Thomas, I., Tulkens, H. (2007). Quelles frontières pour Bruxelles ? Une mise à jour. *Reflets et perspectives de la vie économique*, XLVI(2), 155.

Duprez, C. and Nautet, M. (2019). Economic Flows Between Regions in Belgium. Report, National Bank of Belgium.

Ester, M., Kriegel, H.-P., Xu, X. (1996). A density-based algorithm for discovering clusters in large spatial databases with noise. *Proceedings of the Second International Conference on Knowledge Discovery and Data Mining (KDD-96)*, AAAI Press, 240, 226–231.

Ewing, R., Hamidi, S., Tian, G., Proffitt, D., Tonin, S., Fregolent, L. (2018). Testing Newman and Kenworthy's theory of density and automobile dependence. *Journal of Planning Education and Research*, 38(2), 167–182.

Ferreira, J.A., Condessa, B., Castroe Almeida, J., Pinto, P. (2010). Urban settlements delimitation in low-density areas – An application to the municipality of Tomar (Portugal). *Landscape and Urban Planning*, 97(3), 156–167.

Fujishima, S., Fujiwara, N., Akiyama, Y., Shibasaki, R., Sakuramachi, R. (2020). The size distribution of 'cities' delineated with a network theory-based method and mobile phone GPS data. *International Journal of Economic Theory*, 16(1), 38–50.

Goodchild, M. (2016). Laws in geography. *Conférence au Collège International des Sciences du Territoire*, Paris.

Guérois, M., Bretagnolle, A., Mathian, H., Pavard, A. (2014). Functional Urban Areas (FUA) and European Harmonization. A Feasibility Study from the Comparison of Two Approaches: Commuting Flows and Accessibility Isochrones. Technical Report, Espon 2013 Database, Paris, European Union.

Haisch, T. and Müller, U. (2015). Defining and measuring urban regions: A sensitivity analysis. *Papers in Regional Science*, 94(1), 219–226.

Han, H.-Y., Lai, S.-K., Dang, A.-R., Tan, Z.-B., Wu, C.-F. (2009). Effectiveness of urban construction boundaries in Beijing: An assessment. *Journal of Zhejiang University-SCIENCE A*, 10(9), 1285–1295.

Harig, O., Hecht, R., Burghardt, D., Meinel, G. (2021). Automatic delineation of urban growth boundaries based on topographic data using Germany as a case study. *ISPRS International Journal of Geo-Information*, 10(5), 353.

Henderson, M., Yeh, E.T., Gong, P., Elvidge, C., Baugh, K. (2003). Validation of urban boundaries derived from global night-time satellite imagery. *International Journal of Remote Sensing*, 24(3), 595–609.

Hu, S., Tong, L., Frazier, A.E., Liu, Y. (2015). Urban boundary extraction and sprawl analysis using Landsat images: A case study in Wuhan, China. *Habitat International*, 47, 183–195.

Jayasinghe, A., Kalpana, L.D.C.H.N., Abenayake, C.C., Mahanama, P.K.S. (2021). Urban boundary demarcation – An iCN model approach. *Social Sciences*, 10(12), 448.

Jia, T. and Jiang, B. (2010). Measuring urban sprawl based on massive street nodes and the novel concept of natural cities. arXiv:1010.0541 [physics].

Jiang, B. (2013). Head/tail breaks: A new classification scheme for data with a heavy-tailed distribution. *The Professional Geographer*, 65(3), 482–494.

Jiang, B. (2015). Head/tail breaks for visualization of city structure and dynamics. *Cities*, 43, 69–77.

Jiang, B. (2019). A recursive definition of goodness of space for bridging the concepts of space and place for sustainability. *Sustainability*, 11(15), 4091.

Jiang, B. and Jia, T. (2011). Zipf's law for all the natural cities in the United States: A geospatial perspective. *International Journal of Geographical Information Science*, 25(8), 1269–1281.

Jiang, B. and Liu, X. (2012). Scaling of geographic space from the perspective of city and field blocks and using volunteered geographic information. *International Journal of Geographical Information Science*, 26, 215–229.

Jiang, B. and Miao, Y. (2015). The evolution of natural cities from the perspective of location- based social media. *The Professional Geographer*, 67(2), 295–306.

Jiang, P., Cheng, Q., Gong, Y., Wang, L., Zhang, Y., Cheng, L., Chen, D. (2016). Using urban development boundaries to constrain uncontrolled urban sprawl in China. *Annals of the American Association of Geographers*, 106(6), 1321–1343.

Kauffmann, A. (2012). Delineation of city regions based on commuting interrelations: The example of large cities in Germany. IWH Discussion Paper No. 4. Halle Institute for Economic Research.

Lam, N.S.-N. and Quattrochi, D.A. (1992). On the issues of scale, resolution, and fractal analysis in the mapping sciences. *The Professional Geographer*, 44(1), 88–98.

Liu, F. and Huang, Q. (2019). An approach to determining the spatially contiguous zone of a self-organized urban agglomeration. *Sustainability*, 11(12), 3490.

Liu, Q., Deng, M., Shi, Y., Wang, J. (2012). A density-based spatial clustering algorithm considering both spatial proximity and attribute similarity. *Computers and Geosciences*, 46, 296–309.

Liu, Q., Deng, M., Shi, Y. (2013). Adaptive spatial clustering in the presence of obstacles and facilitators. *Computers and Geosciences*, 56, 104–118.

Liu, L., Xia, B., Wu, H., Zhao, J., Peng, Z., Yu, Y. (2019). Delimitating the natural city with points of interests based on service area and maximum entropy method. *Entropy*, 21(5), 458.

Long, Y. (2016). Redefining Chinese city system with emerging new data. *Applied Geography*, 75, 36–48.

Ma, S., Li, X., Cai, Y. (2017). Delimiting the urban growth boundaries with a modified ant colony optimization model. *Computers, Environment and Urban Systems*, 62, 146–155.

Maisonobe, M., Jégou, L., Eckert, D. (2018). Delineating urban agglomerations across the world: A dataset for studying the spatial distribution of academic research at city level. *Cybergeo: European Journal of Geography*, 871 [Online]. Available at: https://journals.openedition.org/cybergeo/29637.

Masucci, A.P., Arcaute, E., Hatna, E., Stanilov, K., Batty, M. (2015). On the problem of boundaries and scaling for urban street networks. *Journal of the Royal Society Interface*, 12(111), 20150763.

Medda, F., Nijkamp, P., Rietveld, P. (1998). Recognition and classification of urban shapes. *Geographical Analysis*, 30(4), 304–314.

Minkowski, H. (1903). Volumen und Oberfläche. *Mathematische Annalen*, 57, 447–495.

Montero, G., Tannier, C., Thomas, I. (2021). Delineation of cities based on scaling properties of urban patterns: A comparison of three methods. *International Journal of Geographical Information Science*, 35(5), 919–947.

Montero, G., Caruso, G., Hilal, M., Thomas, I. (2022). A partition-free spatial clustering that preserves topology: Application to built-up density. *Journal of Geographical Systems*, 25(1), 5–35.

Moudon, A.V. (1997). Urban morphology as an emerging interdisciplinary field. *Urban Morphology*, 1(1), 3–10.

Nicolau, R. and Cavaco, C. (2018). Automated delimitation of urban areas comprising small-sized towns – Comparison of two methodologies applied to mainland Portugal. *Environment and Planning B: Urban Analytics and City Science*, 45(1), 180–201.

de Olde, C. and Oosterlynck, S. (2021). Taking implementation seriously in the evaluation of urban growth management strategies: "Safeguarding the Future" of the Antwerp City-Region. *Land*, 10(2), 159.

Oliveira, E.A., Andrade, J.S., Makse, H.A. (2015). Large cities are less green. *Scientific Reports*, 4(1), 4235.

Openshaw, S. (1983). *The Modifiable Areal Unit Problem. Concepts and Techniques in Modern Geography*. Geo Books, Norwich.

Ren, Z. (2018). Tutorials for Extracting Street Nodes from OSM Data and Creating Natural Cities from Street Nodes Using ArcGIS Model. A Workshop on Big Data Analytics: Topological and scaling perspective for better understanding and (re-)making sustainable cities, 12 June, Lund.

Rozenfeld, H.D., Rybski, D., Andrade, J.S., Batty, M., Stanley, H.E., Makse, H.A. (2008). Laws of population growth. *Proceedings of the National Academy of Sciences*, 105(48), 18702–18707.

Rozenfeld, H.D., Rybski, D., Gabaix, X., Makse, H.A. (2011). The area and population of cities: New insights from a different perspective on cities. *American Economic Review*, 101(5), 2205–2225.

Rybnikova, N., Portnov, B., Charney, I., Rybnikov, S. (2021). Delineating functional urban areas using a multi-step analysis of artificial light-at-night data. *Remote Sensing*, 13(18), 3714.

Schoning, P., Dysterud, M.V., Engelien, E. (1999). Computerised delimitation of urban settlements: A method based on the use of administrative registers and digital maps. Documents 1999/17, Statistics Norway, November.

Sémécurbe, F. (2020). Analyse de la distribution spatiale des implantations humaines : apports et limites d'indicateurs multi-échelles et trans-échelles. PhD Thesis, Université de Franche-Comté.

Sémécurbe, F., Tannier, C., Roux, S.G. (2019). Applying two fractal methods to characterise the local and global deviations from scale invariance of built patterns throughout mainland France. *Journal of Geographical Systems*, 21(2), 271–293.

Sotomayor-Gómez, B. and Samaniego, H. (2020). City limits in the age of smartphones and urban scaling. *Computers, Environment and Urban Systems*, 79, 101423.

Tan, X. and Huang, B. (2022). Identifying urban agglomerations in China based on density–density correlation functions. *Annals of the American Association of Geographers*, 112(6), 1666–1684.

Tannier, C. (2017). Analyse et simulation de la concentration et de la dispersion des implantations humaines de l'échelle micro-locale à l'échelle régionale - Modèles multi-échelles et trans-échelles. HDR Thesis, Université de Franche-Comté.

Tannier, C. and Thomas, I. (2013). Defining and characterizing urban boundaries: A fractal analysis of theoretical cities and Belgian cities. *Computers, Environment and Urban Systems*, 41, 234–248.

Tannier, C., Vuidel, G., Frankhauser, P. (2007). Délimitation d'ensembles morphologiques par une approche multi-échelle. *Huitièmes Rencontres de Théo Quant*, ThéMA, CNRS-Université de Franche-Comté.

Tannier, C., Thomas, I., Vuidel, G., Frankhauser, P. (2011). A fractal approach to identifying urban boundaries. *Geographical Analysis*, 43(2), 211–227.

Tayyebi, A., Pijanowski, B.C., Tayyebi, A.H. (2011). An urban growth boundary model using neural networks, GIS and radial parameterization: An application to Tehran, Iran. *Landscape and Urban Planning*, 100(1–2), 35–44.

Thomas, I., Frankhauser, P., De Keersmaecker, M. (2007). Fractal dimension versus density of the built-up surfaces in the periphery of Brussels. *Papers in Regional Science*, 86(2), 287–307.

Thomas, I., Cotteels, C., Jones, J., Peeters, D. (2012). Revisiting the extension of the Brussels urban agglomeration: New methods, new data… new results? *Belgeo*, 1–2, 2–13.

Thomas, I., Jones, J., Caruso, G., Gerber, P. (2018). City delineation in European applications of LUTI models: Review and tests. *Transport Reviews*, 38(1), 6–32.

Uchiyama, Y. and Mori, K. (2017). Methods for specifying spatial boundaries of cities in the world: The impacts of delineation methods on city sustainability indices. *Science of The Total Environment*, 592, 345–356.

Usui, H. (2019). A bottom-up approach for delineating urban areas minimizing the connection cost of built clusters: Comparison with top-down-based densely inhabited districts. *Computers, Environment and Urban Systems*, 77, 101363.

Vandermotten, C. (2016). Adéquations et inadéquations du découpage territorial aux contextes politiques et économiques : le cas de la Belgique. *EchoGéo*, 35 [Online]. Available at: http://journals.openedition.org/echogeo/14490.

Vandermotten, C., Halbert, L., Roelandts, M., Cornut, P. (2008). European planning and the Polycentric consensus: Wishful thinking? *Regional Studies*, 42(8), 1205–1217.

Vanderstraeten, L. and Van Hecke, E. (2019). Les régions urbaines en Belgique. *Belgeo*, 1 [Online]. Available at: http://journals.openedition.org/belgeo/32246.

Van Hecke, E., Halleux, J.-M., Decroly, J.-M., Mérenne-Schoumaker, B. (2009). *Noyaux d'habitat et Régions urbaines dans une Belgique urbanisée*. Monographie 9: L'Urbanisation. SPF Economie, P.M.E., Classes moyennes et Energie, Brussels.

Vanneste, D., Thomas, I., Vanderstraeten, L. (2008). The spatial structure(s) of the Belgian housing stock. *Journal of Housing and the Built Environment*, 23(3), 173–198.

Wang, S., Wang, D., Li, C., Li, Y. (2015). Comment on "Clustering by fast search and find of density peaks". arXiv:1501.04267.

Wang, Y., Hu, Z., Zhang, Y. (2021). Delineating the future boundaries of urban development using the FLUS Model: A case study of Zhaoyuan City, China. *IOP Conference Series: Earth and Environmental Science*, 783(1), 012088.

Weber, C. (2001). Urban agglomeration delimitation using remote sensing. In *Remote Sensing and Urban Analysis*, Donnay, J., Barnsley, M., Longley, P. (eds). Taylor & Francis, London, 145–159.

Wu, B., Yu, B., Wu, Q., Chen, Z., Yao, S., Huang, Y., Wu, J. (2018). An extended minimum spanning tree method for characterizing local urban patterns. *International Journal of Geographical Information Science*, 32(3), 450–475.

Xu, Z., Jiao, L., Lan, T., Zhou, Z., Cui, H., Li, C., Xu, G., Liu, Y. (2021). Mapping hierarchical urban boundaries for global urban settlements. *International Journal of Applied Earth Observation and Geoinformation*, 103, 102480.

Yu, B., Shu, S., Liu, H., Song, W., Wu, J., Wang, L., Chen, Z. (2014). Object-based spatial cluster analysis of urban landscape pattern using nighttime light satellite images: A case study of China. *International Journal of Geographical Information Science*, 28(11), 2328–2355.

Zahn, C. (1971). Graph-theoretical methods for detecting and describing gestalt clusters. *IEEE Transactions on Computers*, C-20(1), 68–86.

Zhang, Y., Yu, J., Fan, W. (2008). Fractal features of urban morphology and simulation of urban boundary. *Geo-spatial Information Science*, 11(2), 121–126.

Zhong, X. and Duckham, M. (2016). Characterizing the shapes of noisy, non-uniform, and disconnected point clusters in the plane. *Computers, Environment and Urban Systems*, 57, 48–58.

Zhou, Q. (2015). Comparative study of approaches to delineating built-up areas using road network data: Delineating built-up areas using road network data. *Transactions in GIS*, 19(6), 848–876.

Zhou, Q. and Guo, L. (2018). Empirical approach to threshold determination for the delineation of built-up areas with road network data. *PLOS ONE*, 13(3), e0194806.

Zhou, Y., Tu, M., Wang, S., Liu, W. (2018). A novel approach for identifying urban built-up area boundaries using high-resolution remote-sensing data based on the scale effect. *ISPRS International Journal of Geo-Information*, 7(4), 135.

6

Multifractal Analyses of Population Distributions

François SÉMÉCURBE and Cécile TANNIER

Laboratoire ThéMA, CNRS-Université de Franche-Comté, Besançon, France

6.1. Introduction

In order to study the spatial distribution of a population, geographers generally resort to a division of space into spatial units. The population densities calculated within these units describe the unequal spatial distribution of the population. This approach has a puzzling intrinsic property: the results vary depending on the size and shape of the spatial units under consideration. For example, in mainland France in 2020, the average density was 120 inhabitants per km^2. However, this overall figure masks a great heterogeneity of densities. The Île-de-France region, which includes the French capital, Paris, has a density exceeding 1,000 inhabitants per km^2, while the least dense region, Corsica, has a density of just over 40 inhabitants per km^2. Within the Île-de-France region, the department of Paris, which corresponds to the municipality of Paris, has a density of more than 20,000 inhabitants per km^2, while the department of Seine-et-Marne has a density barely twice as high as the national average. Regardless of the scale of analysis, some areas have a high population concentration and others appear very sparse.

Openshaw (1983) theorized this dependence of analysis on the choice of sample as *the modifiable areal unit problem* (MAUP). The MAUP was initially considered to be a limit to geographical knowledge to be dealt with by choosing an appropriate scale of analysis. Gradually, the idea emerged that no scale is more relevant than another. Geographers now consider the dependence of results on scales of observation as evidence of the multiscalar organization of the phenomena studied (Sanders 2011).

If the MAUP involves a regular and proportional change in the concentration of a given phenomenon across scales (i.e. spatial resolutions or hierarchically nested neighborhoods), and thus if the MAUP has the property of scale invariance, it is possible to quantify it by finding the power law that best describes it. It is then possible to calculate the fractal dimension of the spatial distribution under consideration. However, in a number of cases, the MAUP results not in simple scale invariance, but in a series of interlinked scaling laws. A single fractal dimension cannot then describe them correctly and it can be interesting to carry out multifractal analyses, the fundamental principle of which is that a spatial distribution may present different interlinked self-similar characteristics.

While fractal analysis is applied to data representing the presence or absence of a given phenomenon, multifractal analysis is applied to data representing the *quantity* of the phenomenon in each location of the space under consideration. It consists of a sort of extension of fractal analysis, as it takes into account not only a series of hierarchically nested spatial resolutions (or neighborhoods), as fractal analysis does, but also a *series of points of view* on the quantity contained in each spatial unit.

Probably due to a rather complicated mathematical formalism, multifractal analysis has so far been the subject of relatively few applications in geography, but the theoretical distributions of multifractal geometry have nevertheless been used to enrich the traditional models of geography. Thus, Chen (2014) proposed a multifractal reinterpretation of central place theory using a multifractal growth model. Ozik et al. (2005) used multifractal cascades to represent the spatial distribution of the global population. At a finer scale, Saeedimoghaddam and Stepinski (2021) used multifractal cascades to simulate street networks resembling those of American cities. Cavailhès et al. (2010) combined economic behavioral modeling with multifractal spatial modeling to study the specificities of the land rent gradient and accessibility to urban and rural amenities in a multifractal metropolis. More recently, Chen (2016) used the aggregate proportion of rural and urban populations to model the population distribution in China and the United States using a multifractal cascade. With regard to this article, it should be noted that the absence of consideration of space and of comparison with real population distributions reduces the relevance of the stated conclusions.

In addition, morphological indicators of multifractal analysis have been used to describe and characterize different types of spatial configurations. Haag (1994), on the one hand, and Chen and Zhou (2004), on the other, used multifractal analysis to characterize city systems and the rank-size distribution of systems of settlement. Chen and Wang (2013), Song and Yu (2019) and Song et al. (2021) used it to characterize the urban fabric of Beijing at different dates. Instead of comprehensively describing the entire urban area of Beijing, Tan et al. (2021) explored its local multi-scale configuration. Dealing with other geographical objects, Ariza-Villaverde et al. (2013) and Long and Chen (2021) used multifractal analysis to characterize street networks, Murcio et al. (2015) used it to describe the evolution of London's street network between 1786 and 2010 and Pavón-Domínguez et al. (2017) used it to characterize a public transport network. Finally, Salat et al. (2018) used it to study the spatial distribution of real estate prices in Kyoto (Japan), and Hu et al. (2012) used it to describe changes in the prices of plots for residential use.

In addition to the attempt by Le Bras (1993, 2017) to characterize the multifractal structure of settlement in France, three studies have used multifractal analysis to characterize the spatial distribution of a population (Appleby 1996; Orozco et al. 2015; Sémécurbe et al. 2016). The first article provides an overall description of the populations of the United Kingdom and the United States. The other two consist of an exploration of the spatial heterogeneity of the multifractal characteristics of the populations of Switzerland and France, respectively.

6.2. Theoretical presentation of multifractal analysis

The multifractal analysis of a spatial population distribution aims to characterize the different scaling laws governing the said distribution. Two families of multifractal indicators, generalized dimensions and the singularity spectrum, are generally used for this purpose. The two families of indicators describe the same information but in a different way: the generalized dimensions describe the concentration and dispersion of the population globally, while the singularity spectrum describes these locally. Because of their global character, the estimation of the generalized dimensions is simpler and more precise than that of the spectrum (Arneodo et al. 1995), and the spectrum is obtained using a mathematical transformation of the generalized dimensions.

6.2.1. *Measures and probabilities: mathematical modeling of population distributions*

The multifractal analysis described in this chapter is applied to mathematical objects called "measures", which are generalizations of the concepts of length, area or volume into 1-, 2- or 3-dimensional spaces, respectively. A measure is a mathematical relationship that associates relative weights to different parts of a set (Arneodo et al. 1995). A population distribution in space is a measure of the set under consideration. The parts, in this case, are the different spatial units considered, and the weight of a spatial unit corresponds to the population it contains.

The theoretical definition of a measure is based on mathematical concepts that it is not necessary to know to understand the applications of multifractal geometry in geography. As a first approximation, a measure P is a function that associates a numerical quantity with the different parts of a set E and which has the following properties:

– The measure of the empty part, $\emptyset$, is null, $P(\emptyset) = 0$.

– The measure of the union of two disjoint parts, A and B ($A \cap B = \emptyset$), is the sum of the measures of these two parts $P(A \cup B) = P(A) + P(B)$.

– The measure of the whole E is bounded: $P(E) < +\infty$. In the case of multifractal analysis, it is assumed that P is a probability: $P(E) = 1$.

Multifractal analysis also requires the support of a measure to be defined, namely the set of points whose neighborhood, however small, has a non-zero measure. To use an image, the support of a measure represents its spatial footprint considering a given study area.

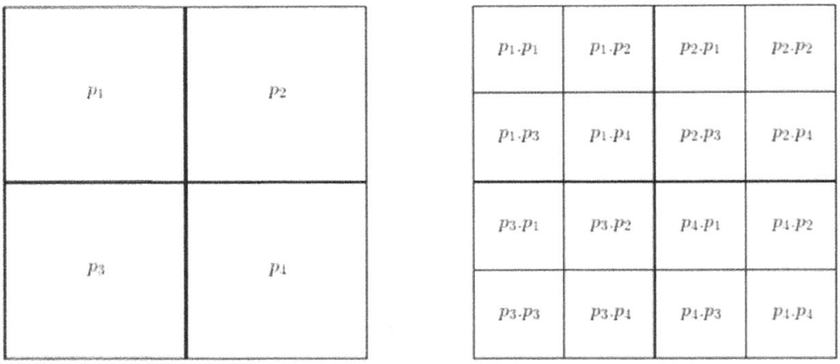

Figure 6.1. *Procedure for constructing a multifractal cascade*

Multifractal cascades are scale-invariant measures useful for modeling theoretical population distributions (Ozik et al. 2005). They are defined as a process of progressive spatial disaggregation of the population. In Figure 6.1, we have shown the generator of a multifractal cascade containing four cells of size δ whose population fractions are p_1, p_2, p_3 and p_4.

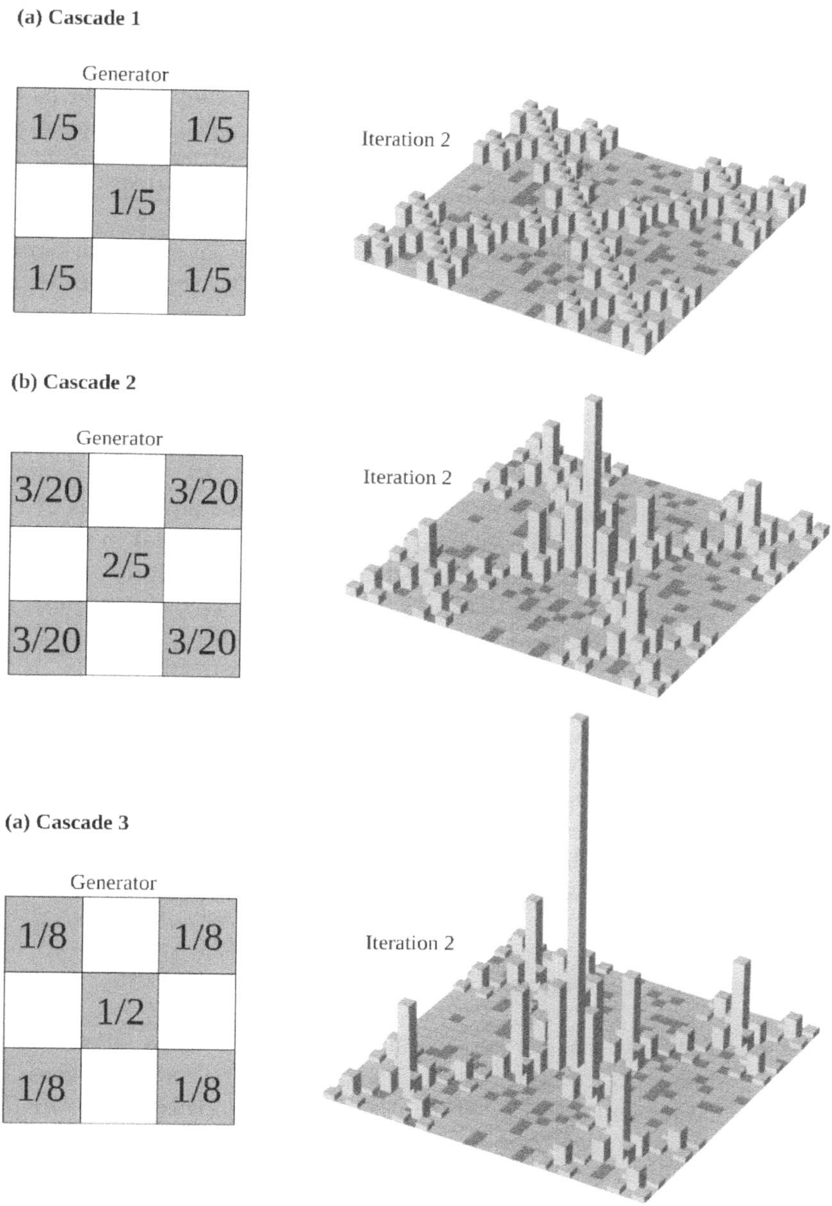

Figure 6.2. *Three multifractal cascades. NB: the cells of the support colored dark gray correspond to shading effects resulting from the 3D view. They should therefore not be taken into account when reading the figure*

These four cells are then divided into smaller cells of size $\delta/2$. A chosen multiplicative process calculates the population in each newly divided cell, given the population in the initial cell (Appleby 1996). For example, $p_1.p_1$, $p_1.p_2$, $p_1.p_3$ and $p_1.p_4$ are the population fractions of the subcells of the initial cell in the upper right of Figure 6.1. A theoretical multifractal cascade is obtained by reproducing this operation for an infinity of successive divisions of cells into smaller cells.

Three different cascades are shown in Figure 6.2. They all have the same support, namely a Vicsek (box) fractal of dimension $log\,(5)/log\,(3)$ (see Chapter 3). The cells of the generator of cascade 1 have the same probability $1/5$. This cascade is called monofractal: the distribution of the population in space depends solely on the location of inhabited places, that is, on the support of the measure, which in this case follows a simple fractal logic. The probabilities of the cells of the generators of cascades 2 and 3 are heterogeneous. The generator of cascade 2 has a central probability of $2/5$ and four peripheral probabilities of $3/20$. The generator of cascade 3 has a central probability of $1/2$ and four peripheral probabilities of $1/8$. Cascades 2 and 3 are said to be multifractal. The initial differences in population concentration within the two generators are accentuated across scales due to the multiplicative process producing each cascade. As the iterations progress, the population of cascade 3 becomes more concentrated than that of cascade 2. The multifractality of cascade 3 is therefore stronger than that of cascade 2.

NOTE.– Saeedimoghaddam et al. (2021) used multifractal cascades to model intersections of the road network of American cities. The probabilities of the generators were chosen so that the generalized dimensions of the cascades resemble those of the cities studied as much as possible. Combining modeling and morphological analysis, their approach made it possible to propose and test different processes generating street grids.

6.2.2. Generalized dimensions

Generalized dimensions can be considered as an extension of the fractal box-counting dimension (see Chapter 4). Calculating generalized dimensions, like the box-counting dimension, relies on the use of regular grids called $\delta - grilles$, composed of squares $c_{i,\delta}$ of size δ indexed by i. On this basis, a generalized dimension D_q is defined as follows:

$$D_q = lim_{\delta\to 0}\frac{I_{q,\delta}}{-log\delta} \qquad [6.1]$$

where $I_{q,\delta} = \frac{1}{1-q}log\left(\sum_i P\left(c_{i,\delta}\right)^q\right)$ for $q \neq 1$ [6.2] and $I_{1,\delta} = -\sum_i P\left(c_{i,\delta}\right) log\left(P\left(c_{i,\delta}\right)\right)$ for $q = 1$ with the writing convention $0^0 = 0$. [6.3]

$I_{q,\delta}$ is the Rényi entropy. It is equal to the Shannon entropy for $q = 1$. The Shannon entropy measures the amount of information needed to locate δ an event at a spatial resolution whose location is random and governed by the probability distribution P.

$\sum_i P\left(c_{i,\delta}\right)^q$ is the moment of order q. For real population distributions, when the moments of order q follow scaling laws $\sum_i P\left(c_{i,\delta}\right)^q \sim \delta^{-(1-q)D_q}$, the generalized dimensions are obtained using their exponent $-(1-q)D_q$.

D_0 is the box-counting dimension of the support of the measure; it characterizes the geometry, the spatial footprint of inhabited places. D_1 is called the information dimension. It is similar to marginal information across the scales. The quantity $log(\delta_1/\delta_2)D_1$ represents the additional information we must know to locate a random event at a spatial resolution δ_2, given that we know its location at resolution δ_1. D_2 is called the correlation dimension and describes the probability that two independent random events (e.g. two inhabitants) are close to each other across the scales. The more D_1 and D_2 are different from D_0, the more the population is concentrated across scales and deviates from a monofractal spatial distribution. More broadly, for any q integer greater than or equal to 2, D_q can be defined as the joint location of q random events of law P across scales (Arneodo et al. 1995).

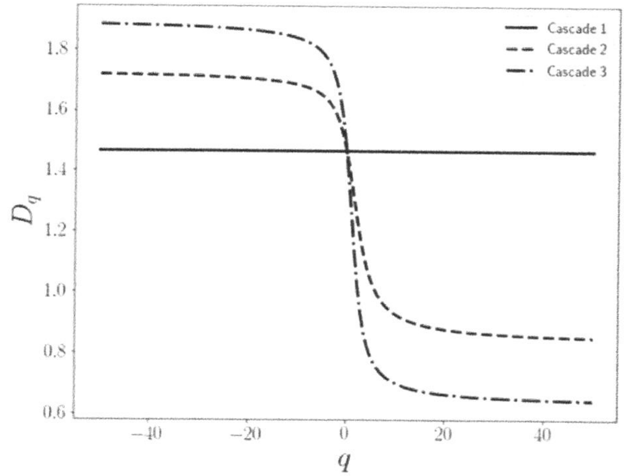

Figure 6.3. *Generalized dimensions of the three multifractal cascades in Figure 6.2*

The generalized dimensions of the three multifractal cascades in Figure 6.2 correspond to the values of the Rényi entropy of their generator:

$$D_q = \frac{1}{1-q}\frac{log(\sum p_i^q)}{log3} \qquad [6.4]$$

In Figure 6.3, we see that the generalized dimensions are decreasing as a function of q (Orozco et al. 2015). The rate of the decrease testifies to the more or less multifractal character of population distributions. Since cascade 1 is monofractal, all its dimensions D_q are identical and have the value of the box-counting dimension of the support of the cascade, in this case a Vicsek (box) fractal that follows a single scaling law of exponent $log\,(5)/log\,(3) \approx 1.46$. Cascades 2 and 3 are multifractal and have strictly decreasing generalized dimensions. An infinity of scaling laws govern their organization across scales. The greater rate of decrease of the generalized dimensions of cascade 3 with respect to cascade 2 reveals its stronger multifractality.

There is another method of multifractal analysis, the sandbox method, which is particularly well suited to the study of networks, such as road networks and public transport networks. Unlike box-counting methods that use fixed grids, it is based on disks of variable radius centered on the analyzed spatial configuration. The fractal dimension of correlation proposed by Grassberger and Procaccia (1983) is a particular case of generalized dimension calculated using the sandbox method and corresponds to D_2. If we consider a set U of N points, this dimension is determined by examining the average number of neighbors that each point has across the scales:

$$\frac{1}{N}\sum_i \sum_j 1_{d(i,j)\leq r} \sim r^{D_2} \qquad [6.5]$$

where:

– i and j are points in the set U;

– $d(i,j)$ is the distance between i and j;

– $\sum_j 1_{d(i,j)\leq r}$ is the number of points contained in the disk centered on the point i of radius r.

The other generalized dimensions are obtained from the exponents $(1-q)D_q$ of the scaling laws followed by the following quantities:

$$\frac{1}{N}\sum_i \left(\sum_j 1_{d(i,j)\leq r}\right)^{q-1} \sim r^{(1-q)D_q} \qquad [6.6]$$

Compared to the box-counting approach, the use of disks improves the accuracy of estimating multifractal indicators. Disks are more naturally adapted than grids to each spatial configuration.

6.2.3. *The singularity spectrum (or multifractal spectrum)*

The multifractal spectrum of a population distribution makes it possible to locally characterize the population concentration. The local concentration is measured using the Hölder exponent (also called the singularity exponent) calculated at any point x on the support of the population distribution:

$$\alpha_H(x) = lim_{r \to 0} \frac{log\left(P(B_{x,r})\right)}{log(r)} \qquad [6.7]$$

where B_x is a ball centered on x with radius r.

The evolution of the local density around the point x as a function of the radius r follows a scaling law of exponent $d_H(x) - 2$:

$$Density(x, r) = \frac{P(B_x)}{\pi r^2} \sim r^{d_H(x)-2} \qquad [6.8]$$

This formula describes the concentration of the population near x:

– When $d_H(x) = 2$, the local density at the point x converges asymptotically towards a finite value and shows very little sensitivity to the value of the radius r. The population is evenly distributed near x.

– When $d_H(x) < 2$, the local density diverges to infinity when the radius tends to 0. x is a singularity that locally concentrates the population. The concentration is strongest when $d_H(x)$ is small and close to 0.

The singularity spectrum is obtained by measuring the Hausdorff fractal dimension of sets of points having the same Hölder coefficient:

$$f(\alpha) = d_H(\{x_0 : d_H(x_0) = \alpha\}) \qquad [6.9]$$

For the three multifractal cascades in Figure 6.2, the values of their spectrum as a function of q are known (Figure 6.4) and depend on the probabilities defined in each cell of the generators:

$$\alpha(q) = \frac{-\Sigma p_i log p_{i,q}}{log 3} \text{ and } f(\alpha(q)) = -\frac{\Sigma p_{i,q} log p_{i,q}}{log(3)} \text{ with } p_{i,q}(i) = \frac{p_i^q}{\Sigma p_i^q} \quad [6.10]$$

where q is a real number.

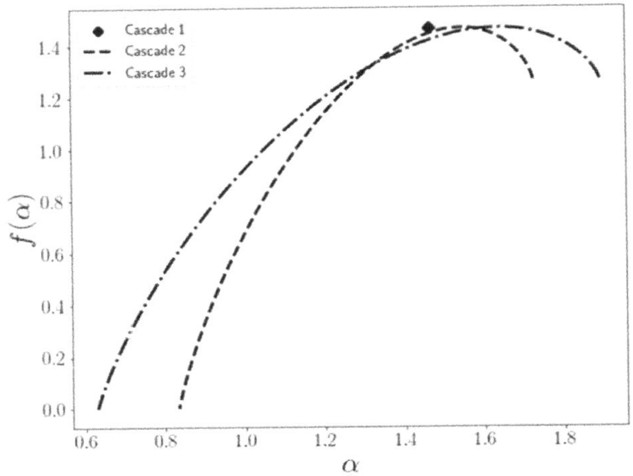

Figure 6.4. *Singularity spectrum of the three multifractal cascades in Figure 6.2*

The numerator of $f(\alpha(q))$ is the Shannon entropy of the distribution $p_{i,q}$. This entropy is at its maximum when all probabilities are equal, which implies that the maximum of the spectra $f(\alpha(q))$ of the multifractal cascades is reached for $q = 0$ and has the box-counting dimension of the support as its value.

The values of the spectrum $f(\alpha)$ are constrained by α: $0 < f(\alpha) < \alpha$. The left-hand side of the inequality is trivial and related to the fact that $f(\alpha)$ is a dimension. The right part is more interesting because it implies that the strength of the singularities constrains their distribution: a strong singularity cannot be found simultaneously in many places.

Since the three cascades of Figure 6.2 have the same support, their spectra have the same maximum $log(5)/log(3)$ (Figure 6.4). The range of the spectrum $(\alpha_{max} - \alpha_{min})$ gives an indication of the more or less definite multifractal character of the distributions. For monofractal cascade 1, the spectrum reduces to a coordinate point (D_0, D_0): all the Hölder coefficients are identical. We find here the fact that the organization of the population distribution is characterized by a single scaling law.

Conversely, as cascades 2 and 3 are multifractal, their spectrum is a curve and not a point. The spectrum of cascade 3 is more extensive than that of cascade 2: at all scales, cascade 3 contains local densities that are more heterogeneous than those of cascade 2. Indeed, as mentioned above, the multifractality of cascade 3 is more pronounced than that of cascade 2.

6.2.4. *Multifractal formalism*

Calculation of the Hölder coefficient is valid only for theoretical population distributions where the radius r may be infinitely small. For real distributions, the singularity spectrum is deduced from the generalized dimensions. To this end, the moments of order q are rewritten in the form of a sum involving the two components of the spectrum:

$$\Sigma_i P\left(c_{i,\delta}\right)^q \sim \Sigma\, \delta^{f(\alpha)}\,(\delta^{\alpha})^q \sim \delta^{f(\alpha)-q\alpha} \qquad [6.11]$$

where $\delta^{f(\alpha)}$ represents the number of squares of measure δ^{α} (see equations [6.2] and [6.3]).

When the size δ of the squares is small enough, the term $\delta^{f(\alpha)-q\alpha}$ is dominated by $\delta^{min_\alpha(q\alpha - f(\alpha))}$. Using equation [6.1], we get the following relationship:

$$\delta^{(1-q)D_q} \sim \delta^{min_\alpha(q\alpha - f(\alpha))} \qquad [6.12]$$

Through moving to the logarithm of the two members of the equation, we deduce the following relation between the generalized dimensions and the multifractal spectrum:

$$(1-q)D_q = min_\alpha\big(q\alpha - f(\alpha)\big) \qquad [6.13]$$

This relationship is a Legendre transform. If the spectrum is differentiable, it simplifies as follows:

$$\alpha(q) = \frac{d}{dq}\Big((1-q)D_q\Big) \qquad [6.14]$$

$$f\big(\alpha(q)\big) = q\alpha(q) - (1-q)D_q \qquad [6.15]$$

Estimation of the spectrum using this Legendre transform is called multifractal formalism. In theory, multifractal formalism is not verified for all measures. In practice, the class of measures that comply with multifractal formalism is large enough that it can be used in many cases without too much risk. The properties of the spectrum described above are general and therefore apply to measures that verify multifractal formalism.

Chhabra and Jensen (1989) proposed a reformulation of the Legendre transform that avoids the need to calculate derivatives:

$$\alpha(q) = lim_{\delta\to 0} \sum_i \frac{P_{\delta,q}(i) log\left(P(c_{i,\delta})\right)}{log(\delta)} \quad [6.16]$$

$$\text{and } f\big(\alpha(q)\big) = lim_{\delta\to 0} \sum_i \frac{P_{\delta,q}(i) log\left(P_{\delta,q}(i)\right)}{log(\delta)} \quad [6.17]$$

$$\text{with } P_{\delta,q}(i) = \frac{P_\delta(i)^q}{\sum_{P_\delta(i)>0} P_\delta(i)^q} \quad [6.18]$$

Here, we find formulas analogous to those used previously for directly calculating the spectra of multifractal cascades.

Recently, Tan et al. (2021) questioned the idea that the multifractal spectrum is a local description of a measure because, due to its indirect calculation which is based on a transformation of generalized dimensions, the multifractal spectrum remains a global mean. Moreover, in the case of real population distributions for which the estimates are obtained considering a restricted range of scales, the value of the Hölder exponent at a point may vary depending on the scales considered. Tan et al. (2021) used these multiscalar variations to produce a spatial typology of the local built-up configuration of the city of Beijing. This method is similar to the multi-radial analysis proposed by Frankhauser (2015).

6.3. Application to the study of the spatial distribution of the population

The implementation of a multifractal analysis requires a database describing the distribution of the population studied. It also requires defining a range of scales to calculate the indicators and a range of values for the parameter q. The results of the analyses sometimes differ significantly depending on the choices made, so these are worth discussing.

The choice of the range of scales results from a trade-off between two constraints:

– Scale invariance is the minimum criterion that a range of scales must verify in order to guarantee the robustness of the results of multifractal analyses. It is verified when the indicators plotted in bi-logarithmic coordinates are aligned for a wide range of scales. The width of a range of scales is satisfactory when the ratio $\delta_{max}/\delta_{min}$ between the coarsest scale δ_{max} and the finest scale δ_{min} is very large (Seuront 2009). In practice, this property is not always verified in published articles (Malcai et al. 1997; Avnir et al. 1998).

– Not all scales are associated with the same phenomenon (Orozco et al. 2015). The same spatial distribution may be monofractal with $D_0 < 2$ for an initial range of scales, then become multifractal for coarser scales and occupy the whole plane ($D_0 = 2$) according to the predominant geographical and socio-economic phenomena (Ariza-Villaverde et al. 2013). Choosing a relevant range of scales is a central element of any geographical analysis.

6.3.1. *Exploration of generalized dimensions*

Here, we present the study by Sémécurbe et al. (2016), which consists of a multifractal analysis of population data gridded at 200 m provided by INSEE for the year 2010 for the entire territory of mainland France. The side length of each cell of the δ-grids is between 800 and 6,400 m. The size of a spatial unit (grid) is fixed at 25 km per side (Figure 6.5). In doing so, mainland France is cut into 992 grid squares, each about four times larger than the cells of the δ-grid at the spatial resolution of 6,400 m. Five values of q are taken into account: $\{0; 0.5; 1; 1.5; 2\}$.

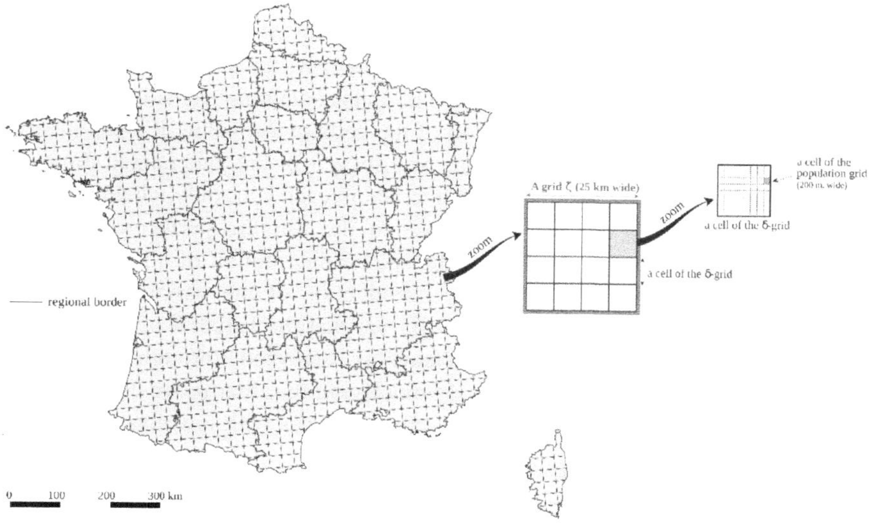

Figure 6.5. *Hierarchical nesting of spatial units of interest defined for the multifractal characterization of the spatial distribution of the population in mainland France*

Note that, in another study, Appleby (1996) used incomplete data to analyze the multifractal organization of the population of the United States and Great Britain; only the population of the largest cities (>2,500 inhabitants in United States and > 5,000 inhabitants in Great Britain) was taken into account. The sensitivity analysis conducted by Appleby measured the effect of this exclusion of data on the calculated multifractal indicators; the lower the population threshold of exclusion, the smaller the fractal dimensions. Orozco et al. (2015) and Sémécurbe et al. (2016), meanwhile, carried out their analyses on exhaustive grid data. In this case, it is errors of population location that may cause estimation biases for multifractal indicators.

For the same value of δ, $\left(\frac{\delta}{200}\right)^2$, δ-grids can be created by varying their origin along the x-axis and y-axis. In theory, the position of the grids does not influence the value of the multifractal indices. However, Appleby (1996) found in practice that the calculated entropies depend on which grids are chosen. He suggests that, for each value of δ, the grid chosen for the study of the spatial distribution of the population is that for which the minimum entropies are obtained. This selection is made for all pairs (δ, q). The use of the minimum criterion is consistent with the fact that entropy describes the minimum amount of information required to locate a random event.

The coefficient q acts as a filter in the calculation of the dimensions D_q. For $q \rightarrow +\infty$; only the cells comprising a large quantity of the phenomenon under consideration contribute to the value of D_q. Conversely, for $q \rightarrow -\infty$, the generalized dimensions describe the spatial distribution of cells containing a small amount of the phenomenon. For $q = 1$, the contribution of the cells to the value of D_q is invariant with regard to the quantity they contain. And for $q = 0$, the generalized dimension does not take into account the quantity contained in each cell; only whether the cells are populated or not is considered. Thanks to the filtering performed by q, the generalized dimensions D_q characterize the spatial distribution of a population in a manner that is both multiscale (via the introduction of the various δ-grids) and multipoint of view (via the coefficient q). By varying the length δ of the grid cells, the spatial distribution of the inhabitants is analyzed through a series of hierarchically nested spatial partitions. By modifying the filter q, we consider either places with high concentrations of inhabitants, or places with low concentrations of inhabitants.

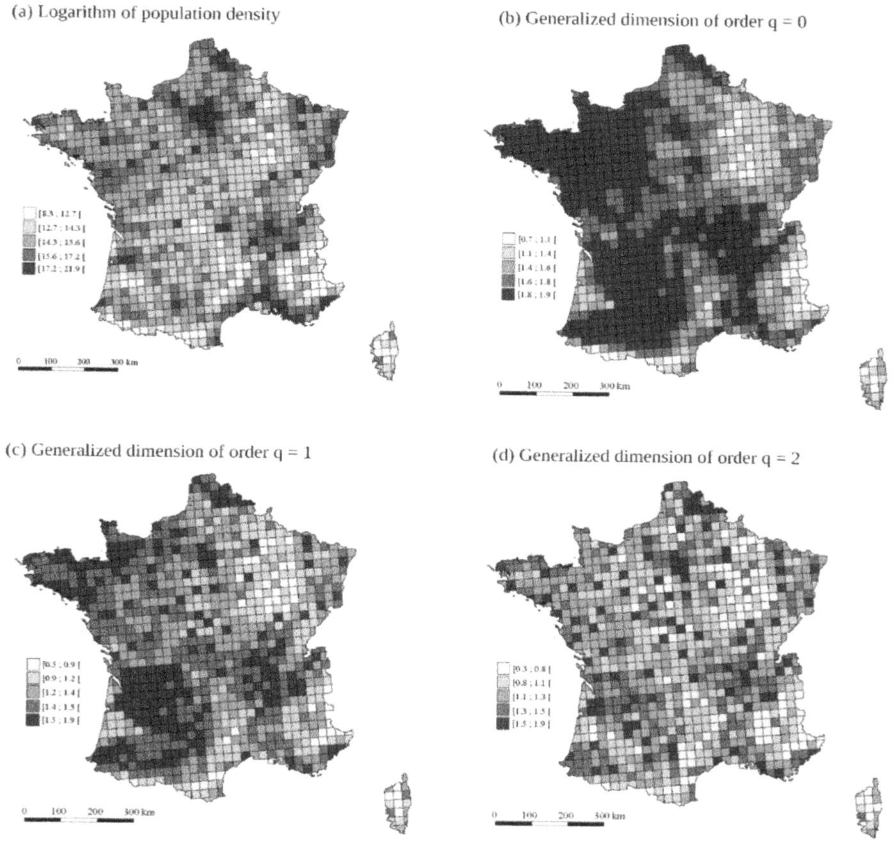

Figure 6.6. *Population densities and generalized multifractal dimensions D_q in mainland France. Source: population data gridded at 200 m, INSEE 2010*

Estimations of generalized dimensions for mainland France show that the dimension D_0 is high in western France and in major cities (Figure 6.6). The indifference of this dimension to density explains why very different spaces may have comparable values. On the contrary, the dimension D_2 separates large metropolises, which have high D_2 dimensions, from rural areas, which have low D_2 dimensions. D_2 is therefore very different from D_0 in the rural areas of western France; these areas are multifractal. Conversely, D_2 is close to D_0 in the rural areas of the north-eastern quarter of France, in the Alps, in the Pyrenees and on the western Atlantic coast (Landes): in these areas, the multifractality of the settlement structure is much less pronounced. As for the metropolises, they have high values of D_0 and D_2, and their multifractality is even less pronounced. Because of the decrease in the generalized dimensions, the map D_1 has an intermediate configuration with respect to the maps D_0 and D_2.

6.3.2. *Typology of spatial distributions according to their multifractal spectrum*

The singularity spectrum $f(\alpha)$ makes it possible to quantitatively characterize the MAUP (Sémécurbe et al. 2016). When reduced to a coordinate point (2, 2), the population is spatially distributed uniformly; the population densities are almost constant locally. In this case, the effect of the MAUP is weak or even absent. When the spectrum is not reduced to the coordinate point (2, 2), the maximum value of $f(\alpha)$ is always less than 2 and the spatial distribution of the population is subject to the MAUP. Two phenomena may occur:

– *Shape of the container (spatial distribution of cells containing inhabitants)*: the lower the maximum value of $f(\alpha)$, the more irregular the shape of the support of the measure (i.e. of the set of "inhabited" cells) and the larger the number of empty cells in the δ-grid for small values of δ. As a result, the MAUP relative to the shape of the container is strong.

– *Content heterogeneity*: two measures with the same maximum value of $f(\alpha)$ may have more or less intense multifractality. It is weak when the singularity spectrum is concentrated around α_0: only the support of the measure (i.e. the container) creates the MAUP, but the population densities vary little between the cells of the δ-grid. In this case, the spatial distribution of the population is almost monofractal. Conversely, a wide singularity spectrum is a sign of strong multifractality, and the MAUP results from both the shape of the container and the heterogeneity of the content.

The 992 spatial units (grid-squares) of mainland France (Figure 6.5) were classified according to their multifractal spectrum $f(\alpha)$. To do this, we used a hierarchical ascending classification (HAC) based on Ward's algorithm. This algorithm minimizes intra-group deviations and maximizes inter-group deviations. The realization of this classification required the measure of the difference between the multifractal spectra of each pair of spatial units as a prerequisite. Since the spectra do not all have the same extent on the *y*-axis, we chose to measure this difference using the Hausdorff geometric distance (see the article by Sémécurbe et al. (2016) for further explanation on this).

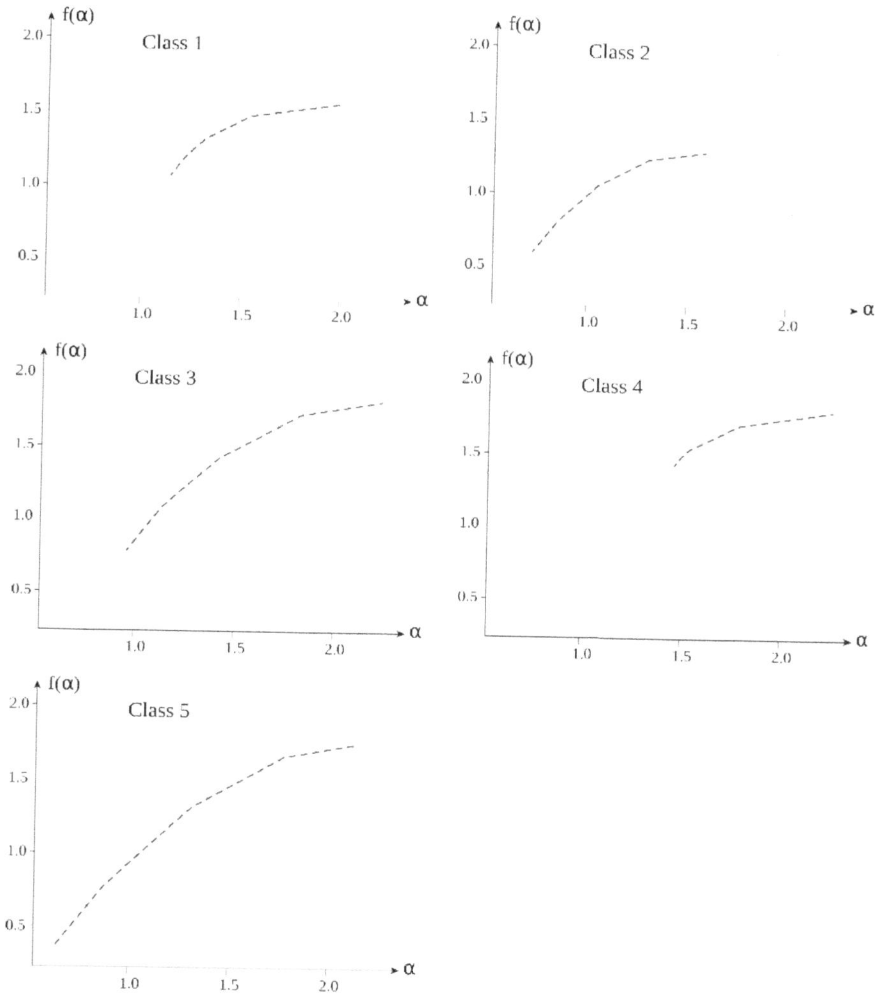

Figure 6.7. *Mean multifractal spectrum characterizing each of the classes*

Five classes of spectrum were identified. Figure 6.7 shows the mean singularity spectrum of each of the classes. Class 2 has the lowest maximum value of f(α), which denotes a strong MAUP relative to the shape of the container (variable alternation of populated cells and unpopulated cells depending on the level of analysis). Since, moreover, the spectrum is quite narrow, the spatial distribution of the population tends towards a monofractal distribution. Class 1 has a maximum value of f(α) almost as low as that of class 2, but the spectrum is wider; the spatial distribution of the population is multifractal, but the MAUP resulting from the shape of the container is stronger than that resulting from the shape of the contents. We observe the opposite trend for class 3: the maximum value of f(α) is very high and the spectrum is very wide; the MAUP resulting from the shape of the container is almost non-existent (almost all the cells are occupied at all levels of analysis), while that resulting from the shape of the content is strong. For class 4, the spectrum combines very high values of α and f(α) and is very narrow; the spatial distribution of the population is uniform (not fractal), with almost all cells populated and having approximately the same number of inhabitants. Finally, the class 5 spectrum is a combination of classes 2 and 3. It reaches values of f(α) less high than the class 3 spectrum but it is very wide; the cells are not all occupied at the different levels of analysis, but the MAUP of the container is relatively weak, while the MAUP of the content is very strong.

The comparison of the map representing the spatial units as a function of their multifractal spectrum class (Figure 6.8) with the maps of population density and generalized fractal dimensions D_0, D_1 and D_2 (Figure 6.6) allows each of the classes to be described in terms of the spatial distribution of the population. In class 1, the population is moderately concentrated locally. In class 2, the population is highly concentrated locally. These two classes of settlement are mainly found in the north-eastern quarter of France (Champagne, Alsace, Franche-Comté), in the Alps, on the Mediterranean coast, in the Pyrenean massif and on the western Atlantic coast (Landes). In class 3, the population is dispersed (the spatial units belonging to this class are located in the outskirts of large cities, as well as in rural areas of western France). In class 4, the population is uniformly dense: these are intra-urban spatial units corresponding to large French cities. Finally, class 5 corresponds to an intermediate spatial configuration, combining the characteristics of classes 2 and 3; villages are surrounded by areas of low population density in which the population is evenly dispersed.

Chen and Wang (2013) used the singularity spectrum to study the evolution of Beijing's built-up fabric between 1988 and 2009. They worked from remote sensing images at a resolution of 30 m for six dates (1988, 1992, 1999, 2006 and 2009), by counting the quantity of built-up pixels present in each cell of the δ-grid. The range of values for δ is not specified by the authors. The series of calculated values for q is {−1; −0.5; 0; 1; 2; 5; 10; 15; 20}, some values being much lower and others much higher than the values of q considered for analysis of the population distribution in mainland France. On this basis, Chen and Wang highlighted three abnormalities of the Beijing multifractal spectrum: (1) the values of D_q and $\alpha(q)$ are greater than 2 for q less than 0; (2) the values of $\alpha(q)$ diverge while those of f(α) converge immediately when q becomes less than 0; (3) when q tends towards +∞, f(α) should approach 0 or a fixed value, which is not the case. From these abnormalities, Chen and Wang conclude that the agglomeration of Beijing is gradually losing its multifractal character. This conclusion seems erroneous to us, because the abnormalities of the values of D_q and f(α) result in fact from problems of estimation of the multifractal parameters, and not from the loss of multifractality of the built-up fabric of Beijing. On the one hand, the high values of f(α) in 2009 (close to 2) can easily be explained; we see on the corresponding image of Beijing that almost all the squares of the grid are occupied. On the other hand, Appleby (1996) pointed out that when q is negative, estimation of the multifractal parameters is very unstable because we focus on cells with very low populations (or a very small quantity of the phenomenon studied) and the generalized entropies are then very sensitive to the positioning of the δ-grid. Finally, Wendt et al. (2009) showed that multifractality may be studied considering q values greater than 2 only for mathematical theoretical constructions and not for real-world objects.

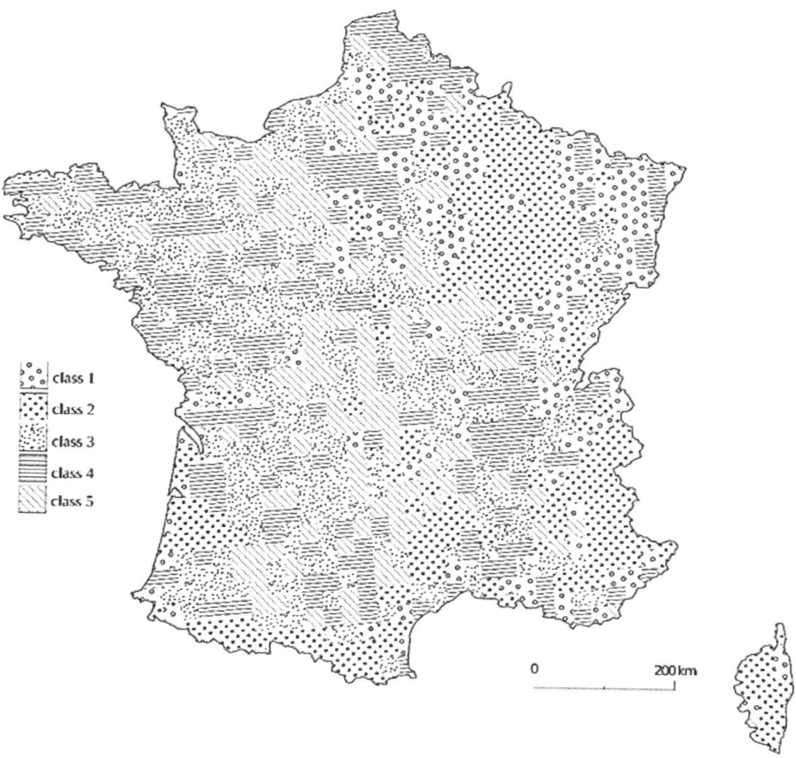

Figure 6.8. *Classification of the multifractal spectra of the spatial distribution of the population in France. 992 spatial units 25 km wide*

The method of spatial analysis based on the classification of multifractal spectra could be applied to characterize and compare the spatial concentration of different types of population, for example, inhabitants with high incomes or low incomes, or inhabitants over 65 years old or under 30 years old. In any case, performing this type of analysis would require considering a fairly large study area, as the spatial units must be large enough to contain different cells of the δ-grid at the coarsest spatial resolution. Furthermore, the size of the cells of the δ-grid at the finest spatial resolution should be higher than the spatial resolution of the analyzed data.

6.4. Conclusion and perspectives

The use of multifractal analysis is motivated by the existence of subparts of heterogeneous fractal dimensions in many spatial configurations of interest to geographers (Batty and Longley 1994; Frankhauser 2008). However, this finding alone is not sufficient to guarantee the multifractality of the configurations studied. Indeed, for a spatial configuration to be multifractal, it is necessary that the configurations of its subparts are themselves fractal. The article by Salat et al. (2018), which studies the spatial distribution of real estate prices in Kyoto (Japan) using multifractal analysis, illustrates the difficulty of separating the analytical instrument from the analyzed objects. Their maps of local fractal dimensions are structured and follow a center–periphery gradient. This juxtaposition of fractal structures of heterogeneous dimensions effectively invalidates the multifractality hypothesis formulated by the authors, but does not call into question the results obtained. Indeed, it seems quite legitimate to apply multifractal analysis to objects that are not multifractal as long as they are not considered to be multifractal. On the other hand, when multifractal analysis is applied automatically and without taking a wider perspective on the nature of the configurations studied, the risk is to tend towards normative discourses. Thus, Huang and Chen (2018) claim that urban built configurations that deviate from a multifractal configuration are "pathological". However, no current theory allows us to explain the multifractality of urban built-up fabrics and therefore to justify such an assertion.

The multifractal analyses presented in this chapter are based on the examination of a single quantity, the population. *Joint* multifractal analysis generalizes multifractal analysis to bivariate data (Meneveau et al. 1990). Ariza-Villaverde et al. (2014) used it to explore relationships across scales between the morphology of the road network and ambient noise. The rarity of use of this method in geography is probably explained by a mathematical formalism even more complicated than that of simple multifractal analysis and by some problems in estimating multifractal indicators (Jaffard et al. 2019). The bivariate multifractal cascades proposed by Meneveau et al. (1990) could allow modeling across scales of social phenomena, such as segregation or social mixing, in which two sub-populations are involved.

6.5. References

Appleby, S. (1996). Multifractal characterization of the distribution pattern of the human population. *Geographical Analysis*, 28(2), 147–160.

Ariza-Villaverde, A.B., Jiménez-Hornero, F.J., De Ravé, E.G. (2013). Multifractal analysis of axial maps applied to the study of urban morphology. *Computers, Environment and Urban Systems*, 38, 1–10.

Ariza-Villaverde, A.B., Jiménez-Hornero, F.J., De Ravé, E.G. (2014). Influence of urban morphology on total noise pollution: Multifractal description. *Science of the Total Environment*, 472, 1–8.

Arneodo, A., Argoul, F., Bacry, E., Elezgaray, J., Muzy, J.-F. (1995). *Ondelettes, multifractales et turbulence : de l'ADN aux croissances cristallines*. Diderot éditeur, Arts et Sciences, Paris.

Avnir, D., Biham, O., Lidar, D., Malcai, O. (1998). Is the geometry of nature fractal? *Science*, 279(5347), 39–40.

Batty, M. and Longley, P.A. (1994). *Fractal Cities: A Geometry of Form and Function*. Academic Press, London.

Cavailhès, J., Frankhauser, P., Peeters, D., Thomas, I. (2010). Residential equilibrium in a multifractal metropolitan area. *The Annals of Regional Science*, 45(3), 681–704.

Chen, Y. (2014). Multifractals of central place systems: Models, dimension spectrums, and empirical analysis. *Physica A*, 402, 266–282.

Chen, Y. (2016). Defining urban and rural regions by multifractal spectrums of urbanization. *Fractals*, 24(01), 1650004.

Chen, Y. and Wang, J. (2013). Multifractal characterization of urban form and growth: The case of Beijing. *Environment and Planning B: Planning and Design*, 40(5), 884–904.

Chen, Y. and Zhou, Y. (2004). Multi-fractal measures of city-size distributions based on the three-parameter Zipf model. *Chaos, Solitons & Fractals*, 22(4), 793–805.

Chhabra, A. and Jensen, R.V. (1989). Direct determination of the f(α) singularity spectrum. *Physical Review Letters*, 62(12), 1327.

Frankhauser, P. (2008). Fractal geometry for measuring and modelling urban patterns. *The Dynamics of Complex Urban Systems*. Physica-Verlag HD, 213–243.

Frankhauser, P. (2015). From fractal urban pattern analysis to fractal urban planning concepts. In *Computational Approaches for Urban Environments, Geotechnologies and the Environment*, Helbich, M., Jokar Arsanjani, J., Leitner, M. (eds). Springer International Publishing, Switzerland, 13–48.

Grassberger, P. and Procaccia, I. (1983). Characterization of strange attractors. *Physical Review Letters*, 50(5), 346.

Haag, G. (1994). The rank-size distribution of settlements as a dynamic multifractal phenomenon. *Chaos, Solitons and Fractals*, 4(4), 519–534.

Hu, S., Cheng, Q., Wang, L., Xie, S. (2012). Multifractal characterization of urban residential land price in space and time. *Applied Geography*, 34, 161–170.

Huang, L.S. and Chen, Y.G. (2018). A comparison between two OLS-based approaches to estimating urban multifractal parameters. *Fractals*, 26(01), 1850019.

Jaffard, S., Seuret, S., Wendt, H., Leonarduzzi, R., Roux, S., Abry, P. (2019). Multivariate multifractal analysis. *Applied and Computational Harmonic Analysis*, 46(3), 653–663.

Le Bras, H. (1993). *La planète au village*. Editions de l'Aube, Paris.

Le Bras, H. (2017). La structure multifractale du peuplement. In *Villes, réseaux et transport. Le défi fractal*, Dupuy, G. (ed.). Economica, Paris, 41–66.

Long, Y. and Chen, Y. (2021). Multifractal scaling analyses of urban street network structure: The cases of twelve megacities in China. *PLoS ONE*, 16(2), e0246925.

Malcai, O., Lidar, D.A., Biham, O., Avnir, D. (1997). Scaling range and cutoffs in empirical fractals. *Physical Review E*, 56(3), 2817.

Meneveau, C., Sreenivasan, K.R., Kailasnath, P., Fan, M.S. (1990). Joint multifractal measures: Theory and applications to turbulence. *Physical Review A*, 41(2), 894.

Murcio, R., Masucci, A.P., Arcaute, E., Batty, M. (2015). Multifractal to monofractal evolution of the London street network. *Physical Review E: Statistical, Nonlinear, and Soft Matter Physics*, 92(6), 062130.

Openshaw, S. (1983). *The Modifiable Areal Unit Problem. Concepts and Techniques in Modern Geography*. Geo Books, Norwich.

Orozco, C.D.V., Golay, J., Kanevski, M. (2015). Multifractal portrayal of the Swiss population. *Cybergeo: European Journal of Geography*, (714) [Online]. Available at: http://cybergeo.revues.org/26829.

Ozik, J., Hunt, B.R., Ott, E. (2005). Formation of multifractal population patterns from reproductive growth and local resettlement. *Physical Review E*, 72(4), 046213.

Pavón-Domínguez, P., Ariza-Villaverde, A.B., Rincón-Casado, A., de Ravé, E.G., Jiménez-Hornero, F.J. (2017). Fractal and multifractal characterization of the scaling geometry of an urban bus-transport network. *Computers, Environment Urban Systems*, 64, 229–238.

Saeedimoghaddam, M. and Stepinski, T.F. (2021). Multiplicative random cascade models of multifractal urban structures. *Physica A: Statistical Mechanics and its Applications*, 569, 125767.

Salat, H., Murcio, R., Yano, K., Arcaute, E. (2018). Uncovering inequality through multifractality of land prices: 1912 and contemporary Kyoto. *PloS ONE*, 13(4), e0196737.

Sanders, L. (2011). Géographie quantitative et analyse spatiale : quelles formes de scientificités ? In *Les sciences humaines sont-elles des sciences ?* Martin, T. (ed.). Vuibert, 71–91.

Sémécurbe, F., Tannier, C., Roux, S.G. (2016). Spatial distribution of human population in France: Exploring the modifiable areal unit problem using multifractal analysis. *Geographical Analysis*, 48(3), 292–313.

Seuront, L. (2009). *Fractals and Multifractals in Ecology and Aquatic Science*. CRC Press, Taylor & Francis, Boca Raton, London, New York.

Song, Z. and Yu, L. (2019). Multifractal features of spatial variation in construction land in Beijing (1985–2015). *Palgrave Communications*, 5(1), 1–15.

Song, Z., Jin, W., Jiang, G., Li, S., Ma, W. (2021). Typical and atypical multifractal systems of urban spaces–using construction land in Zhengzhou from 1988 to 2015 as an example. *Chaos, Solitons & Fractals*, 145, 110732.

Tan, X., Huang, B., Batty, M., Li, J. (2021). Urban spatial organization, multifractals, and evolutionary patterns in large cities. *Annals of the American Association of Geographers*, 111(5), 1539–1558.

Wendt, H., Abry, P., Jaffard, S., Ji, H., Shen, Z. (2009). Wavelet leader multifractal analysis for texture classification. *Image Processing (ICIP), 16th IEEE International Conference*, 3829–3832.

7

Characterizing Deviations from Scale Invariance Using Cross-Scale Signatures

François SÉMÉCURBE and Cécile TANNIER
Laboratoire ThéMA, CNRS-Université de Franche-Comté, Besançon, France

The fundamental assumption underlying fractal analysis is that the relevant information for describing an object is not found at a given scale, but in the way that the local and global forms of the object are linked together across scales. Implementing this idea is based on examining the object at different scales, and each scale corresponding to a level of resolution of its details. If, when moving from a coarse scale to a finer scale, the quantity of emergent details follows a scaling law, the object respects a regularity across scales called scale invariance (or self-similarity)[1]. Fractal dimensions calculated from the exponents of the scaling laws characterize this regularity.

Unlike theoretical figures for which fractal dimensions can be calculated for infinitely precise scales and infinite ranges of scales, the range of scales for which real-world objects are scale-invariant is limited, regardless of the fractal dimension estimation method adopted. "[...] Any empirical measurement of fractal dimension thus highlights a domain of validity of the identified structure, generally between a lower threshold and an upper threshold below or beyond which the fractal dimension changes or can no longer be established" (Pumain 2010). The level of resolution of the source data also interferes with the determination of the domain of validity of the

1. On this subject, see section 2.6 of Chapter 2.

fractality of the structures. "Indeed, since the measure of fractal dimension consists in counting the sites occupied by a given type of object in boxes of increasingly smaller size, this counting becomes impossible when the scale of the supporting cartographic document does not allow discernment of full and empty boxes or the network segments at a very large scale" (Pumain 2010). From the late 1990s, Malcai et al. (1997) and Avnir et al. (1998) found that the vast majority of fractal analyses published in the journals *Physical Review* and *Physical Review Letters* were calculated over ranges of scales that were too narrow for the objects to be considered as fractals. However, even though being able to calculate a fractal dimension does not guarantee that the objects analyzed are fractal, a fractal dimension is nonetheless useful for characterizing geographical objects:

> Since the fractal dimension is determined from counts performed at different levels of resolution of a spatial distribution, this calculation highlights systematic variations in the distribution of a given geographical fact across scales. These variations are non-proportional, nonlinear, they would not appear if we were satisfied with simpler measures of distributions such as that of densities, so often used in classical geography [...] (François et al. 1995).

Early on, however, pioneering works in geography insisted on the fact that the fractal dimension is not assumed to be constant in reality (Goodchild 1980); most often, it is constant for a certain scale interval but varies across successive scale intervals (Lam 1990; Lam and Quattrochi 1992; White and Engelen 1994; Tannier and Pumain 2005). Thus, for a given built-up fabric, which may be that of a neighborhood or an entire city, even though the fit of a count curve to a power law is statistically correct, the difference between two scales in the number of elements counted in neighborhoods of increasing or decreasing size may vary significantly between scales. This has been demonstrated using scaling behavior curves drawn for different cities or neighborhoods (Frankhauser 1998a, 1998b, 2004; Tannier and Pumain 2005; Thomas et al. 2010).

Several methods have been proposed to quantify this type of deviation from scale invariance. The *MorphoLim* method (Tannier et al. 2011)[2], for example, makes it possible to determine the characteristic scale of a built-up fabric, which is that where the deviation from scale invariance is the greatest. The multi-scale signatures introduced by Batty (1991) make it possible to highlight deviations from scale invariance. For Frankhauser (1998a), these deviations are a by-product of fractal analysis; they make it possible to identify ruptures in the spatial organization of built-up fabrics in space and time. Sémécurbe et al. (2019) used the measure of

2. On this subject, see Chapter 5.

deviations from scale invariance to characterize patterns of buildings in mainland France, and Tan et al. (2021) used it to analyze the evolutions of the built environment of the city of Beijing. Outside of the field of geography, Palmer (1988) used cross-scale signatures in ecology to summarize how plants organize themselves in space. More recently, Raghavendra and Dutt (2010) have implemented cross-scale signatures to distinguish letter forms, Florindo et al. (2012) to describe biological phenomena, and Abry et al. (2015) to differentiate photographic paper media.

7.1. Cross-scale signatures: a tool for exploring deviations from scale invariance

Analyzing using cross-scale signatures takes up the concepts of fractal analysis but does not assume any regularity between the scales of observation. Purely descriptive, it offers a framework in which deviation from scale invariance is no longer a handicap but the very object of analysis, so that any objects, fractals or not, may be studied with this method.

7.1.1. *Limitations of fractal dimension for characterizing built patterns*

Self-similarity is a property of an object for which each part shares a resemblance with the whole. This property may be formalized by means of scaling laws (or power laws). For example, if we overlay a distribution of points with a grid of regular squares of size *s*, the number of squares needed to overlay the distribution, *M(s)*, varies according to a power function of *s*:

$$M(s) \approx s^{-D} \quad [7.1]$$

where the exponent *D* is the fractal box-counting dimension. For sets of points such as the centroids of buildings, estimation of the box-counting dimension is imprecise and it is preferable to favor another method of fractal analysis: the sandbox method, which is a generalization of the correlation analysis (see Chapter 4). The squares are replaced by disks centered on each point of the set being analyzed. For a point e, $M_r(e)$ designates the number of points contained in the disk centered on e with radius r, in other words, the number of neighbors of e located at a distance of r. An estimate of the sandbox fractal dimension is obtained by examining the average behavior of $M_r(e)$ across scales:

$$M(r) = \sum M_r\,(e)^{-1} \approx r^{-D} \quad [7.2]$$

In practice, the fractal dimension D is determined by estimating, using a linear regression, the slope of the line formed by $log(M(r))$ as a function of $log(r)$:

$$log(M(r)) = a - Dlog(r) \quad [7.3]$$

Let us apply the sandbox method to the centroids of buildings in two classic configurations of rural geography: a bocage area where buildings are scattered, in Normandy, and an open-field area where buildings are concentrated, in eastern France (Figure 7.1). The range of scales adopted to calculate the fractal dimension is 50, 100, 200, 400 and 800 m. This choice is based on the observation that beyond a radius of 800 m, all buildings have at least one neighbor (the fractal dimension is then equal to 2), and that below 50 m, the buildings are isolated from each other (the fractal dimension is then close to 0). Distances of 50 and 800 m limit the domain of validity of the fractal analysis. The fractal dimensions obtained with the sandbox method are almost identical for the two configurations, respectively, 1.26 for bocage and 1.22 for open-field, while visually these appear very different. Indeed, the fractal dimension, which is the slope of the straight line that best fits the empirical curve, does not measure scale invariance but an average behavior across scales that masks the complexity of the underlying cross-scale organization. However, scale invariance is not respected for the bocage configuration. It can be seen in Figure 7.2 that, for this configuration, the points of the empirical count curve of the average number of neighbors are not perfectly aligned on a straight line. Conversely, the alignment of the points of the curve of the open-field configuration shows the existence of a scale invariance which is, by the way, not visually obvious in Figure 7.1.

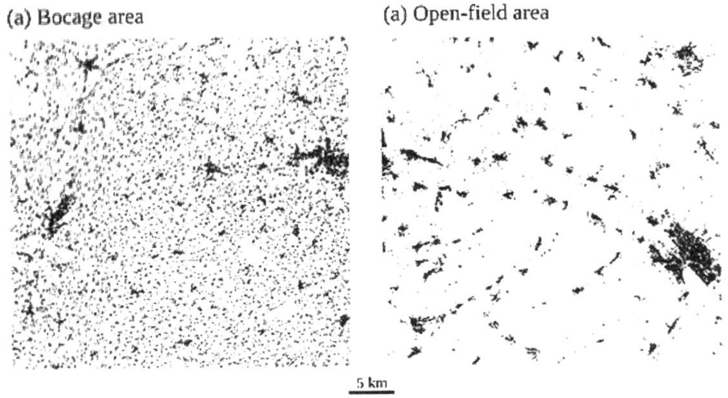

Figure 7.1. *The two built configurations studied. (a) West of France, between Coutances and Saint-Lô. (b) East of France, near Chalons-en-Champagne (data sources: IGN BD Topo (2012))*

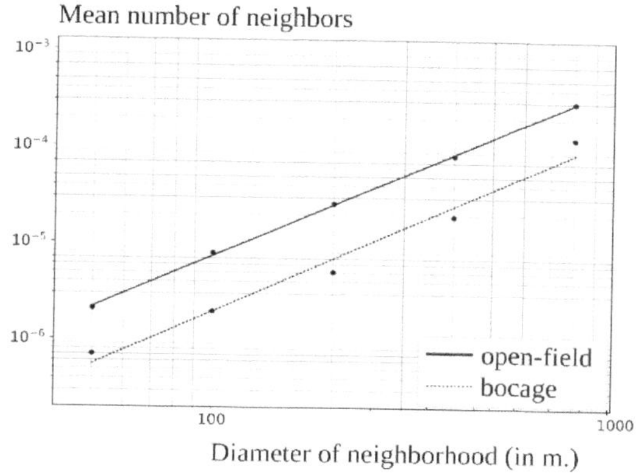

Figure 7.2. *Sandbox fractal analysis of the spatial distribution of building centroids in the images of bocage and open-field areas shown in Figure 7.1*

Let us now apply a regionalized version of the sandbox method, the *Geographically Weighted Fractal Analysis* (GWFA) (see Chapter 4), to all of the centroids of buildings in mainland France. The territory of France has been cut into squares of 2 km on each side or 145,178 squares in total. For each square, all of the buildings located less than 8 km from its center are extracted. The sandbox method is applied over the same series of radii of neighborhoods as before: 50, 100, 200, 400 and 800 m. The fractal dimensions obtained range between 0 and 2 depending on whether a rural or urban space is involved (Figure 7.3a). In very dense spaces, the vast majority of the territory is covered with buildings, so the fractal dimension is close to 2. However, in very sparse rural areas, buildings are isolated from each other and the fractal dimension is almost zero. These fractal dimensions are in fact highly correlated with the logarithm of the density of buildings (correlation coefficient equal to 0.9), which is a limit for describing the heterogeneity of the forms of settlement in mainland France.

The linear coefficients of determination (R^2), showing the quality of the statistical estimation of the fractal dimension, describe a different geography (Figure 7.3b). The R^2s divide mainland France roughly into two parts along a North–East South–West axis. To the northeast of this axis, they are close to 1, which means that the quality of the statistical estimate is very good. This is generally not the case to the southwest of the axis. The fact that the R^2s deviate from 1 indicates a deviation from scale invariance, which is invisible when only considering the fractal dimensions. The scale invariance corresponding to high R^2 is therefore only verified for certain areas of mainland France. Outside of these areas, the fractal dimension only imperfectly describes the cross-scale organization of built-up fabrics.

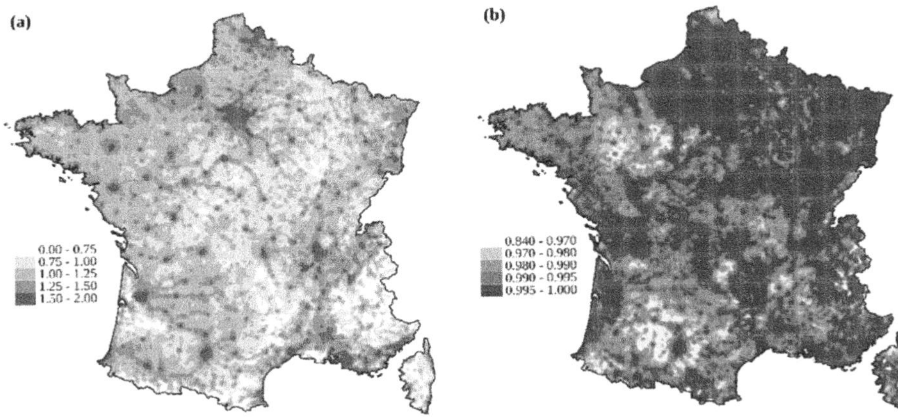

Figure 7.3. *Fractal analysis of 145,178 built-up fabrics in mainland France (squares of 2 km side): (a) Fractal dimensions estimated using the GWFA for a range of scales from 50 to 800 m. (b) Statistical quality of the estimation of fractal dimensions measured by means of the coefficient of determination R^2. Based on Sémécurbe et al. (2019)*

7.1.2. Calculation of cross-scale signatures

Cross-scale signatures make it possible to precisely describe the way in which the objects studied, namely spatial distributions of the centroids of buildings here, deviate from a hypothesis of scale invariance. They are calculated from the same material as fractal dimensions: the bi-logarithmic count curves of the number of elements in a neighborhood of increasing or decreasing size. But instead of summarizing these curves by an overall rate of change, namely, the fractal dimension corresponding to the slope of the straight line fitted on the count curve, we estimate specific variations, namely, local slopes, which from a theoretical point of view amounts to calculating a derived function. We approximate the derivative by calculating the rates of change between two scales that we deem to be sufficiently close. Thus, instead of calculating an average slope considering a given range of scales (fractal dimension), the cross-scale signatures consist of the successive rates of change of $log(M(r))$ as a function of $log(r)$:

$$Sc(r_i, r_{i+1}) = \frac{log(M(r_i)) - log(M(r_{i+1}))}{log(r_i) - log(r_{i+1})} \quad [7.4]$$

For scale-invariant objects, the local slopes merge with the overall slope; the rates of change are all equal to the fractal dimension. In this case, the cross-scale signatures provide no additional information on the spatial organization of the objects studied. However, when the distribution is not scale-invariant, the curve formed by $log(M(r))$ as a function of $log(r)$ is not a line and the rates of change of $log(M(r))$ are not constant. Cross-scale signatures $Sc(r_i, r_{i+1})$ measure the deviation from scale invariance and describe oscillations around an average value, the fractal dimension, scale by scale.

The cross-scale signatures from the sandbox analysis applied to the two bocage and open-field configurations shown in Figure 7.1 consist of four rates of change (Sc100–50, Sc200–100, Sc400–200 and Sc800–400) calculated for each scale (Figure 7.4). The cross-scale signatures highlight a difference in organization of these two configurations that was invisible with fractal dimensions; while the signature of the open-field configuration is almost flat (hence almost scale-invariant), the signature of the bocage increases starting from the second indicator (therefore not scale-invariant).

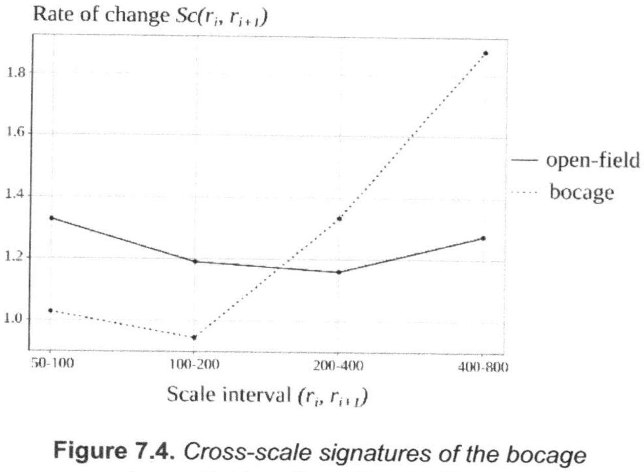

Figure 7.4. *Cross-scale signatures of the bocage and open-field configurations in Figure 7.1*

In the same way, the cross-scale signatures calculated from the results of the GWFA for the 145,178 built-up fabrics of mainland France locally highlight deviations from scale invariance. The decrease in signatures in the open-field territories of north-eastern France contrasts with their increase in the bocage landscapes of western and south-western France (Figure 7.5).

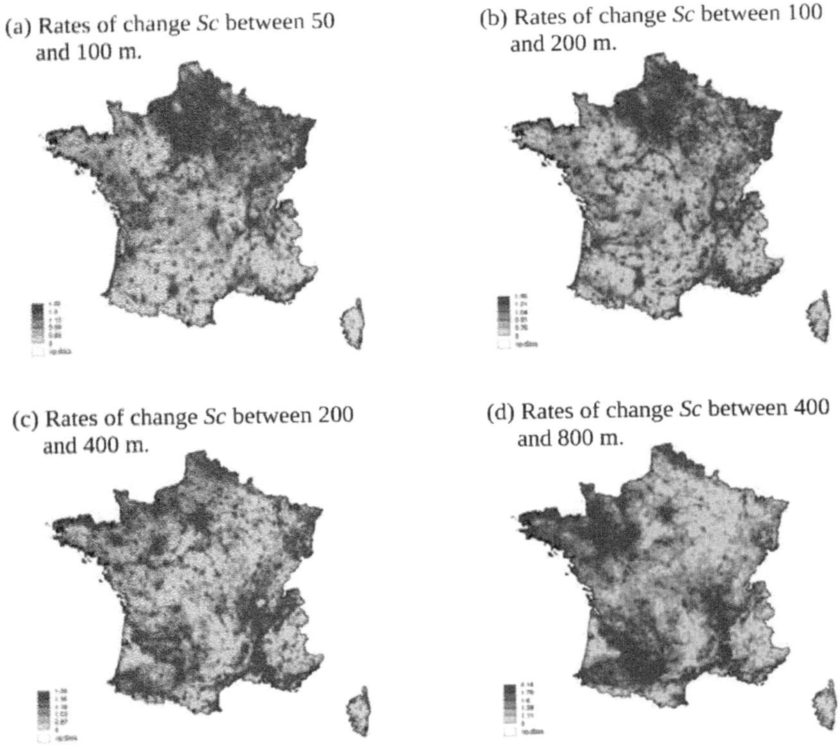

Figure 7.5. *Cross-scale indicators calculated from the GWFA for the built-up fabrics of mainland France in squares of 2 km side. Source: Sémécurbe (2020, p. 121)*

7.2. Synthesizing the information provided by cross-scale signatures

By construction, the information resulting from a cross-scale analysis is very rich and requires the use of statistical methods of information reduction to make it clear and legible. Two complementary statistical methods are presented in this chapter: an a posteriori typology of cross-scale signatures and an a priori decomposition of them.

7.2.1. *Typology of cross-scale signatures using principal component analysis*

Principal component analysis (PCA) is a data analysis method that aims to synthesize the relationships between variables relating to a set of statistical individuals. The principal components, the results of the PCA, are linear combinations of variables decorrelated from each other. They are obtained by searching for the axes, or privileged directions, which best preserve the positioning of individuals with respect to the variables considered (inertia of the whole). This method was applied to the four rates of change (Sc50–100, Sc100–200, Sc200–400, Sc400–800) calculated locally for the 145,178 built-up fabrics of mainland France.

The four rates of change are highly correlated with each other (Figure 7.6); here, we can see that the maps corresponding to the two extreme indicators of signatures are almost mirrors of each other (Sc100–50 and Sc800–400, Figures 7.5a and d). The first two axes of the PCA summarize nearly 90% of the information provided by the cross-scale signatures of the built-up fabrics of mainland France. Considering only the first two axes of the PCA therefore guarantees satisfactory representation of the signatures while considerably reducing the complexity of the phenomenon studied.

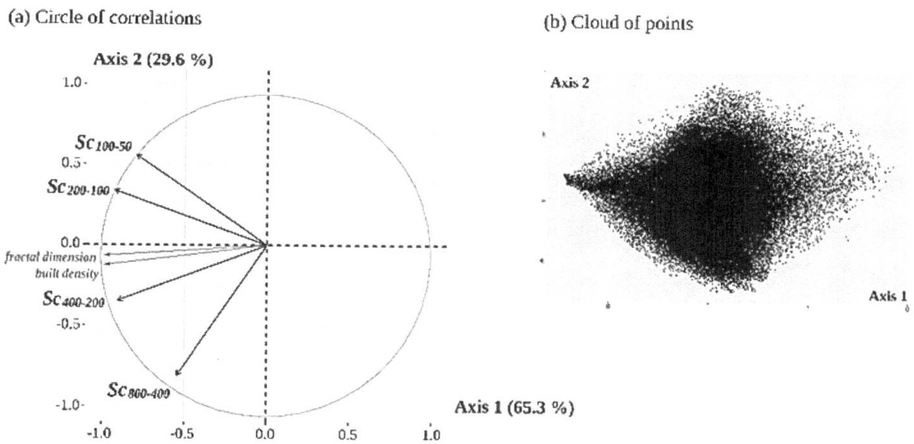

Figure 7.6. *Results of the principal component analysis of the cross-scale signatures of the 145,178 built-up fabrics of mainland France*

The first axis (65% of inertia) is highly correlated positively with the four cross-scale signatures and characterizes what statisticians call a "size effect": it orders entities according to the value of their sizes. Logically, it is highly correlated with the fractal dimension calculated over all scales and with the density of buildings. We visually find this correlation on the map of the first axis (Figure 7.7a) which reproduces the classic division between dense urban spaces and sparse rural, forest and mountain spaces.

The second axis (30% of inertia) is more interesting and describes the way in which the built-up fabrics deviate from a situation of scale invariance. It opposes the indicators Sc100–50, Sc200–100 to the indicators Sc400–200, Sc800–400. In other words, if we control for the "size effect" by taking spaces of the same fractal dimension, high Sc100–50, Sc200–100 and low Sc400–200, Sc800–400 correspond to a high value on the second axis and vice versa. Thus, this axis synthesizes the opposition between the signatures calculated for the extreme scale ranges (100–50 and 800–400). In the open-field zones of the northeastern quarter of France, the values on axis 2 are low (Figure 7.7b). Indeed, scale invariance is approximately verified for the scale ranges under consideration (between 50 and 800 m). However, the amplitude of the values on axis 2 is large in the bocage areas of western France.

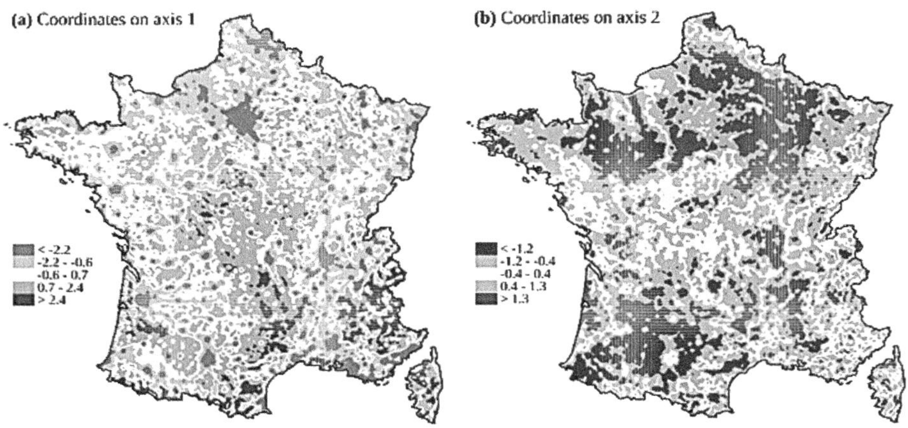

Figure 7.7. *Coordinates of the 145,178 built-up fabrics of mainland France on the first two axes of the principal component analysis of the cross-scale signatures. For a color version of this figure, see www.iste.co.uk/tannier/fractal.zip*

Finally, the first axis orders the territories according to a "mean fractal dimension", while the second axis roughly separates the built-up fabrics according to the increase or decrease in their cross-scale signatures. This clear distinction detected by the PCA results from the existence, in mainland France, of very different forms of settlement in comparable proportions.

In order to further simplify observation of the built-up fabrics and thus better understand their shape, a typology of the information contained in the two axes of the PCA was made using the dynamic clouds algorithm. This classification algorithm is iterative. At the first step, we randomly select as many centers as the number of classes we want to obtain. Each point to be classified is assigned to the nearest center. We then recalculate new class centers corresponding to the barycenters of the classified points. We iterate the process as long as points change classes. At the end of this process, the classes of points are identified and their centers describe their average characteristics.

The choice of the number of classes is made by arbitrating between, on the one hand, the maximum heterogeneity authorized within the classes, that is, the minimum degree of resemblance for elements of the same class, and on the other hand, the variability between the classes that we consider relevant, that is, the level of detail that we want to preserve. For this purpose, we study the scree plot, which represents the sum of the squared distances of each point from its assigned center as a function of the number of classes. In our case, the scree plot suggests a division into six classes (see Sémécurbe et al. 2019).

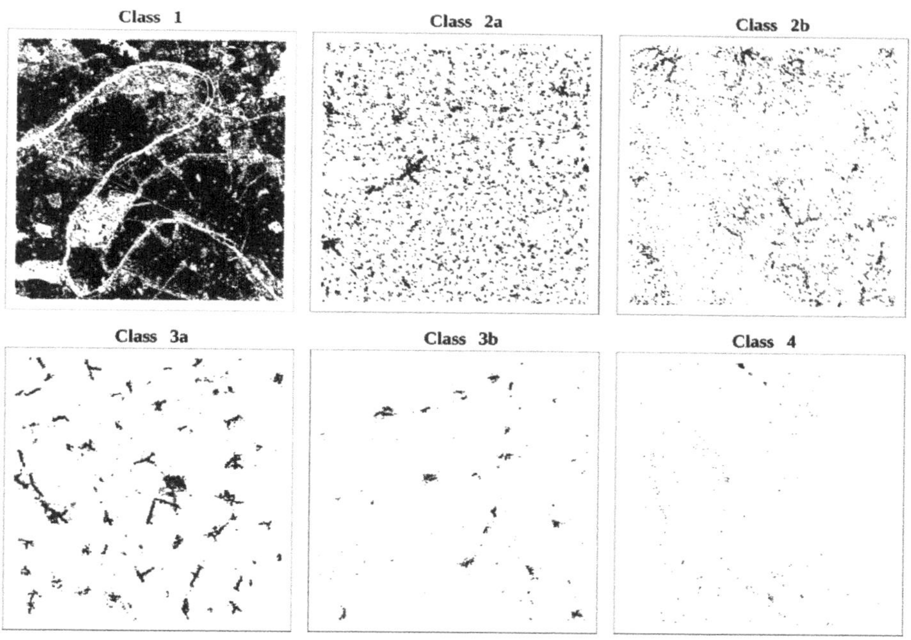

Figure 7.8. *Characteristic built-up fabrics of each class. Source: Sémécurbe et al. (2019)*

Class 1 groups together spatial units with high fractal dimensions that correspond approximately to cities (Figure 7.8). Classes 2a and 2b correspond to organizations of scattered buildings; many buildings are located outside of village spaces. Class 2a buildings occupy space more evenly than class 2b buildings. In contrast, classes 3a and 3b include built-up fabrics where buildings are concentrated mainly in village spaces. Villages in class 3a are larger and more numerous than in class 3b. The differences between elements (a) and (b) of the nomenclature reflect a specific positioning on axis 1; the elements of classes (a) have higher fractal dimensions than those belonging to classes (b). Class 4, characterized by low fractal dimensions, corresponds to very sparse spaces such as forests and mountain ranges.

The map of the typology drawn from this classification is very well structured even though no continuity constraint has been introduced into the classification (Figure 7.9). Spatial units belonging to the same class form related sets, reflecting the great coherence of local and regional processes of structuring of settlements over the long history of their establishment. The intermediate classes 2a, 2b, 3a and 3b are located on the periphery of urban spaces and reflect a wide variability of suburban and rural forms.

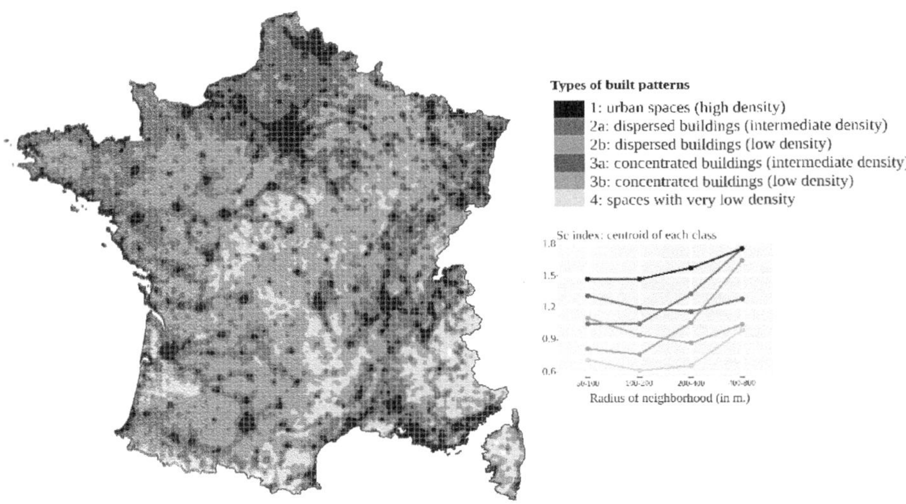

Figure 7.9. *Typology of 145,178 built fabrics in mainland France based on the characterization of how the spatial distribution of buildings deviates locally from a scale-invariant distribution. Based on Sémécurbe et al. (2019). For a color version of this figure, see www.iste.co.uk/tannier/fractal.zip*

The growth of signatures in the bocage areas is easily explained. At large scales, between 50 and 100 m, the probability of finding many neighboring buildings around each building is low because the houses are mostly isolated from each other. Therefore, the rates of change are low. On the contrary, for small scales

(400 and 800 m), each building is surrounded by many buildings: the rates of change are then high. In open-field spaces, where the buildings are agglomerated in villages of more or less modest size, each building has many neighbors considering the small neighborhood distances; rates are high. Thus, the signatures tend to decrease gradually up to a certain distance threshold. Indeed, each time we move from a fine scale to a coarser scale (respectively small and large neighborhood radii), the relative proportion of empty space increases around each building and the rates decrease. Once this distance threshold is exceeded, each building becomes a neighbor to those in the other villages, which explains why Sc800–400 is then higher than Sc400–200 in areas of concentrated population.

One limitation to the statistical approach adopted is that the results obtained may vary considerably depending on the study area under consideration. Indeed, applying a PCA to the cross-scale signatures of the built-up fabrics of the whole of Europe, or even to a small morphologically homogeneous subset of the territory of France, would have produced potentially very different typologies.

7.2.2. *Decomposition of cross-scale signatures using orthogonal polynomials*

Here, we are no longer synthesizing a posteriori the information provided by the cross-scale signatures, but decomposing them a priori into previously defined components whose thematic significance has been determined. The objective is to produce indicators that are comparable, regardless of the study areas under consideration.

Forriez and Nottale (2010) model the scale function $M_r(i)$ in the form of a polynomial of degree 2:

$$M_r(i) = a_0 - a_1 \times log(r) - a_2 \times \left(log(r)\right)^2 + \epsilon \qquad [7.5]$$

This formula amounts to introducing an additional term $log(r)^2$ into the usual methods of estimating fractal dimensions. This new term captures the deviation from a situation of scale invariance. It mechanically improves the fit of the estimated scale functions for the configurations in Figure 7.1, at least for the bocage configuration (Table 7.1). In the case of the open-field configuration, the scale invariance property being approximately verified, the improvement in the quality of the estimation of fractal dimension is marginal.

The differences in the value of a_1 between the classical polynomial model of degree 1 and the polynomial model of degree 2 (0.669 vs. 1.265 in bocage and 1.264 vs. 1.226 in open-field, Table 7.1) are explained by the collinearity between the vectors log(r) = (log(50), log(100), log(200), log(400), log(800)) and $\log(r)^2$ = ($\log(50)^2$), $\log(100)^2$), $\log(200)^2$), $\log(400)^2$), $\log(800)^2$). Their correlation coefficient is greater than 0.95. In doing so, regression does not enable us to separate the effects attributable to each of them: $\log(r)^2$ "captures" part of the explanatory power provided by log(r). As a result, a_1 can no longer be considered as a fractal dimension.

Parameter	Polynomial of degree 1		Polynomial of degree 2		Orthogonal polynomials	
	Bocage	*Open-field*	*Bocage*	*Open-field*	*Bocage*	*Open-field*
a_1	1.265	1.226	0.669	1.264	1.265	1.226
a_2			1.968	-0.124	0.558	-0.035
R^2	*0.976*	*0.999*	*0.997*	*0.999*	*9.997*	*0.999*

Table 7.1. *Estimating the sandbox fractal dimension according to different statistical models for the bocage and open-field configurations in Figure 7.1. Source: Sémécurbe (2020, p. 130)*

In order to estimate the parameters of a statistical model independently of the degree of the polynomials introduced into it, Kennedy and Gentle (1980) suggest the use of orthogonal regression. The idea is to replace the two terms log(r) and $\log(r)^2$ in equation [7.5] by linear combinations of these, called orthogonal polynomials:

$$M_r(i) = a_0\vec{1} + a_1P_1(log(\vec{r})) + a_2P_2(log(\vec{r})) + \epsilon \qquad [7.6]$$

The first combination is a polynomial of degree 1 in $log(\vec{r})$: $P_1(log(\vec{r})) = b_{1,0}\vec{1} + log(\vec{r})$.

The second combination is a polynomial of degree 2: $P_2(log(\vec{r})) = b_{2,0}\vec{1} + b_{2,1}log(\vec{r}) + log(\vec{r})^2$, where $b_{1,0}, b_{2,0}$ and $b_{2,1}$ are constants determined by the following orthogonality equations:

$$P_1(log(\vec{r})) \cdot 1 = 0$$

$$P_2(log(\vec{r})) \cdot 1 = 0$$

$$P_1(log(\vec{r})) \cdot P_2(log(\vec{r})) = 0$$

The notation · designates the usual scalar product between two vectors. Here, we consider five scales r, for memory 50, 100, 200, 400 and 800 m. In doing so, the scalar product $P_1(\log(r)) \cdot P_2(\log(r))$ is equal to $P_1(\log(50)) \times P_2(\log(50)) + P_1(\log(100)) \times P_2(\log(100)) + P_1(\log(200)) \times P_2(\log(200)) + P_1(\log(400)) \times P_2(\log(400)) + P_1(\log(800)) \times P_2(\log(800))$.

The Gram–Schmidt algorithm is generally used to solve the system of orthogonality equations discussed above. It gives the values of $P_1(log(\vec{r}))$ = (-2, -1, 0, 1, 2) and $P_2(log(\vec{r}))$ = (2, -1, -2, -1, 2) (Figure 7.10a). The curve $P_1(log(\vec{r}))$ is just a y-axis shift of the curve $log(\vec{r})$ while $P_2(log(\vec{r}))$ is a parabola centered on the median scale of 200 m.

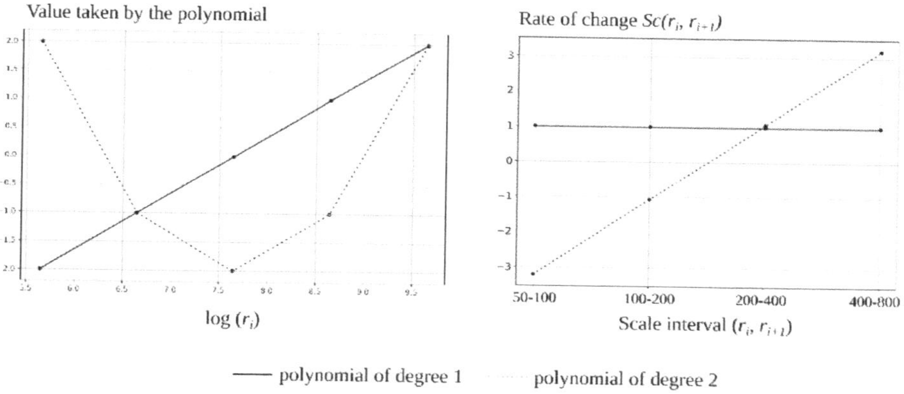

Figure 7.10. *Decomposition of fractal dimension estimation using orthogonal polynomials*

The scaling functions estimated using orthogonal polynomials (equation [7.6]) are identical to those obtained by introducing a polynomial of degree 2 (equation [7.5]). However, the significance of the coefficients a_1 and a_2 differs. In the case of modeling with orthogonal polynomials, the coefficient a_1 characterizes the average cross-scale behavior of the object. When the object is quasi-scale invariant, a_1 corresponds exactly to the fractal dimension. Moreover, since $a_1 \times P_1(\log(r))$ is a polynomial of degree 1, its rates of change between two scales are constant and its representation is a horizontal line (Figure 7.10). As for the coefficient a_2, it describes the variation in the rates of change. Since $a_2 \times P_2(\log(r))$ is a polynomial of degree 2, its rates of change between two scales vary and its representation is a rising or falling line. a_2 indicates how the fractal dimension changes or does not change across scales. The sign of a_2 describes the overall direction of evolution of signatures across scales; a positive sign is associated with a growth of signatures and a negative sign with a decrease. The absolute value of a_2 reflects the intensity of the deviation from a situation of scale invariance. In the case where the cross-scale signatures present regular variations (signatures whose points are perfectly aligned on a rising or falling straight line), a_2 is proportional to the variation between two consecutive rates of change. It therefore completely summarizes the information contained in this type of signature. In the case where the signatures are non-regular, for example, that of the bocage in Figure 7.11a, a_2 quantifies the mean variation between the rates.

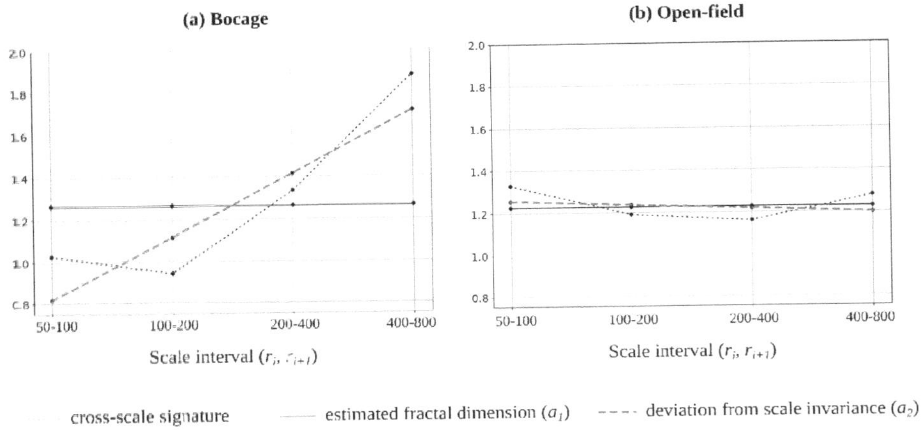

Figure 7.11. *Modeling of cross-scale signatures using the coefficients a_1 and a_2 of the orthogonal polynomials calculated for the bocage and open-field configurations in Figure 7.1*

Modeling the cross-scale signatures of the bocage and open-field configurations of Figure 7.1 using orthogonal polynomials makes it possible to obtain the estimated fractal dimension, namely, the coefficient a_1, and an indicator of deviation from scale invariance a_2 graphically centered around a_1 (Figure 7.11). In the case of the bocage configuration, the value of $a_2 = 0.558$ is high and reflects the growth of signatures relatively well. In the case of the open-field configuration, $a_2 = -0.035$ is almost zero because this configuration is, for the radii of analysis considered, almost perfectly fractal. Adding a polynomial of order 2 therefore contributes almost nothing.

Modeling the cross-scale signatures of the 145,178 built-up fabrics of mainland France with orthogonal polynomials makes it possible to map the spatial distribution of the coefficients of the second polynomial a_2 (Figure 7.12). The absolute value of a_2 is very large in the west (bocage areas) and the south of France, where scale invariance is not verified. It is almost zero in most of the open-field territories of eastern France as well as in some areas of the center and southwest, on the coastal side. In these areas, the fractal dimension is correctly estimated.

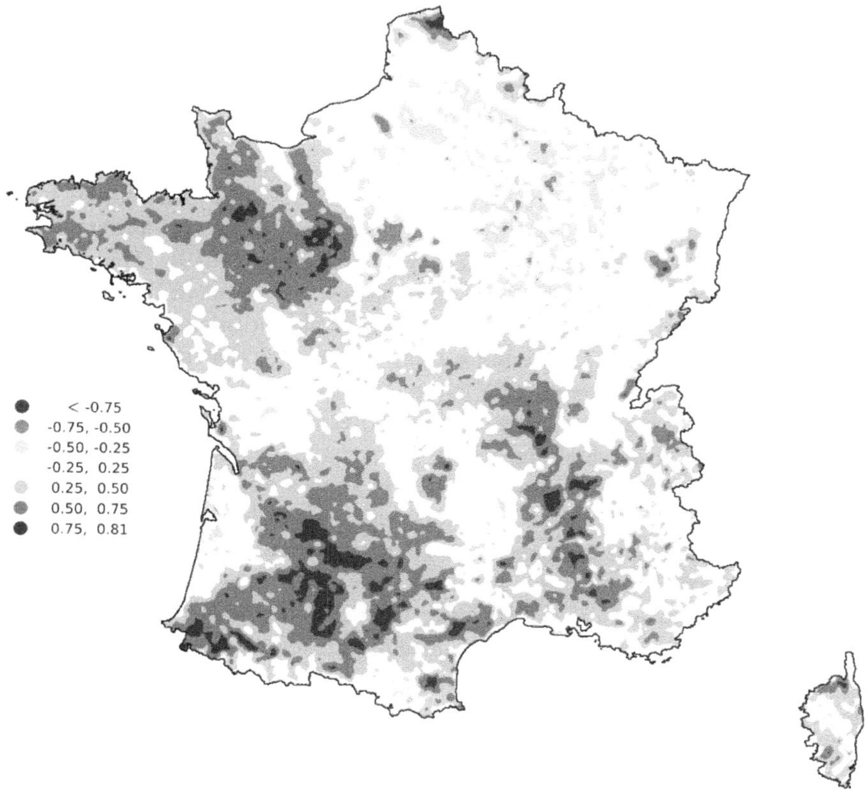

Figure 7.12. *Deviations from scale invariance measured for 145,178 built-up fabrics in mainland France (as a result of the decomposition into orthogonal polynomials of the cross-scale signatures calculated with the GWFA over the range of scales 50–800 m for q = 0: coefficient value a_2). For a color version of this figure, see www.iste.co.uk/tannier/fractal.zip*

The map of Figure 7.12 is visually very similar to that of Figure 7.7b which represents the coordinates on axis 2 of the PCA calculated for the cross-scale signatures. On both maps, the areas shaded in red are identical (same shape and same location), and represent the areas where scale invariance is not verified. However, in Figure 7.12, we detect other small additional non-scale-invariant areas, represented in shades of blue. In these areas, the cross-scale signatures decrease instead of increasing, as is the case in the red-shaded areas. The information provided by the coefficient a_2 of the orthogonal polynomial modeling is therefore more precise than that provided by the PCA of the cross-scale signatures. Moreover, the significance of the values of a_2 is much clearer than that of the coordinates of points on the axes of a PCA.

7.3. Conclusion

Although the statistical self-similarity of geographical shapes is spatially circumscribed and applies to a restricted range of scales, measurements of fractal dimension make it possible to differentiate between them, the differences or similarities identified having visual coherence (Tannier 2019). The study of deviations from scale invariance allows us to deepen the results of fractal analyses: we hypothesize that a spatial distribution is scale invariant (or statistically self-similar), and we study deviations from this model. Deviations from scale invariance may appear for some scale ranges and not others. Deviations from scale invariance may also vary in space; spatial differentiations are thus highlighted (Tannier 2019). Cross-scale signatures are a good tool for exploring and characterizing deviations from scale invariance. Their a priori decomposition using orthogonal polynomials produces indicators with a clear thematic significance: the coefficient a_1 corresponds to the estimated fractal dimension and the coefficient a_2 measures deviations from scale invariance. The interest of a decomposition by orthogonal polynomials lies less in their statistical properties per se than in the interpretability of the indicators thus obtained (coefficients a_1 and a_2) and in the fact that the values of these two indicators are comparable regardless of the study areas under consideration.

7.4. References

Abry, P., Jaffard, S., Wendt, H. (2015). Irregularities and scaling in signal and image processing: Multifractal analysis. In *Benoit Mandelbrot: A Life in Many Dimensions*, Frame, M. and Cohen, N. (eds). World Scientific, Singapore.

Avnir, D., Biham, O., Lidar, D., Malcai, O. (1998). Is the geometry of nature fractal? *Science New Series*, 279(5347), 39–40.

Batty, M. (1991). Cities as fractals: Simulating growth and form. In *Fractals and Chaos*, Crilly, A.J., Earnshow, R.A., Jones, H. (eds). Springer, New York.

Florindo, J.B., Backes, A.R., de Castro, M., Bruno, O.M. (2012). A comparative study on multiscale fractal dimension descriptors. *Pattern Recognition Letters*, 33(6), 798–806.

Forriez, M. and Nottale, L. (2010). Lois d'échelle et transitions fractal – non fractal en géographie. *L'Espace géographique*, 39(2), 97–112.

François, N., Frankhauser, P., Pumain, D. (1995). Villes, densité et fractalité. Nouvelles représentations de la répartition de la population. *Les Annales de la recherche urbaine*, 67, 55–64.

Frankhauser, P. (1998a). The fractal approach: A new tool for the spatial analysis of urban agglomerations. *Population*, 4, 205–240.

Frankhauser P. (1998b). Fractals geometry of urban patterns and their morphogenesis. *Discrete Dynamics in Nature and Society*, 2, 127–145.

Frankhauser, P. (2004). Comparing the morphology of urban patterns in Europe. A fractal approach. In *European Cities: Insights on Outskirts, Structures*, Borsdorf, A. and Zembri, P. (eds). ESF COST Office, Brussels.

Goodchild, M. (1980). Fractals and the accuracy of geographical measures. *Mathematical Geology*, 12(2), 85–98.

Kennedy, W.J. and Gentle, J.E. (1980). *Statistical Computing*. Marcel Dekker, New York.

Lam, N. (1990). Description and measurement of landsat TM images using fractals. *Photogrammetric Engineering and Remote Sensing*, 56(2), 187–195.

Lam, N. and Quattrochi, D. (1992). On the issues of scale, resolution, and fractal analysis in the mapping sciences. *Professional Geographer*, 44(1), 88–98.

Malcai, O., Lidar, D.A., Biham, O., Avnir, D. (1997). Scaling range and cutoffs in empirical fractals. *Physical Review E*, 56(3), 2817.

Palmer, M.W. (1988). Fractal geometry: A tool for describing spatial patterns of plant communities. *Vegetatio*, 75(1–2), 91–102.

Pumain, D. (2010). Commentaire : transition fractal – non fractal et objets géographiques. *L'Espace géographique*, 39(2), 113–116.

Raghavendra, B.S. and Narayana Dutt, D. (2010). Signal characterization using fractal dimension. *Fractals*, 18(3), 287–292

Sémécurbe, F. (2020). Analyse de la distribution spatiale des implantations humaines – apports et limites d'indicateurs multi-échelles et trans-échelles. PhD Thesis, Université de Franche-Comté, Besançon.

Sémécurbe, F., Tannier, C., Roux, S.G. (2019). Applying two fractal methods to characterise the local and global deviations from scale-invariance of built patterns throughout mainland France. *Journal of Geographical Systems*, 21(2), 271–293.

Seuront, L (2010). *Fractals and Multifractals in Ecology and Aquatic Science*. CRC Press, Routlege Taylor & Francis, London.

Tan, X., Huang, B., Batty, M., Li, J. (2021). Urban spatial organization, multifractals, and evolutionary patterns in large cities. *Annals of the American Association of Geographers*, 111(5), 1539–1558.

Tannier, C. (2019). Effects of distance and scale dependence in geographical models of cities and territories. In *Geographical Modeling: Cities and Territories*, Pumain, D. (ed.). ISTE Ltd, London and John Wiley & Sons, New York.

Tannier, C. and Pumain, D. (2005). Fractals in urban geography: A theoretical outline and an empirical example. *Cybergeo: European Journal of Geography*, 307. doi: 10.4000/cybergeo.3275.

Tannier, C., Thomas, I., Vuidel, G., Frankhauser, P. (2011). A fractal approach to identifying urban boundaries. *Geographical Analysis*, 43(2), 211–227.

Thomas, I., Frankhauser, P., Frenay, B., Verleysen, M. (2010). Clustering patterns of urban built-up areas with curves of fractal scaling behaviour. *Environment and Planning B: Planning and Design*, 37, 942–954.

White, R. and Engelen, G. (1994). Urban systems dynamics and cellular automata: Fractal structures between order and chaos. *Chaos, Solitons and Fractals*, 4(4), 563–583.

PART 3

Urban Forms and Fractal Planning

8

Principles of Fractal Planning and Urban Design

Cécile TANNIER
Laboratoire ThéMA, CNRS-Université de Franche-Comté, Besançon, France

A number of researchers have applied fractal geometry to answer questions in urban and regional planning and urban design. This area of research covers a variety of scientific disciplines, including geography, physics, architecture, psychology and ecology. However, Jahanmiri and Parker (2022) have shown that very few scientific articles on this topic are published in planning and urban design journals (16% of the 428 documents studied by the two aforementioned authors). According to them, the relatively high level of mathematical knowledge needed to understand the applications of fractals proposed by researchers explains their near-total lack of diffusion in the world of planning and urban design.

Scientific publications on the application of fractals to planning and urban design address the cities and territories involved at different levels of analysis. At the finest level, studies have dealt with the architecture of buildings (Bovill 1996; Ostwald and Vaughan 2008; Vaughan and Ostwald 2008) and the shape of parks or squares (Lu 2010); at mesoscopic levels, they have dealt with the shape of blocks or neighborhoods (Cooper 2003, 2005; Cooper and Oskrochi 2008; Thomas et al. 2010a, 2010b), the system of parks within a city (Li et al. 2021; You et al. 2022) or the overall shape of cities (Batty and Longley 1994; Frankhauser 1998; Benguigui et al. 2000; Stamps 2002; Hagerhall et al. 2004); at the macroscopic level,

Fractal Geometry in Human Geography and Planning,
coordinated by Cécile TANNIER. © ISTE Ltd 2024.

they have dealt with systems of cities or regional settlement (Arlinghaus 1985; Wong and Fotheringham 1990; Frankhauser et al. 2007; Chen 2011; Chen 2014).

In this chapter, based on studies relating to the application of fractals to planning and urban design, we will first expound various hypotheses relating to the functional advantages of fractal built and non-built forms (section 8.1). We will then discuss the realism of the fractal planning standards proposed by some researchers (section 8.2), and finally argue in favor of introducing fractal geometric rules into geographical models of urban and regional planning (section 8.3).

8.1. Hypotheses about the functional advantages of fractal built and non-built forms

The comparative functional advantages of fractal forms compared to non-fractal forms have been relatively little studied so far, and the value of fractal forms for urban and regional planning is not proven. However, proving that fractal built forms (blocks, neighborhoods, cities) and non-built forms (intra-urban parks, green and natural spaces) could help cities and territories to function sustainably seems very difficult; not because of the intrinsic complexity of these forms, but because assessing the sustainability of the functioning of a district, a city or a region implies taking account of many diverse phenomena, linked together by complex relationships.

We offer here a non-exhaustive overview of research dealing with the functional implications of the fractality of built and non-built forms. Some research aims to prove the interest of fractal forms, while other research simply starts from this hypothesis.

8.1.1. *Response to public expectations for residential choice and satisfaction*

The aesthetic and landscape quality of fractal urban forms is promoted in a few articles (Stamps 2002; Cooper 2003, 2005; Hagerhall et al. 2004; Cooper and Oskrochi 2008). Cooper et al. (2010) found a positive statistical relationship between fractal dimension and the landscape quality of streets perceived by pedestrians who frequent these areas. Cooper et al. (2013) showed that street scenes dominated by vegetation have a higher fractal dimension, and therefore better visual quality (Jahanmiri and Parker 2022).

Advocacy for the implementation of fractal architecture or urban design is not new. Its spokespersons, the best known among them being Bovill (1996), Salingaros and West (1999), Joye (2007), and Salingaros (2011, 2012), most often appeal to the ideas of Alexander (1964). Living structures, especially fractal structures, show great variations in the repetition of their constituent elements, which means that monotony is entirely avoided. They are perceived by human beings as physiologically and psychologically positive; they favorably influence the well-being, mental health and comfort (especially visual) of individuals. For scholars primarily in the fields of architecture, environmental psychology and neuroscience, biophilia or "love of nature" (Wilson 1984) explains the "natural" appetite of human beings for fractal forms: "We are 'wired' to look at fractals and not the Euclidean catalogue of circles, squares, and triangles that most buildings are composed of" (Taylor 2021); "[...] our neurological system was created (evolved) to respond directly and exquisitely to complex, fractal, hierarchical geometric environments" (Salingaros 2011). Neuroscience experiments have shown fractal shapes to have a positive impact on attention, visual performance (pattern recognition, navigation), aesthetic appeal and stress reduction in individuals (Taylor 2021). Conversely, the lack of fractal aesthetics creates pressure on the visual system, resulting in undesirable effects such as headaches (Le et al. 2017). Brielmann et al. (2022) report the results of experiments based on the use of electroencephalograms and show that the human cognitive system, in particular with regard to visual tasks, is more efficient when people's attention is engaged in contemplation of fractal figures with a dimension between 1.3 and 1.5 (Hagerhall et al. 2008). This is seen to be consistent from the point of view of biophilia, knowing that this range of values of fractal dimension is the one that prevails in natural fractal forms.

Researchers from other scientific fields, including geography and economics, have also suggested that a fractal urban form would better meet public expectations than a non-fractal form. For Niang and Soboyejo (2006), fractal residential development is beneficial for the construction of affordable housing. According to Thomas et al. (2008), there is a positive statistical relationship between residential satisfaction of the inhabitants and the fractal dimension of the built-up fabric in the case of the Walloon municipalities in Belgium. Frankhauser and Genre-Grandpierre (1998) and Cavailhès et al. (2004) showed that a city that is the shape of a fractal box (or Vicsek fractal, see Chapter 3) is more interesting than a compact, square-shaped city for individuals who often frequent medium-sized secondary centers, small neighborhood centers and green and natural spaces, but who visit the main center less frequently. For other researchers, however, neither green amenities nor landscape quality have any real influence on household residential choices (see, for example, Vogt and Marans (2004)). Yet, we can hypothesize that the fractal city does not correspond to a model of residential choice at a given moment, but to a model of residential satisfaction over time; in a fractal city, the quality of the residential environment would be preserved or even improved, and consequently so would the satisfaction of individuals.

Distance of access from places of residence to stores, services and green and natural spaces

A series of studies, conducted on a few suburban municipalities located north of the city of Besançon (eastern France), aimed to compare the respective interest of fractal and non-fractal residential developments in terms of residents' distance from stores and services, as well as green and natural spaces used on a daily and weekly basis. First, theoretical scenarios of fractal and non-fractal residential development were evaluated. The intensity of simulated residential development varied greatly across scenarios.

A first study consisted of comparing 16 theoretical scenarios, eight fractals (of fractal dimension equal to 1.26, 1.46, 1.63 or 1.77) and eight non-fractals (Figure 8.1) (Tannier et al. 2010). Three evaluation criteria were considered: the number of contacts between built and non-built cells, the average distance (as the crow flies) to the nearest store or service used on a daily basis and the average distance to the nearest store or service used on a weekly basis. The results obtained showed that the relative gain in accessibility of unbuilt spaces in the fractal scenarios is higher than the relative loss in accessibility of stores and services, compared to the non-fractal scenarios.

A second study consisted of comparing 50 theoretical residential development scenarios, 25 fractal scenarios (five different values of fractal dimension) and 25 non-fractal scenarios (Tannier et al. 2012). The locations of stores and services varied according to the scenarios. This study showed that fractal scenarios required shorter trips to access different types of green and natural spaces than non-fractal scenarios, but longer trips to visit different types of stores and services. It was possible to improve the accessibility of stores and services in fractal scenarios by changing the location of stores and services, but it was impossible to improve the accessibility of green spaces in non-fractal scenarios. Consequently, fractal residential development combined with a proactive commercial development policy could allow good access for the population to both green and natural spaces and to stores and services.

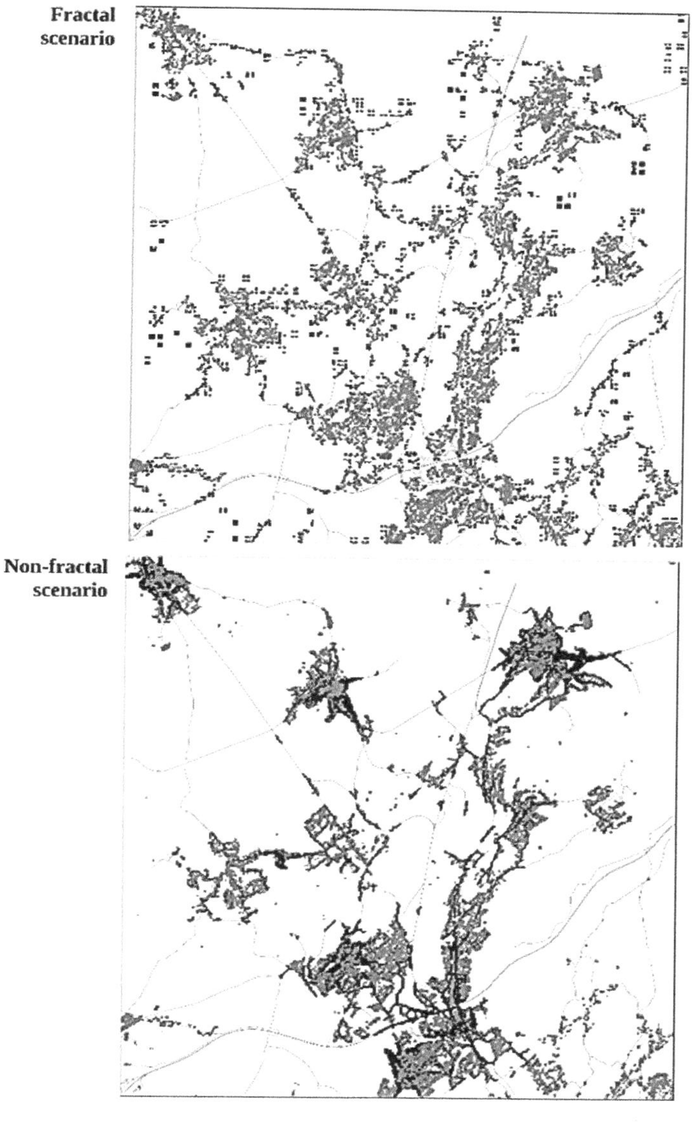

Figure 8.1. *Two theoretical scenarios of residential development. Northern sector of the metropolitan area of Besançon (eastern France). 14,928 previously urbanized cells (in gray); 7,765 newly urbanized cells (in black). Red lines: road network. Adapted from: Tannier et al. (2010). For a color version of this figure, see www.iste.co.uk/tannier/fractal.zip*

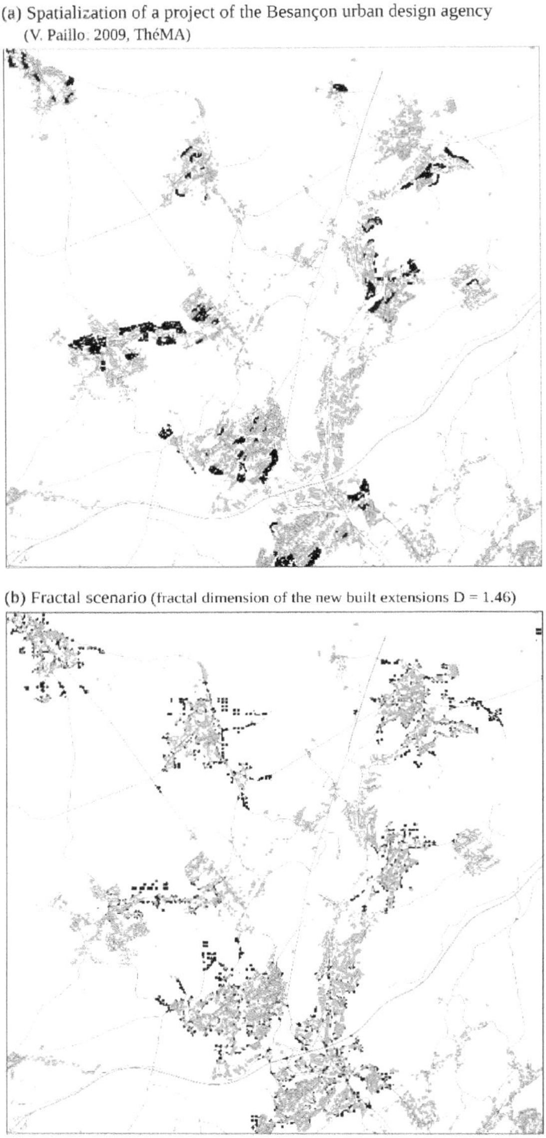

Figure 8.2. *Two realistic residential development scenarios. Northern sector of the metropolitan area of Besançon (eastern France). 14,928 previously urbanized cells (in gray); 2,353 newly urbanized cells (in black). Red lines: road network. Adapted from: Frankhauser et al. (2010). For a color version of this figure, see www.iste.co.uk/tannier/fractal.zip*

Two other studies were devoted to comparing realistic, rather than theoretical, residential development scenarios (Figure 8.2). The realism of these scenarios resulted from the introduction of additional planning constraints, particularly in terms of the intensity of the desired residential development and its location (only the areas closest to stores and services could be urbanized). The first study compared five scenarios (Frankhauser et al. 2010). Two scenarios represent fractal forms of residential development. Two other scenarios represent non-fractal and very compact forms of residential development. The fifth scenario corresponds to spatialization of an urbanization project of the Urban Design Agency of the metropolitan area of Besançon (AUDAB). This scenario is not fractal, but the simulated residential development preserves free spaces within the built-up fabrics. Comparison of the scenarios showed that, in all cases, the average distance to the nearest store and service, calculated over the network, was lower compared to the initial situation. The two non-fractal scenarios improved the average accessibility of stores and services more than the others. On the other hand, they degraded the average accessibility of unbuilt spaces to residents, while the two fractal scenarios and the spatialization of the project of the urban design agency allowed improvements to it. The fractal scenarios showed very dendritic forms of residential development, which was perceived a priori as negative by planning practitioners. However, they maintained an acceptable distance not only to unbuilt spaces, but also to stores and services. A second study, more precise in terms of measuring the accessibility of stores and services, confirmed these results (Houot et al. 2009). Twenty-five fractal residential development scenarios were compared. The results showed that the accessibility of stores and services in the different scenarios was always better than the accessibility calculated for the initial situation. Fractal urbanization does not necessarily degrade the accessibility of stores and services, but on the contrary, may improve it. The study also showed that the fractal dimension of the simulated residential developments influenced accessibility for residents to stores and services more than the spatial configuration of the shopping options.

8.1.2. *Spatial structuring of intra-urban parks and green spaces*

A series of studies on the structure of urban parks and green spaces, mainly conducted in China and using the tools of fractal geometry, have supported the idea that each park or urban system of parks and squares must be organized according to a spatial hierarchy, in accordance with a fractal law of nesting scales. Liang et al. (2013) studied four parks, one in the USA, one in Argentina and two in China, adopting Bovill's (1996) box-counting method applied to a range of very fine scales (5 m, 10 m, 20 m, 40 m and 80 m). These scales correspond to the scales of local human interactions (interpersonal distances) that take place outdoors. Based on this same range of scales, Liang et al. (2013) also studied the hierarchical nesting of empty spaces within each park. In the conclusion of their study, the authors develop an argument in favor of the feasibility and interest of applying fractal geometry to the structure of urban parks following a hierarchical logic. Gong et al. (2020) studied the fractality of the urban park system in the city of Dalian (China). They showed that the perimeters of parks are more complex than those of built-up fabrics, while the fractal dimension of their surface area is lower. The gradient of the prevalence of parks from the center to the outskirts of the city of Dalian shows an excessive concentration in the urban center, while the pericentral and peripheral spaces are relatively devoid of them. Li et al. (2021) showed that the rank-size distribution of parks in major cities in China (Beijing, Shanghai, Guangzhou and Shenzhen) does not follow a scaling law which is supposed to be the "right" structure for proper functioning. According to these authors, the higher the fractal dimension of the system of parks, the more it is multi-scale and the more it is able to adapt to new functional needs in a process of dynamic evolution. You et al. (2022) suggest that the park system in Fuzhou (China) is hierarchical, but that the self-organizing process that is supposed to lead to an optimal configuration is not complete. However, they specify that their model is at a preliminary stage of design that does not allow quantification of the threshold of spatial extension of the self-organization process. They also note that "there is still a long way to go before the theory can be applied to actual park system planning".

8.1.3. *Preservation of biodiversity*

From an ecological point of view, urban environments are often very dynamic (Savard et al. 2000; Angold et al. 2006). For example, they may harbor a remarkable variety of pollinating insects (Baldock et al. 2015). However, urbanization also leads to a significant loss of biodiversity, resulting from both the reduction and fragmentation of habitats of animal and plant species (Fahrig 1997; Hanski 2011) and the barrier effect of urbanized areas (Marull and Mallarach 2005) and transport infrastructure (Forman and Alexander 1998; Gurrutxaga et al. 2010; Girardet et al. 2013). These impacts vary according to the form and intensity of urbanization (Alberti 2005). Considering an urban area as a whole, urban sprawl increases traffic along roads and railways and, consequently, the barrier effect of transport networks. At a more local scale, dense urban development may avoid the loss of ecological habitat (Conway 2009), but, concomitantly, increase the barrier effect of urbanized areas (Aguilera et al. 2011). Conversely, with diffuse urban development, the barrier effect of urbanized areas is reduced, but the fragmentation of ecological habitats is increased (Czamanski et al. 2008).

These observations lead to three very general questions concerning the relationship between urban forms and the preservation of biodiversity: at the scale of an urban area, is it more desirable to concentrate urban developments in the main urban center or to distribute growth across several centers and sub-centers (polycentric urban development)? Should the urbanized edge (perimeters of cities) be smooth and clear or tortuous and more indistinct? On the scale of neighborhoods or blocks, should urban development be very dense or moderately dense? These questions arise in much the same terms when considering the quality of life of humans and the environmental impacts due to their movements (including air and noise pollution) and housing (e.g. household energy consumption for heating and cooling). These questions are extremely difficult to answer because of the many interacting variables that must be accounted for. But when we consider biodiversity as a whole, the level of complexity of the questions asked still increases enormously due to the existence of thousands of different animal and plant species in interaction, each with more or less specific needs, some adapting better than others to urban environments and/or tolerating better than others the anthropization of their places of life.

It is therefore clear that the interest of fractal forms of urban development for the preservation of biodiversity is a field of research that has remained very little explored until now. A study based on the simulation of theoretical scenarios of fractal and non-fractal residential developments showed that reduction in the connectivity of ecological habitats is less marked for fractal than non-fractal scenarios when the urbanization process is less intense and when the species considered have short dispersal distances. When the intensity of the process of urbanization increases, non-fractal scenarios are in all cases more advantageous than fractal scenarios (Tannier et al. 2012).

Another study, based on realistic scenarios of fractal-shaped residential development at the local level, considered the impact of these developments on three forest animal species (lynx, wildcat and red squirrel) with, respectively, long, medium and short dispersal distances (Tannier et al. 2016). The results show that residential development concentrated in the main built-up cluster is more likely to preserve the functional ecological connectivity of the habitats of the three species than residential development distributed over all built-up clusters (except the smallest). For all but one scenario, the decline in connectivity is mainly due to the barrier effect of transport networks (increased traffic) rather than the barrier effect of urbanized areas. Depending on the scenarios, the connectivity of the habitats of the three animal species is more or less impacted without establishing a systematic link between the shape parameters of the simulated residential development and the impacts measured.

With regard to the preservation of biodiversity, maintaining or creating interconnected networks of green and natural spaces of varying sizes is a commonly advocated principle of development. Indeed, patches of ecological habitats, which are of small size and/or separated from each other by long distances, generally constitute poor conditions for the preservation of biodiversity[1] (Forman 1995; Collinge 1996). From this perspective, it seems interesting to promote certain fractal forms of urban development that make it possible to avoid fragmentation of unbuilt spaces. This is particularly the case for certain multifractal built configurations in which unbuilt spaces may have very diverse sizes, their rank-size distribution following a scaling law (Cavailhès et al. 2010; Frankhauser 2015; Frankhauser et al. 2018).

8.1.4. *Intra- and inter-urban spatial hierarchies*

The rank-size distribution of built-up clusters follows a scaling law in the case of certain mono- or multifractal figures that are likely to correspond to city models (see Chapter 3). However, there is currently no indisputable argument as to whether this property should represent an objective of development to be pursued. As Tannier and Thomas (2013) point out, it is currently unclear whether, to ensure "virtuous" urban functioning, a city must have a main built-up cluster that is significantly larger than the others, or have the shape of a hybrid fractal carpet rather than some other shape. Thus, for example, if all the buildings in a city were part of the same built-up cluster, fragmentation of unbuilt spaces would be avoided. But planning strategies aimed at achieving this goal (containment strategies such as Urban Growth Boundaries and green belts) have the side effect of causing discontinuous residential extensions to appear beyond the boundaries fixed by the planning (*leapfrogging*) (Vyn 2012; Peeters et al. 2015).

Schweitzer and Steinbrink (2002) hypothesized that the rank-size distribution of built-up clusters in cities evolved over time until it approached a scale-invariant distribution. If this were the case, we may conclude that the exponent of the corresponding scaling law, which could be a fractal dimension, would be an

1. Oliveira et al. (2011), however, stress that the interpenetration of ecological habitats and urban areas is not desirable in all cases. Indeed, it may sometimes be necessary to disrupt ecological connectivity in order to avoid the spread of invasive species, threatening intra-urban biodiversity, or wild species liable to disrupt human activities.

indicator of the stage of development of an urban fabric, and that deviations from a rank-size distribution of the built-up clusters to a scale-invariant distribution would indicate potential for future developments. However, other empirical research findings contradict this hypothesis. Notably, Benguigui et al. (2006) showed that the rank-size distribution of the built-up clusters of Tel Aviv followed a Zipf law between 1935 and 1964, and then gradually deviated from it between 1974 and 2000.

The central place models of Christaller (1933) and Lösch (1940) offer functional explanations for the emergence of a hierarchical system of centers and sub-centers, based on a supply of stores and services that are differentiated in terms of the frequency of use, associated with catchment areas that are also differentiated in terms of size. Howard's Garden City model (1898), New Urbanism, decentralized concentration (Urban Task Force 1999) and transit-oriented development (TOD) (Calthorpe 1993) also provide for a hierarchy of centers and sub-centers of fixed sizes (Yamu and van Nes 2017). However, empirical case studies that have sought to relate the more-or-less asserted polycentrism of certain urban spaces to the organization of flows and movements in these spaces have not clearly highlighted the advantages of such a configuration (Schwanen et al. 2001, 2004; Aguilera 2005). In doing so, promoting hierarchical polycentric urban development is currently a partisan planning approach based on a number of arguments, but whose advantages and disadvantages have not yet been clearly demonstrated.

8.1.5. *Shape of the perimeter of urban agglomerations*

Researchers have noted that the relationship between the perimeter of cities and their surface area resembles that of some fractal objects (Longley and Batty 1989; Frankhauser 1994): their perimeter increases enormously when the spatial resolution at which they are observed is refined, while their surface area increases relatively little and tends towards an identifiable limit. Based on this observation, some researchers have had the idea to lengthen the outline of the built-up clusters, which does not greatly increase their surface area, in order to maximize direct contact between the interior and the exterior (Arlinghaus and Nystuen 1990). Following this logic, Frankhauser (2000) compared the distances to the center and the urbanized border (i.e. the boundary between the city and the countryside) of two theoretical forms of cities with the same surface area, but with a smooth outline in one case (a square city) and a dendritic one in another (a city in the form of a fractal figure named a teragon). With this second form of city, the average distance to the urbanized border, from anywhere in the city, is much less than that of the square city, while the average distance to the center is almost equivalent. The study of six other theoretical cities mapped at building scale, some fractals and others non-fractals, confirmed that a very dendritic fractal border makes it possible to significantly increase the number of buildings located less than 5 m from it (Tannier and Thomas 2013).

8.2. Fractal planning standards: between myths and realities

We have seen previously that some research aims to prove the functional interest of fractal urban forms, while other research relies on this hypothesis to state rules or propose models of fractal planning that are supposed to be relevant. The two questions we will discuss in this section of the chapter refer to this second stance. We will first question the existence of a "right" fractal dimension for the planning of cities, then the optimality of fractal city shapes.

8.2.1. *Is there a "right" fractal dimension for city planning?*

Some authors argue that fractal dimension is a relevant tool for optimizing the occupation of space (Chen 2017; Cheng et al. 2021). Chen and Luo (2006) even go so far as stating that the ideal fractal dimension tends towards 1.71. However, the same fractal dimension may characterize a wide variety of shapes. For example, Figure 8.3 shows two fractal generators of the same fractal dimension $D = 1.46$. Generator (a) creates a fractal figure in which all elements are connected to each other, while generator (b) creates a fractal figure that mixes clusters of connected and non-connected elements. Indeed, the fractal dimension says little about the orientation and overall shape of the ultimate figure (Batty and Longley 1994). Moreover, regardless of the fractal dimension, the size ratio between the largest morphological cluster and the second, third, etc., largest clusters may vary considerably from one figure to another. Finally, the same value of fractal dimension can be produced quantitatively and qualitatively by very different processes (Pumain 2017).

Since the same fractal dimension value may characterize very different shapes, the question of the "right" fractal dimension, the one that would be optimal for a city, lacks meaning (Genre-Grandpierre 2017). No fractal dimension value can serve as a universal standard for urban planning. Nevertheless, as Jahanmiri and Parker (2022) suggest, fractal dimension can be used for guiding urban planning and evaluating outcomes. When combined with other metrics and an understanding of the urban context, fractal dimension can reflect human well-being outcomes and urban aesthetics. Fractal values between a certain range may be more functional or desirable.

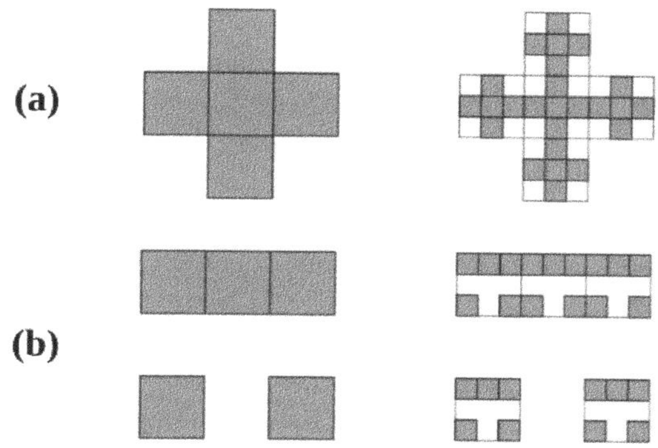

Figure 8.3. *Two fractal generators of the same dimension D = 1.46, from which two very different figures result*

In fact, fractal dimensions present, like all measurements, limits in terms of both calculation and interpretation. They are nonetheless interesting because they enable characterization of very heterogeneous spatial distributions. In particular, it is possible to set a fractal dimension to be achieved for the development of a given built-up fabric (through extension and/or densification), adapting and justifying the choice of this dimension in light of the objectives pursued, the specificities of the built-up fabric under consideration and the nature of the data representing it. For example, in Tannier et al. (2012), the authors estimated the fractal dimension of self-similarity of a suburban built-up fabric in the metropolitan area of Besançon (eastern France) (equal to 1.43). They also calculated its fractal dimension of correlation (equal to 1.5)[2]. Based on this, they may have argued that a residential development should have a target fractal dimension value greater than approximately 1.46. However, they wanted to test other scenarios with lower fractal dimensions (equal to 1.26 or 1), in order to develop the sectors of the study area that were characterized by dimensions less than 1.46, or to accentuate the fractality of the built-up fabric for scale intervals for which the tissue was not statistically self-similar. In any case, if we want to maintain a certain hierarchy of unbuilt spaces within a built-up fabric, the fractal dimension should not exceed 1.8 (Frankhauser 2004).

2. For a definition of these two fractal dimensions, see Chapters 2 and 4.

8.2.2. *Are fractal city shapes, by nature, optimal?*

In a self-organized system, a meso- or macroscopic order emerges from interactions at the microscopic level (and, possibly, from interactions between levels), which in turn constrains the future evolution of the system[3]. Starting from the premise that a fractal order may emerge from self-organizing processes, fractality (or multifractality) is sometimes seen as a desirable state of equilibrium. Deviations from fractality can then be considered as a sign of dysfunction. For Chen and Wang (2013), for example, deviations of the built-up fabric of the city of Beijing from a multifractal structure indicate the decline and degeneration of the central city. Chen (2016) asserts that multifractality represents an optimal structure for human geographical systems since a fractal object can occupy its space in the most efficient way. Thus, using the ideas from multifractals to design or plan urban and rural developments, we can make the best of human geographical space. This idea that fractal or multifractal figures are optimal or, at least, particularly efficient for city planning has been taken up by a number of researchers, for example, Gong et al. (2020) and You et al. (2022). Yamu and van Nes (2017) point out that fractal dusts and multifractal Sierpinski gaskets (Vicsek figures) are particularly useful at the regional level (Yamu and Frankhauser 2015). At the urban level, monofractal figures may be more relevant when a higher built density is expected (Murcio et al. 2015). Referring to Frankhauser (2015), Cheng et al. (2021) also suggest that a spatial structure that is too simple, that is, non-fractal, may be evidence of either over-planning or under-structuring. Over-simplicity of a spatial structure may affect its morphological evolution and compromise its proper functioning.

However, built-up fabrics often show deviations from statistical self-similarity (Frankhauser 1998, 2004; Tannier and Pumain 2005; Thomas et al. 2010b; Sémécurbe et al. 2019). The self-similarity of cities is indeed spatially circumscribed and concerns restricted ranges of scales. Yet, not all cities or parts of cities that are not self-similar are dysfunctional or in decline.

Adopting an organicist perspective, other authors argue that urban planning should aim to create fractal forms because they are observed in nature (in which they

---

3. Some authors add to the idea of self-organization that of a limiting factor, often corresponding to a maximum size of city that is impossible to exceed (Schweitzer and Steinbrink 2002) or to a maximum rate of urbanization (Chen 2016). Chen (2016) goes even further by hypothesizing that an urbanization process should ideally stop at the point at which the urbanization rate is equal to 0.618 and the corresponding urban–rural ratio, which he calls the *golden ratio*, is equal to 1.618.

emerge spontaneously) and are therefore "naturally" virtuous and optimal. Among the often-cited examples of natural fractal shapes that meet certain optimizing principles, we can note the veins of leaves and lungs (tracheobronchial tree). On this basis, fractals are established as a universal aesthetic principle (Salingaros 2003; Jiang and Sui 2014). Fractal forms are considered as a "natural" alternative to the urban forms traditionally used as a reference in planning (Nguyen et al. 2022). However, adopting this point of view often leads to more emphasis being placed on emergent forms than on the processes that give rise to them. The resulting models are essentially structural: they take into account neither the real behaviors of individuals (results of a combination of aspirations, constraints and available means), nor the strong emergence that characterizes social systems[4].

Here, we suggest adopting another point of view, considering that a form is not intrinsically optimal: it is optimal in terms of the processes (behaviors, practices, modes of operation) of which it allows optimization. In the case of human settlements in general, and cities in particular, the social, political, cultural, economic, natural, etc., contexts are extremely diverse and change over time. This results in a very great diversity of individual spatial behaviors and practices. This is why possible optima with regard to the shape of cities are necessarily context-dependent and not universal.

Considering a given context, it is possible to evaluate the functional advantages of fractal and non-fractal forms for cities (or settlements) (Tannier et al. 2012a, 2012b). However, such an enterprise is difficult in the sense that the advantages and disadvantages of different city forms are usually not unequivocal, but, on the contrary, more or less contradictory. For example, compact and dense urban development may avoid the loss of ecological habitats, but, at the same time, increase the barrier effects of built-up areas against the movements of animals. Conversely, less dense and more diffuse urban development may increase the fragmentation of ecological habitats but reduce the barrier effects of urbanization.

---

4. Different authors distinguish weak emergence, where macroscopic structures can be observed from the outside as a result of microscopic processes, from strong emergence, where the dynamics of the microscopic levels adapt in turn to the structures of the macroscopic levels generated by them and which they observe from the inside (Livet et al. 2010). Socio-spatial systems are characterized by strong emergence; insofar as they belong, or not, to a social group or a place, individuals participate in the construction of collective references to that group or place (weak emergence). In return, the adoption (or not) of these collective references by the individual or the group influences their behavior (strong emergence).

8.3. Why and how to introduce fractal geometric rules into geographic models of urban and regional planning?

To answer this question, we take the example of two simulation models aimed at providing decision support for urban and regional planning. These two models have the characteristic of imposing self-similarity as a principle that new urban developments must respect. The first model, *MUP-City*[5], allows the simulation of fractal residential development under accessibility constraints (Tannier et al. 2012). The second model, *Fractalopolis*[6] (Frankhauser et al. 2018), is dedicated to the interactive design of multifractal urban development plans. These two models are intended to simulate different possible configurations of urban development starting from an initial spatial configuration (buildings, road and path networks, public transport stations, stores and services, green and leisure infrastructures). They do not incorporate any temporal dynamics; not only is the configuration obtained by simulation not situated in time, but the transition from the initial configuration to the simulated configuration is done in a single step ("one-shot"), rather than the outcome of a trajectory passing successively through a series of intermediate states.

MUP-City and *Fractalopolis* are both based on an explicit spatial scale dependency principle. This scale dependence is relative to the shape of urban developments: an unbuilt space at a macro- or mesoscopic level cannot be a built space at a microscopic level. Thus, *MUP-City* and *Fractalopolis* preserve, at a fine spatial resolution, the unbuilt spaces identified at a coarser spatial resolution. They respect the principle according to which the fractality of urban fabrics essentially originates in the spatial organization of unbuilt spaces and not in that of built spaces. The *Fractalopolis* model also incorporates a functional scale dependency principle: the functional level of each center of urban development, set by the model, determines the catchment areas of the centers, as well as the number and diversity of stores, services and green and leisure infrastructures to be expected in each of them.

In both models, scale dependence is integrated in the form of recursive geometric rules, as a result of which the spatial configurations created are statistically self-similar. In the case of *MUP-City*, this is a fractal rule of urbanization which imposes a certain spacing between the built cells and a certain form of connectivity between the non-built cells. In the case of *Fractalopolis*, it is a system of iterated

5. See: https://sourcesup.renater.fr/www/mupcity/fr.html.

6. See: https://sourcesup.renater.fr/www/fractalopolis/.

functions that jointly defines the spacing between the different centers of planned development (i.e. their relative position) and their surface area, as well as a certain connectivity between spaces located outside the centers.

The simulation results consist of the identification of cells that are potentially interesting for urbanization, in the case of *MUP-City*, and urban or suburban centers in which new housing units could be built, in the case of *Fractalopolis*.

8.3.1. *MUP-City: simulation of fractal residential development under accessibility constraints*

With *MUP-City*, the new simulated residential developments are in the form of square cells with a size of 20 m a side in most applications performed so far. The location of the cells which are of value for urbanizing is determined by a fractal urbanization rule based on a multi-scale decomposition of the existing built-up fabric (Figure 8.4). In addition to choosing a fractal dimension for simulated residential developments (corresponding to the variable N_{max} in Figure 8.4), the fractal rule of urbanization dictates that cells not urbanized at the decomposition level i_1 cannot be urbanized at a finer level of decomposition i_2. The target fractal dimension when N_{max} = 4 is 1.26; 1.46 when N_{max} = 5; 1.63 when N_{max} = 6; 1.77 when N_{max} = 7.

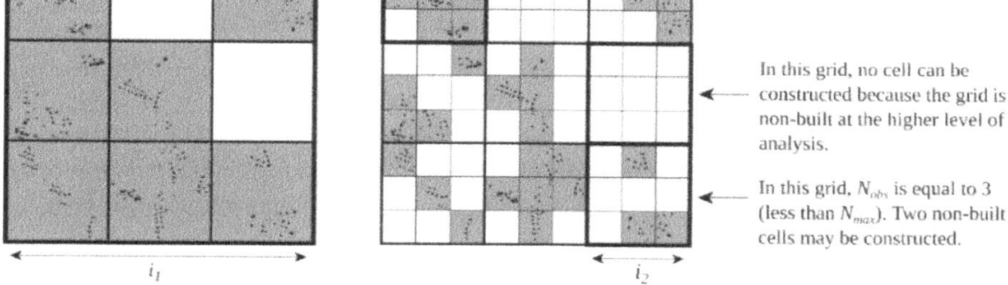

Figure 8.4. *Example of the application of the fractal rule of urbanization of MUP-City for N_{max} equal to 5, that is, a fractal dimension equal to 1.46. Source: Tannier et al. (2017)*

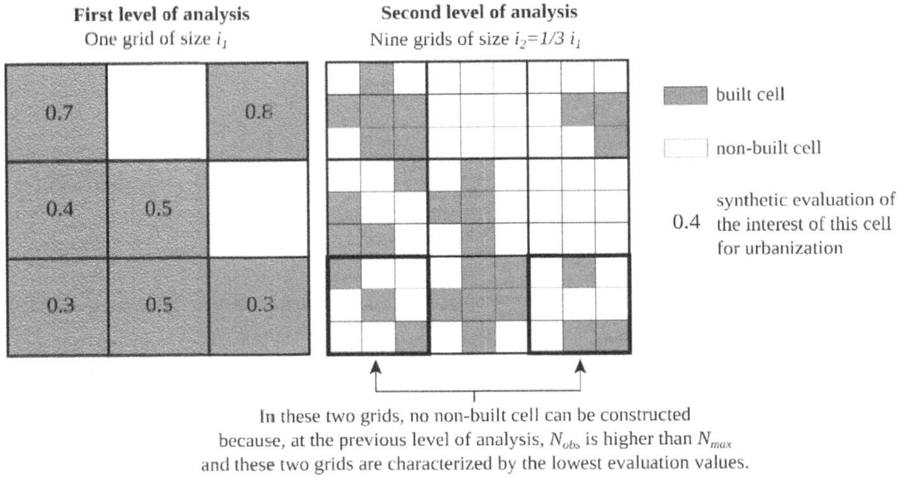

Figure 8.5. *Strict variant of the fractal rule of urbanization for N_{max} equal to 5, that is, a fractal dimension equal to 1.46*

Note that in Figure 8.4, certain grids include more urbanized cells than should be present in a statistically self-similar object. This is due to the fact that the initial built-up fabric is not organized according to a logic of mono-fractal nesting of scales with reduction factor r equal to 3. In order for simulated residential developments to approach more closely to statistical self-similarity, a "strict" variant of the fractal rule of urbanization has been introduced into *MUP-City* (Figure 8.5).

MUP-City allows the simulation of very diverse forms of residential development (Figure 8.6). The simulated developments correspond to a densification of existing built-up fabrics, by increasing their local fractal dimension, and/or by creating new morphologically self-similar residential extensions. Note that these extensions result in an increase in the length of the urbanized border. It should also be noted that simulating fractal residential development makes it possible to obtain, at a local level, visually realistic forms of residential development. Simulated residential developments may take the form of more or less sinuous linear extensions, interstitial infills or more or less dense infills on the periphery of existing built-up areas.

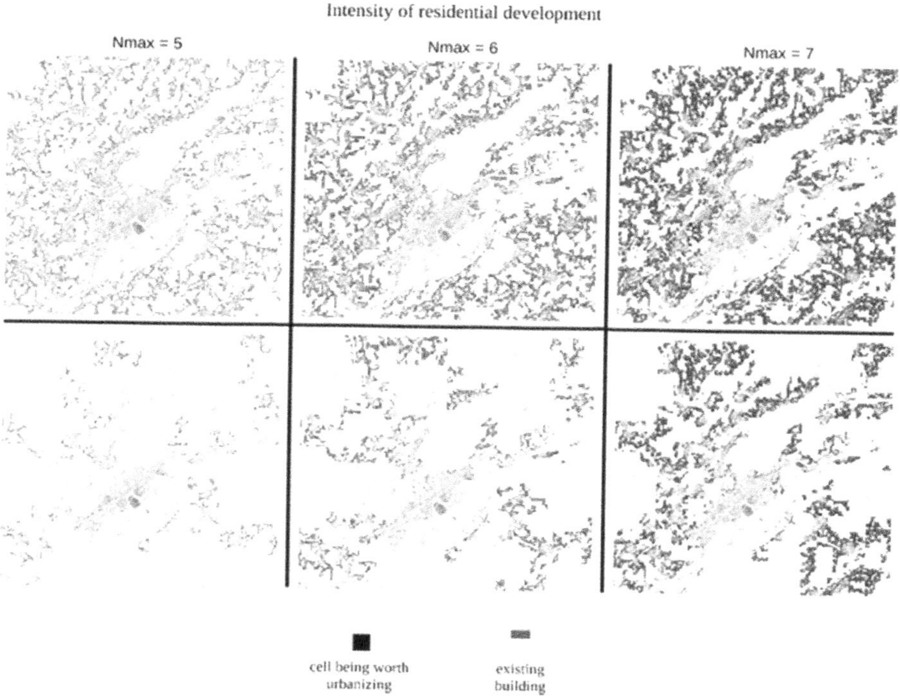

Figure 8.6. *Examples of results obtained with* MUP-City *over the agglomeration of Besançon (eastern France). First line: application of the basic fractal urbanization rule. Second line: application of the strict variant of the fractal urbanization rule. Source: Colomb (2019, p. 114). For a color version of this figure, see www.iste.co.uk/tannier/fractal.zip*

8.3.2. Fractalopolis*: design of multifractal urban development plans*

In *Fractalopolis*, the creation of a multifractal development plan follows an iterative logic based on the definition and application of an iterated function system (IFS) (Barnsley 1988) (see Chapter 2). Two types of areas are distinguished: those favorable to urban development (called "centers of planned development") and those in which little or no development is envisaged. The initiator of a development plan designed with *Fractalopolis* is a square of size L positioned over the study area. The generator of the plan generally contains a central square of the chosen size $S_1 = r_1 \times L$, placed over the main urban center, and N squares of smaller size $S_0 = r_0 \times L$ placed over the secondary development centers. The number of secondary centers, the value of the parameters r_0 and r_1 and the number of functional levels (usually three or four, corresponding to the number of iterations for which an IFS is applied) are defined specifically for each study case. Subsequently, the IFS chosen to create a development plan is applied iteratively until the smallest squares are of approximately the size of urban neighborhoods or blocks. At this iteration step, the length of their side varies between 500 m and 300 m, approximately. A development plan constructed on this basis is said to be multifractal because the generator of the plan includes several reduction factors r (Mandelbrot 1982).

To illustrate our point, we present an example of the application of *Fractalopolis* whose objective was to design a multifractal development plan for the metropolitan area of Besançon (eastern France). This metropolitan area included approximately 200,000 inhabitants in 2011, of whom approximately 117,000 were located in the central municipality of Besançon. The multifractal development plan for the agglomeration, designed with *Fractalopolis* (version 1.0), starts from the initial situation of 2011 and sets targets to be achieved by 2026 regarding the location of new housing constructions (Frankhauser et al. 2018). The generator chosen to design the plan consists of a primary center of planned development (first functional level) and five secondary centers (second functional level) (Figure 8.7). The difference in size between the elements of the generator is small in order to correspond to the spatial organization of the study area. The main center is located in the municipality of Besançon. The secondary development centers are located in suburban municipalities well served by public transport.

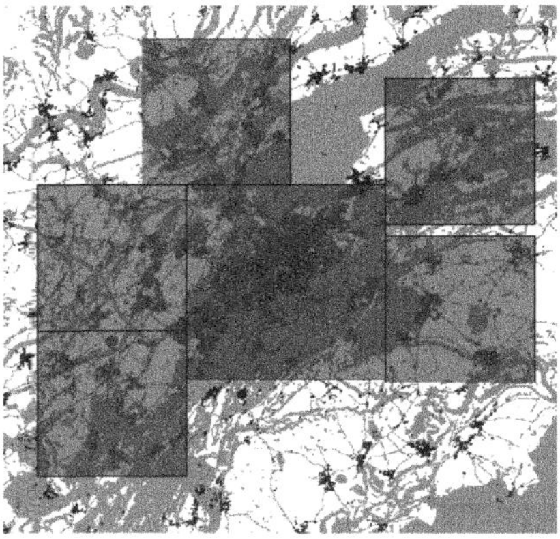

Figure 8.7. *Generator of a multifractal development plan for the metropolitan area of Besançon (eastern France). In gray: non-urbanizable areas. Red dots: public transport stations. Source: Frankhauser et al. (2018). For a color version of this figure, see www.iste.co.uk/tannier/fractal.zip*

(a) Strict application of the generator | (b) Plan obtained after manual movement of centers

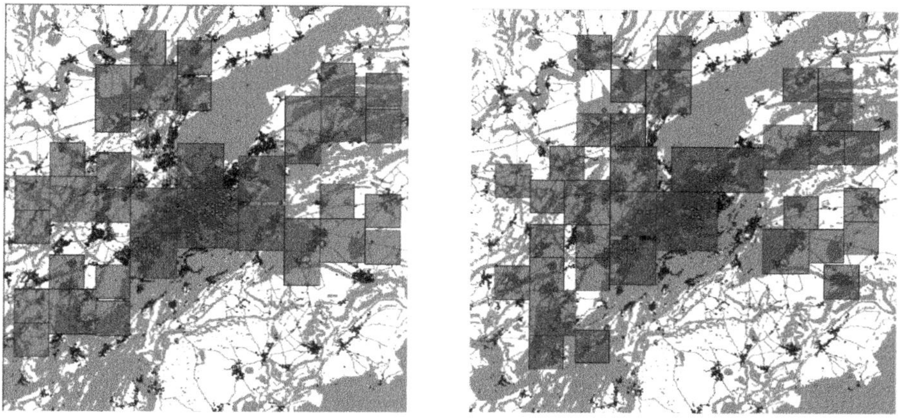

Figure 8.8. *Multifractal configuration resulting from the application of the generator shown in Figure 8.7 to the second step of creating the Besançon development plan. In gray: non-urbanizable areas. Red dots: public transport stations. For a color version of this figure, see www.iste.co.uk/tannier/fractal.zip*

Figure 8.8(a) shows the multifractal configuration resulting from the application of the generator to the second stage of creating the Besançon development plan. This geometrically regular configuration must be modified to correspond to the location of the existing urban and suburban centers. Movement of the squares from their positions initially defined by the generator is performed manually, but *Fractalopolis* helps to follow a certain number of rules governing movement of the squares:

– In order to maintain the statistically self-similar character of the multifractal plan, the squares representing each center of development may not overlap. Moreover, they cannot overlay spaces located outside the centers of development generated in the previous steps of iteration (which are mainly, in the case of Besançon, natural and agricultural spaces). This ensures both the preservation of unbuilt spaces connected to each other and the penetration of unbuilt spaces into built spaces at different scales (green alleys and green corridors).

– At the first and second stages of iteration, since the squares representing the planned centers of development are very large, they may overlay non-urbanizable areas. Then, as the iterations progress, the squares become smaller and it is possible to position them in such a way as to preserve the non-urbanizable areas.

– Centers at the highest functional levels must be served by public transport. This rule is not imposed in *Fractalopolis*, but must be respected by the designer of a development plan.

– Good access to stores and services must be ensured. To do this, the accessibility of each center is represented using a color code in *Fractalopolis* (Figure 8.9). Each time a square is moved, all its accessibility indicators are automatically recalculated.

Figure 8.9. *(a) Multifractal configuration resulting from the application of the generator shown in Figure 8.7 to the third step of creating the Besançon development plan. (b) Accessibility of centers of development. Better accessibility is indicated by centers in darker green. For a color version of this figure, see www.iste.co.uk/tannier/fractal.zip*

Observing these rules for moving the squares when working on the plan in the second iteration step made it possible to place the sub-centers of the third functional level around the city center, close to the public transport stations (see Figure 8.8(b), to be compared with Figure 8.8(a)). Then, at the third step of iteration, the position of the squares was chosen to guarantee the best possible accessibility and preserve the non-urbanizable areas (Figure 8.9). Thus, the squares overlay fewer non-urbanizable areas in Figure 8.9(a) than in Figure 8.8(b).

The multifractal plan thus obtained (Figure 8.9) has the following characteristics: axial development along the public transport axes; the centers of development are spatially related except in an area between Besançon and Saône because of the presence of non-urbanizable areas (steep slopes and protected natural areas); spaces outside the centers of development are also related, thus avoiding fragmentation of green and natural spaces; large areas in which no development is planned (which are, currently or potentially, green and natural spaces) penetrate the agglomeration from the outside to the city center of Besançon.

An important point to note is that due to the presence of two reduction factors in the generator of a multifractal development plan, the surface area of each center depends on its relative position, that is, its distance from the higher-level centers. Therefore, the surface area of centers at different hierarchical levels may be the same, while the surface area of centers of the same hierarchical level may be different. In this respect, the model differs from Christaller's (1933) central place model, but corresponds to the empirical observation that centers of the same functional level are more extensive when located near a large urban center and less extensive when located in sparsely populated areas.

Many other multifractal plans for the metropolitan area of Besançon could be imagined, in particular in response to the modifications of the generator, other goals in terms of accessibility to services or additional planning constraints. Through the application of *Fractalopolis* to the metropolitan area of Besançon, we understand intuitively that a single "good" spatial and functional hierarchy of centers of planned development does not exist. However, accentuation of the initial hierarchy or its inflection in one direction or another may be interesting.

"[...] is it essential for good urban planning to maintain a good adaptability of urban fabrics to future conditions that no one can predict? Have we properly measured the flexibility of adaptation of fractal structures?" (Pumain 2017). To these questions, it is possible to answer that the application of the *Fractalopolis* model to the metropolitan area of Besançon illustrates a certain form of adaptability of a fractal model of urban development; the location, number and characteristics of the different centers (and therefore the distances separating them, including the possibilities of mobility offered by the presence of public transport stations) can be defined according to the specificities of each study area, in its current and possibly its future state. It would also be possible to apply *Fractalopolis* to the planning of *shrinking cities*: *Fractalopolis* would help to determine those centers in which a certain number of dwellings should be destroyed and, possibly, replaced by green or agricultural spaces.

8.3.3. *Fractal geometric rules in geographic models of cities and territories*

In this section, we would like to show that introducing fractal geometric rules into geographic models of urban and regional planning is interesting because it allows effects of distance and scale dependence to be explicitly accounted for in models.

8.3.3.1. *Accounting for scale dependence*

When considering the spatial distribution of human settlements, scale dependence means that observing a state of concentration at a certain scale (i.e. considering a certain level of aggregation or spatial resolution) implies observing a state of dispersion at a finer or coarser scale. A consequence of scale dependence is the *modifiable areal unit problem* (Openshaw 1983): the states of dispersion or concentration measured depend on the size of the elementary spatial entities considered, their shape and their position, as well as the delimitation of the study area. The level of detail of the representation of space (i.e. the scale, or the level of aggregation or spatial resolution) influences the measurement of geographic shapes (Tobler 1969, 1990). In addition, aggregation has a "smoothing" effect in the sense that the concentrations measured are generally higher at coarser levels of aggregation or spatial resolution (Tobler 2004). This is why it is impossible to define a true population density at a point, since any estimate depends explicitly on the scale at which density is measured; remarkably few geographical variables are scale-less (Goodchild 2001).

It turns out that a crucial property of fractal objects is that their size depends on the scale of measurement adopted (Falconer 2013)[7]. In addition, the density of points (or elements) of a fractal object decreases as the scale (i.e. the level of aggregation or spatial resolution) adopted becomes finer. This decrease in density is all the more marked when the fractal dimension is low. The density of points of a fractal object is therefore scale-dependent, and this scale dependence is geometric in nature. Conversely, the density of an object inscribed in the plane (in 2D) with a fractal dimension equal to 2, which is in this case the maximum fractal dimension and which characterizes uniform objects, non-fractal by definition, does not vary between scales.

The application of successive spatial decompositions respecting a hierarchical nesting principle in *MUP-City* and the application of an IFS in *Fractalopolis* are two types of fractal geometric rules that allow the introduction of an explicit scale dependence into both models.

8.3.3.2. *Accounting for distance effects*

There are two types of distance effects. On the one hand, they may manifest in the form of a friction attenuating the mutual influence of two phenomena as they move away from each other (which may be measured in metric distance, cost, time or otherwise). On the other hand, distance effects may be manifested in the form of

7. For further details on this subject, see Chapter 2, section 2.4.

belonging or not to the same territory (the same neighborhood, the same city, the same country). In this case, they partly explain the existence of spatial discontinuities and boundary effects. The main consequence of distance effects is the first law of geography: elements close to each other are more likely to be related or similar than distant elements (Tobler 1970).

Through the introduction of fractal geometric rules into a model, distances between elements are more or less strictly imposed at different scales. Thus, in *MUP-City* and *Fractalopolis*, distances between spatial entities created by simulation (cells or centers) are imposed at different hierarchically nested scales without modeling distance effects, either in terms of friction or belonging to the same territory. These effects are indirectly accounted for through a series of hypotheses by which it is assumed that the spacings between spatial entities (cells or centers), imposed by geometric rules, allow certain mobility behaviors that are desired for a given planning scenario, while preventing others that we wish to avoid in this scenario. The mobility behaviors involved may include individuals traveling to access stores and services, public transport, and green and leisure infrastructure (Frankhauser et al. 2018), or the juvenile dispersal of animal populations between different habitats (Tannier et al. 2016).

In *MUP-City* and *Fractalopolis*, in addition to the geometric rules for locating entities created by simulation (cells or development centers), distance effects in terms of friction and the role of belonging to the same territory are modeled as part of the evaluation of the interest of cells or centers for being subjects of residential development.

8.4. Conclusion

In this chapter, we have shown that there is nothing to prove the functional interest of built and non-built fractal forms, although a series of studies have helped to support this hypothesis. As Jahanmiri and Parker (2022) point out, the role of fractals in planning theory is still not fully established, in contrast to their role in landscape ecology theory (where they reflect processes of landscape fragmentation and biodiversity changes) and in urban growth theories (where they complement agglomerative central place theories by reflecting the influence of dispersive forces).

Some reflections in this chapter lead to contradicting the idea that a scale-invariant or self-similar spatial distribution would represent an absolute optimum. On the one hand, a shape cannot be optimal in itself. It is so only in relation to the processes which it allows to be optimized. Our understanding is still (very) poor concerning the effect of self-similar spatial distributions on geographical processes whose existence is empirically recognized. On the other hand, absolutes do not exist in terms of the spatial distribution of human settlements due to the diversity of social, political, economic and natural contexts, which influence how people function in a space. Therefore, it seems impossible to define planning solutions, in particular in the form of numerical standards (e.g. an absolute threshold defining an acceptable distance to the nearest public transport station or an exponent value of a power law), which could be successfully applied anywhere. This is why the fractal simulation models *MUP-City* and *Fractalopolis* do not contain any rules determining optimum levels of concentration or dispersion (buildings, housing units, stores and services) at each level of analysis. They are only used to define and respect a certain number of constraints, via a general model that must be configured for each specific situation. Jahanmiri and Parker (2022) also point out that fractal theories should encourage planners to see their role as guiding, rather than mandating, development patterns, acknowledging the incremental nature of fractal urban growth processes. According to them, fractal principles argue against urban designs and planning interventions that impose pre-defined and large-scale blueprints of forms. Fractal principles are in line with planning theories that acknowledge complexity and holism, such as those pioneered by Jacobs (1961) and Alexander (1964), and in contrast with simplicity and reductionism ideas of models such as the Garden City (Howard 1898).

The additional argument that we have developed in this chapter is that fractal geographic simulation models are interesting for urban and regional planning because of their geometric and explicitly scale-dependent nature. Scale dependence can be introduced into a fractal simulation model by means of successive spatial decompositions respecting a hierarchical nesting principle (*MUP-City*) or iteratively applied geometric transformations (*Fractalopolis*). It is thus possible to explicitly account for both spatial and functional hierarchies. Concerning distance effects, the spacings between spatial entities created by simulation are imposed more or less strictly via the application of fractal geometric rules. In addition, in *MUP-City* and *Fractalopolis*, the distance effects in terms of friction and of belonging to the same territory are modeled as part of the evaluation of interest in urbanizing cells or centers.

Finally, introducing fractal geometric rules into models allows us to better understand how space is organized across scales, and how it is possible to modify this spatial organization by jointly taking into account distance effects and scale dependence.

8.5. References

Aguilera, A. (2005). Growth in commuting distances in French polycentric metropolitan areas: Paris, Lyon, Marseille. *Urban Studies*, 42(9), 1537–1547.

Aguilera, F., Valenzuela, L.M., Botequilha-Leitão, A. (2011). Landscape metrics in the analysis of urban land use patterns: A case study in a Spanish metropolitan area. *Landscape and Urban Planning*, 99, 226–238.

Alberti, M. (2005). The effects of urban patterns on ecosystem function. *International Regional Science Review*, 28, 168–192.

Alexander, C. (1964). *Notes on the Synthesis of Form*. Harvard University Press, Cambridge, MA.

Angold, P.G., Sadler, J.P., Hill, M.O., Pullin, A., Rushton, S., Austin, K., Small, E., Wood, B., Wadsworth, R., Sanderson, R. et al. (2006). Biodiversity in urban habitat patches. *Science of the Total Environment*, 360, 196–204.

Arlinghaus, S.L. (1985). Fractals take a central place. *Geografiska Annaler, Series B, Human Geography*, 67(2), 83–88.

Arlinghaus, S.L. and Nystuen, J.D. (1990). Geometry of boundary exchanges. *Geographical Review*, 80(1), 21–31.

Baldock, K.C.R., Goddard, M.A., Hicks, D.M., Kunin, W.E., Mitschunas, N., Osgathorpe, L.M., Potts, S.G., Robertson, K.M., Scott, A.V., Stone, G.N. et al. (2015). Where is the UK's pollinator biodiversity? The importance of urban areas for flower-visiting insects. *Proceedings of the Royal Society B (Biological Sciences)*, 282, 20142849.

Barnsley, M. (1988). *Fractals Everywhere*. Academic Press, New York.

Batty, M. and Longley, P.A. (1994). *Fractal Cities: A Geometry of Form and Function*. Academic Press, San Diego, CA and London.

Benguigui, L., Czamanski, D., Marinov, M., Portugali, Y. (2000). When and where is a city fractal? *Environment and Planning B: Planning and Design*, 27, 507–519.

Benguigui, L., Blumenfeld-Lieberthal, E., Czamanski, D. (2006). The dynamics of the Tel Aviv morphology. *Environment and Planning B: Planning and Design*, 33, 269–284.

Bovill, C. (1996). *Fractal Geometry in Architecture and Design*. Boston Birkhauser Publishing, Cambridge, MA.

Brielmann, A.A., Buras, N.H., Salingaros, N.A., Taylor, R.P. (2022). What happens in your brain when you walk down the street? Implications of architectural proportions, Biophilia, and Fractal Geometry for Urban Science. *Urban Science*, 6, 3.

Calthorpe, P. (1993). *The Next American Metropolis: Ecology, Community, and the American Dream*. Princeton Architectural Press, New York.

Cavailhès, J., Frankhauser, P., Peeters, D., Thomas, I. (2004). Where Alonso meets Sierpinski: An urban economic model of a fractal metropolitan area. *Environment and Planning A*, 36(8), 1471–1498.

Cavailhès, J., Frankhauser, P., Peeters, D., Thomas, I. (2010). Residential equilibrium in a multifractal metropolitan area. *The Annals of Regional Science*, 45(3), 681–704.

Chen, Y. (2011). Fractal systems of central places based on intermittency of space-filling. *Chaos Solitons Fractals*, 44(8), 619–632.

Chen, Y. (2014). Multifractals of central place systems: Models, dimension spectrums, and empirical analysis. *Physica A*, 402, 266–282.

Chen, Y. (2016). Defining urban and rural regions by multifractal spectrums of urbanization. *Fractals*, 24(1), 1650004.

Chen, Y. (2017). Multi-scaling allometric analysis for urban and regional development. *Physica A*, 465, 673–689.

Chen, Y.G. and Luo, J. (2006). A tentative theoretical interpretation of fractal dimension of urban form. *Urban Studies*, 13(5), 35–40.

Chen, Y. and Wang, J. (2013). Multifractal characterization of urban form and growth: The case of Beijing. *Environment and Planning B: Planning and Design*, 40(5), 884–904.

Cheng, L., Feng, R., Wang, L. (2021). Fractal characteristic analysis of urban land-cover spatial patterns with spatiotemporal remote sensing images in Shenzhen city (1988–2015). *Remote Sensing*, 13, 4640.

Christaller, W. (1980). *Die zentralen Orte in Süddeutschland. Eine ökonomisch-geographische Untersuchung über die Gesetzmäßigkeit der Verbreitung und Entwicklung der Siedlungen mit städtischer Funktion.* Wissenschaftliche Buchgesellschaft, Darmstadt (Reprint, originally published 1933, Fischer, Jena).

Collinge, S.K. (1996). Ecological consequences of habitat fragmentation: Implications for landscape architecture and planning. *Landscape and Urban Planning*, 36(1), 59–77.

Colomb, M. (2019). Simulation de formes réalistes de développement résidentiel, de l'échelle du bâtiment à celle de l'ensemble d'une région urbaine. PhD Thesis, University Paris-Est.

Conway, T. (2009). Local environmental impacts of alternative forms of residential development. *Environment and Planning B: Planning and Design*, 36, 927–943.

Cooper, J. (2003). Fractal assessment of street-level skylines: A possible means of assessing and comparing character. *Urban Morphology*, 7(2), 73–82.

Cooper, J. (2005). Assessing urban character: the use of fractal analysis of street edges. *Urban Morphology*, 9(2), 96–107.

Cooper, J. and Oskrochi, R. (2008). Fractal analysis of street vistas: A potential tool for assessing levels of visual variety in everyday street scenes. *Environment and Planning B: Planning and Design*, 35, 349–363.

Cooper, J., Watkinson, D., Oskrochi, R. (2010). Fractal analysis and perception of visual quality in everyday street vistas. *Environment and Planning B: Planning and Design*, 37, 808–822.

Cooper, J., Su, M.-L., Oskrochi, R. (2013). The influence of fractal dimension and vegetation on the perceptions of streetscape quality in Taipei: With comparative comments made in relation to two British case studies. *Environment and Planning B: Planning and Design*, 40, 43–62.

Czamanski, D., Benenson, I., Malkinson, D., Marinov, M., Roth, R., Wittenberg, L. (2008). Urban sprawl and ecosystems – Can nature survive? *International Review of Environmental and Resource Economics*, 2, 321–366.

Fahrig, L. (1997). Relative effects of habitat loss and fragmentation on population extinction. *Journal of Wildlife Management*, 61, 603–610.

Falconer, K.J. (2013). *Fractals. A Very Short Introduction*. Oxford University Press, Oxford.

Forman, R.T.T. (1995). *Land Mosaics: The Ecology of Landscapes and Regions*. University Press, Cambridge.

Forman, R.T.T. and Alexander, L.E. (1998). Roads and their major ecological effects. *Annual Review of Ecology and Systematics*, 29, 207–231.

Frankhauser, P. (1994). *La fractalité des structures urbaines*. Anthropos, Paris.

Frankhauser, P. (1998). The fractal approach: A new tool for the spatial analysis of urban agglomerations. *Population*, 4, 205–240.

Frankhauser, P. (2000). La fragmentation des espaces urbains et périurbains – une approche fractale. In *Structures des villes, entreprise et marchés urbains*, Derycke, P.H. (ed.). L'Harmattan, Paris, 25–54.

Frankhauser, P. (2004). Comparing the morphology of urban patterns in Europe. A fractal approach. In *European Cities: Insights on Outskirts, Structures*, Borsdorf, A. and Zembri, P. (eds). ESF COST Office, Brussels, 79–105.

Frankhauser, P. (2015). From fractal urban pattern analysis to fractal urban planning concepts. In *Computational Approaches for Urban Environments, Geotechnologies and the Environment*, Helbich, M., Jokar Arsanjani, J., Leitner, M. (eds). Springer International Publishing, Switzerland, 13–48.

Frankhauser, P. and Genre-Grandpierre, C. (1998). La géométrie fractale – un nouvel outil pour évaluer le rôle de la morphologie des réseaux de transport public dans l'organisation spatiale des agglomérations. *Les Cahiers Scientifiques du Transport*, 33, 41–78.

Frankhauser, P., Tannier, C., Vuidel, G., Houot, H. (2007). Approche fractale de l'urbanisation – Méthodes d'analyse d'accessibilité et simulations multi-échelles. *11th World Conference on Transportation Research*, Berkeley, 24–28 June.

Frankhauser, P., Tannier, C., Vuidel, G., Houot, H. (2010). Une approche multi-échelle pour le développement résidentiel des nouveaux espaces urbains. In *Modéliser la ville. Forme urbaine et politiques de transport*, Antoni, J.-P. (ed). Economica, Paris, 306–332.

Frankhauser, P., Tannier, C., Vuidel, G., Houot, H. (2018). An integrated multifractal modelling to urban and regional planning. *Computers, Environment and Urban Systems*, 67, 132–146.

Genre-Grandpierre, C. (2017). Commentaire sur le chapitre 1. In *Villes, réseaux et transport. Le défi fractal*, Dupuy, G. (ed.). Economica, Paris, 31–39.

Girardet, X., Foltête, J.-C., Clauzel, C. (2013). Designing a graph-based approach to landscape ecological assessment of linear infrastructures. *Environmental Impact Assessment Review*, 42, 10–17.

Gong, Y., Li, X., Cong, X., Liu, H. (2020). Research on the complexity of forms and structures of urban green spaces based on fractal models. *Complexity*, 4213412.

Goodchild, M. (2001). Models of scale and scales of modelling. In *Modelling Scale in Geographical Information Science*, Tate, N. and Atkinson, P.M. (eds). John Wiley & Sons Ltd, Chichester, 3–10.

Gurrutxaga, M., Lozano, P.J., Del Barrio, G. (2010). Assessing highway permeability for the restoration of landscape connectivity between protected areas in the Basque Country, Northern Spain. *Landscape Research*, 35, 529–550.

Hagerhall, C.M., Purcell, T., Taylor, R. (2004). Fractal dimension of landscape silhouette outlines as a predictor of landscape preference. *Journal of Environmental Psychology*, 24, 247–255.

Hagerhall, C.M., Laike, T., Taylor, R.P., Küller, M., Küller, R., Martin, T.P. (2008). Investigation of EEG response to fractal patterns. *Perception*, 37, 1488–1494.

Hanski, I. (2011). Habitat loss, the dynamics of biodiversity, and a perspective on conservation. *Ambio*, 40, 248–255.

Houot, H., Tannier, C., Vuidel, G., Frankhauser, P. (2009). Multi-scale scenarios of urban residential development and accessibility to retail stores and facilities. *16th European Colloquium on Theoretical and Quantitative Geography*, Maynooth-Ireland, 4–8 September.

Howard, E. (1898). *To-Morrow: A Peaceful Path to Real Reform*. Swan Sonnenschein & Co, London.

Jacobs, J. (1961). *The Death and Life of Great American Cities*. Vintage Books, New York.

Jahanmiri, F. and Parker, D.C. (2022). An overview of fractal geometry applied to urban planning. *Land*, 11, 475.

Jiang, B. and Sui, D. (2014). A new kind of beauty out of the underlying scaling of geographic space. *The Professional Geographer*, 66(4), 676–686.

Joye, Y. (2007). Architectural lessons from environmental psychology: The case of Biophilic architecture. *Review of General Psychology*, 11, 305–328.

Le, A.T.D., Payne, J., Clarke, C., Kelly, M.A., Prudenziati, F., Armsby, E., Penacchio, O., Wilkins, A.J. (2017). Discomfort from urban scenes, metabolic consequences. *Landscape and Urban Planning*, 160, 61–68.

Li, Z., Zhao, W., Nie, M. (2021). Scale characteristics and optimization of park green space in megacities based on the fractal measurement model: A case study of Beijing, Shanghai, Guangzhou, and Shenzhen. *Sustainability*, 13, 8554.

Liang, G., Hu, Y., Sun, H. (2013). The design evaluation of the green space layout of urban squares based on fractal theory. *Nexus Network Journal*, 15, 33–49.

Livet P., Müller J.-P., Phan D., Sanders L. (2010). Ontology, as a mediator for agent-based modeling in social science. *Journal of Artificial societies and Social Simulation*, 13(1), 3.

Longley, P.A. and Batty, M. (1989). Fractal measurement and line generalization. *Computers & Geosciences*, 15(2), 167–183.

Lösch, A. (1940). *Die räumliche Ordnung der Wirtschaft. Eine Untersuchung über Standort, Wirtschaftsgebiete und internationalem Handel*. Fischer, Jena.

Lu, S. (2010). Hidden orders in Chinese gardens: Irregular fractal structure and its generative rules. *Environment and Planning B: Planning and Design*, 37(6), 1076–1094.

Mandelbrot, B. (1982). *The Fractal Geometry of Nature*. W.H. Freeman and Company, New York.

Marull, J. and Mallarach, J.M. (2005). A GIS methodology for assessing ecological connectivity: Application to the Barcelona Metropolitan Area. *Landscape and Urban Planning*, 71, 243–262.

Murcio, R., Masucci, A.P., Arcaute, E., Batty, M. (2015). Multifractal to monofractal evolution of the London street network. *Physical Review*, 92, 062130-1–062130-12.

Nguyen, T.T., Hoffmann, E., Buerkert, A. (2022). Spatial patterns of urbanising landscapes in the North Indian Punjab show features predicted by fractal theory. *Scientific Reports*, 12, 1819.

Niang, M. and Soboyejo, W.O. (2006). On the use of quasi-crystal and fractal concepts in the design of affordable rural/urban landscapes. *Perspectives on Global Development and Technology*, 5(4), 411–421.

Openshaw, S. (1983). *The Modifiable Areal Unit Problem. Concepts and Techniques in Modern Geography*. Geo Books, Kent, and Headley Brothers Ltd., Norwich.

Ostwald, M.J. and Vaughan, J. (2008). Determining the fractal dimension of the architecture of Eileen Gray. In *ANZASCA 2008: Innovation Inspiration and Instruction: New Knowledge in the Architectural Sciences*, Gu, N., Gul, L.F., Ostwald, M.J., Williams, A. (eds). Australian and New Zealand Architectural Science Association, Newcastle, 9–16.

Peeters, D., Caruso, G., Cavailhès, J., Thomas, I., Frankhauser, P., Vuidel, G. (2015). Emergence of leapfrogging from residential choice with endogenous green space: Analytical results. *Journal of Regional Science*, 55(3), 491–512.

Pumain, D. (2017). Commentaire sur le chapitre 3 – Les fractales doivent-elles guider l'aménagement urbain ? In *Villes, réseaux et transport. Le défi fractal*, Dupuy, G. (ed.). Economica, Paris, 113–119.

Salingaros, N.A. (2003). Connecting the fractal city. Keynote speech. *5th Biennial of Towns and Town Planners in Europe, Barcelona.*

Salingaros, N.A. (2011). Why monotonous repetition is unsatisfying. *Meandering Through Mathematics*, 2 September, arXiv:1109.1461v1.

Salingaros, N.A. (2012). Fractal art and architecture reduce physiological stress. *Journal of Biourbanism*, 2, 11–28.

Salingaros, N.A. and West, B.J.A. (1999). Universal rule for distribution of sizes. *Journal of Environment and Planning B: Planning and Design*, 26, 909–923.

Savard, J.-P., Clergeau, P., Mennechez, G. (2000). Biodiversity concepts and urban ecosystems. *Landscape and Urban Planning*, 48, 131–142.

Schwanen, T., Dieleman, F.M., Dijst, M. (2001). Travel behaviour in Dutch monocentric and polycentric urban systems. *Journal of Transport Geography*, 9, 173–186.

Schwanen, T., Dijst M., Dieleman, F.M. (2004). Policies for urban form and their impact on travel: The Netherlands experience. *Urban Studies*, 41(3), 579–603.

Schweitzer, F. and Steinbrink, J. (2002). Analysis and computer simulation of urban cluster distributions. In *Fundamental Principles of Urban Growth*, Humpert, K., Brenner, K., Becker, S. (eds). Müller + Busmann, Wuppertal, 142–157.

Sémécurbe, F., Tannier, C., Roux, S.G. (2019). Applying two fractal methods to characterise the local and global deviations from scale-invariance of built patterns throughout mainland France. *Journal of Geographical Systems*, 21(2), 271–293.

Stamps, A. (2002). Fractals, skylines, nature and beauty. *Landscape and Urban Planning*, 60, 163–184.

Tannier, C. and Pumain, D. (2005). Fractals in urban geography: A theoretical outline and an empirical example. *Cybergeo: European Journal of Geography*, 307.

Tannier, C. and Thomas, I. (2013). Defining and characterizing urban boundaries: A fractal analysis of theoretical cities and Belgian cities. *Computers, Environment and Urban Systems*, 41, 234–248.

Tannier, C., Vuidel, G., Frankhauser, P., Houot, H. (2010). Simulation fractale d'urbanisation – MUP-city, un modèle multi-échelle pour localiser de nouvelles implantations résidentielles. *Revue internationale de géomatique*, 20(3), 303–329.

Tannier, C., Foltête, J.-C., Girardet, X. (2012a). Assessing the capacity of different urban forms to preserve the connectivity of ecological habitats. *Landscape and Urban Planning*, 105(1–2), 128–139.

Tannier, C., Vuidel, G., Houot, H., Frankhauser, P. (2012b). Spatial accessibility to amenities in fractal and non fractal urban patterns. *Environment and Planning B: Planning and Design*, 39(5), 801–819.

Tannier, C., Bourgeois, M., Houot, H., Foltête, J.-C. (2016). Impact of urban developments on the functional connectivity of forested habitats: A joint contribution of advanced urban models and landscape graphs. *Land Use Policy*, 52, 76–91.

Tannier, C., Bourgeois, M., Frémond, M. (2017). Simulation de scénarios fractals d'urbanisation et des mobilités quotidiennes résultantes. In *Villes, réseaux et transport. Le défi fractal*, Dupuy, G. (ed). Economica, Paris, 155–189.

Taylor, R.P. (2021). The potential of Biophilic fractal designs to promote health and performance: A review of experiments and applications. *Sustainability*, 13, 823.

Thomas, I., Frankhauser, P., Biernacki, C. (2008a). The morphology of built-up landscapes in Wallonia (Belgium): A classification using fractal indices. *Landscape and Urban Planning*, 84, 99–115.

Thomas, I., Tannier, C., Frankhauser, P. (2008b). Is there a link between fractal dimensions and other indicators of the built-up environment at a regional level. *Cybergeo: European Journal of Geography*, 413.

Thomas, I., Frankhauser, P., Badariotti, D. (2010a). Comparing the fractality of European urban neighbourhoods: Do national contexts matter? *Journal of Geographical Systems*, 14, 189–208.

Thomas, I., Frankhauser, P., Frenay, B., Verleysen, M. (2010b). Clustering patterns of urban built-up areas with curves of fractal scaling behaviour. *Environment and Planning B: Planning and Design*, 37, 942–954.

Tobler, W. (1969). Geographical filters and their inverses. *Geographical Analysis*, 1(3), 234–253.

Tobler, W. (1970). A computer movie simulating urban growth in the Detroit region. *Economic Geography*, 46(2), 234–240.

Tobler, W. (1990). Frame independent spatial analysis. In *Accuracy of Spatial Databases*, Goodchild, M. and Gopal, S. (eds). Taylor & Francis, London, 115–222.

Tobler, W. (2004). On the first law of geography: A reply. *Annals of the Association of American Geographers*, 94(2), 304–310.

Urban Task Force (1999). *Towards an Urban Renaissance*. E & FN Spon, London.

Vaughan, J. and Ostwald, M.J. (2008). Approaching Euclidean limits: A fractal analysis of the architecture of Kazuyo Sejima. In *ANZASCA 2008: Innovation Inspiration and Instruction: New Knowledge in the Architectural Sciences*, Gu, N., Gul, L.F., Ostwald, M.J., Williams, A. (eds). Australian and New Zealand Architectural Science Association, Newcastle, 285–292.

Vogt, C.A. and Marans, R.W. (2004). Natural resources and open space in the residential decision process: A study of recent movers to fringe counties in Southeast Michigan. *Landscape and Urban Planning*, 69, 255–269.

Vyn, R.J. (2012). Examining for evidence of the leapfrog effect in the context of strict agricultural zoning. *Land Economics*, 88(3), 457–477.

Wilson, E.O. (1984). *Biophilia*. Harvard University Press, Cambridge, MA.

Wong, D.W.S. and Fotheringham, A.S. (1990). Urban systems as examples of bounded chaos: Exploring the relationship between fractal dimension, rank-size, and rural to urban migration. *Geografiska Annaler: Series B, Human Geography*, 72(2–3), 89–99.

Yamu, C. and Frankhauser, P. (2015). Spatial accessibility to amenities, natural areas and urban green spaces: Using a multiscale, multifractal simulation model for managing urban sprawl. *Environment and Planning B: Planning & Design*, 42(6), 1054–1078.

Yamu, C. and van Nes, A. (2017). An integrated modeling approach combining multifractal urban planning with a space syntax perspective. *Urban Science*, 1, 37.

You, M., Guan, C., Lai, R. (2022). Spatial structure of an urban park system based on fractal theory: A case study of Fuzhou, China. *Remote Sensing*, 14, 2144.

9

Multifractal Forward Planning: The *Fractalopolis* Model Applied to the Case of Greater Paris

Pierre FRANKHAUSER[1] and Olivier BONIN[2]

[1] *Laboratoire ThéMA, CNRS-Université de Franche-Comté, Besançon, France*

[2] *Laboratoire Ville Mobilité Transport (LVMT), École des Ponts ParisTech (ENPC), Gustave Eiffel University, Champs-sur-Marne, France*

The objectives of sustainable development impose many constraints on urban areas, in particular environmental constraints: the need to drastically reduce energy consumption (mobility, buildings, infrastructure and services, industry, etc.), the struggle against urban heat islands, resilience to various risks (floods, landslides, storms, etc.), the struggle against soil artificialization, and also the desire for local energy production, development of shorter supply chains, market gardening and urban agriculture. Solutions advocated by public authorities are often drastic: a total halt to the spatial expansion of cities (and therefore densification), eliminating fossil fuel vehicles, installing green roofs, etc. However, they struggle to adapt to the reality of the historical development of cities and provoke resistance from the inhabitants. It therefore seems important to try to avoid monolithic and very dogmatic solutions whose effectiveness is not demonstrated, such as densification at any cost or complete fixation of the built perimeter of the city, to adopt a more pragmatic approach to urban planning. While reducing car use is a reasonable objective, it is necessary to account for the spatial practices of residents, their needs and access to the places they frequent, in already-planned territories, which have a history and are the seat of specific spatial practices.

Fractal Geometry in Human Geography and Planning,
coordinated by Cécile TANNIER.

Accessibility to various services necessary for the well-being of inhabitants seems to us to be the first essential aspect to account for in approaching the planning of an urban space. It directly impacts the kilometers traveled, and therefore energy consumption, greenhouse gas emissions and pollutant emissions, but above all it may help organize the urban space into different centralities, a condition necessary for balanced urban growth. This concerns accessibility to jobs, education and health, and also to stores and services as well as leisure areas and nature. The second essential aspect is that of green amenities, including green and blue infrastructure, which organize the urban space by allowing respiration and proximity to nature. Green amenities, by taking up space, compete with the classic functions of a city (housing, stores, services, jobs, network infrastructure, etc.). Particular attention should therefore be paid to the proper articulation of these two types of functions when proposing plans or principles for urban layout.

We propose here, using *Fractalopolis*, an approach to planning for accessibility that uses multifractal geometry as its basis and formal geometric framework, and that aims to empirically reconcile several constraints: the basic needs of the population and the resulting potential mobility, the preservation of green and blue infrastructure, and the proximity to green and natural spaces. We do not propose a new urban form, the fractal city, which would compete with existing models (dense compact city, finger-plan city, sprawling city, etc.) but a practical planning concept which may help to diagnose an existing territory and shift its trajectory towards more sustainability.

Fractalopolis consists of both a multi-scale urban planning concept and a multifractal formalization of this concept, which are associated with operational software that enables their implementation. Instead of being a support for the strict application of sustainable planning rules, *Fractalopolis* offers a framework for spatial analysis allowing the integration of different planning constraints in a relatively flexible way, adapting to the specificities of each study area. Based on explicit multifractal modeling, *Fractalopolis* aims to reconcile the diversity of built densities and the preservation of green and blue infrastructure, promoting accessibility at different scales. *Fractalopolis* follows the tradition of different approaches to urban design dating back to Howard's Garden City (1898), while integrating concepts of urban hierarchy. It mobilizes two fundamental properties of fractal geometry: its multi-scale character and its capacity to reconcile geometric properties that are a priori incompatible, thus making it possible to propose original forms of occupying space.

The *Fractalopolis* model was developed at the ThéMA laboratory (Besançon, France) by Gilles Vuidel as part of several research contracts funded by the PREDIT program[1]. The software is in open access[2]. A version intended for use in a professional context is under development. The first applications dealt with the French urban centers of Besançon (eastern France) and Lyon within the framework of the PREDIT Vilmodes project (Antoni et al. 2014). In her thesis, Claudia Yamu then applied the concept to the metropolitan area of Greater Vienna (Austria) (Czerkauer-Yamu 2012; Yamu and Frankhauser 2015). A recent application carried out as part of the SOFT research project funded by ADEME[3] focused on the *Est Ensemble* territorial authority, near Paris (Bonin et al. 2020; Frankhauser 2021).

In the remainder of this chapter, we will first develop our approach to accessibility and then to green and blue infrastructure, before describing how these two approaches are linked in *Fractalopolis*. At the end of the chapter, we will present an example of the application of *Fractalopolis* to the *Est Ensemble* sector, located near Paris.

9.1. Planning for accessibility: reducing distances traveled and organizing urban centers

9.1.1. *Accessibility and residential satisfaction*

Households take a number of accessibility criteria into account when making residential choices. Schwanen et al. (2004) indicate that they tend to select places of residence that minimize the distance or access time to jobs (Brun and Fagnani 1994) as well as to stores (Lerman 1976) and popular leisure areas (Guo and Bhat 2002).

Many publications emphasize the importance of the presence of green amenities in individuals' residential environment, which, according to Max-Neef's (1991) classification, meets the needs of *idleness* and *protection*. Several authors have shown that residents prefer a green environment over a built environment (Lamb and

---

1. PREDIT (Research and Innovation Program in Land Transport) was a program for the coordination of French research and innovation policies in land transport (road, rail and inland waterways). Three ministries responsible for research, transport and industry and three agencies (ANR, ADEME, OSEO) were involved in this program to combine their efforts to encourage public support for research.

2. *Fractalopolis* 1.2: https://sourcesup.renater.fr/www/fractalopolis/.

3. Ecological Transition Agency.

Purcell 1990; Kaplan 1995). Others have shown the importance of good accessibility to recreational areas (Guo and Bhat 2002). Property prices confirm these preferences. They decrease with distance from green spaces (Luttik 2000; Hobden et al. 2004). Bolitzer and Netusil (2000), for example, have highlighted the positive impact of proximity to green spaces on land prices. They point out that their results make an important contribution to quantifying the benefits of maintaining access to the open countryside in urban settings. Kestens et al. (2004) suggest differentiating the role of distances from the countryside. Distances below 100 m correspond to a view of the countryside, while distances of 500 m to 600 m instead refer to walking distance.

Accessibility by foot appears to be an important criterion for assessing access to recreational spaces (Hobden et al. 2004). Investigations in Flanders have shown that there is an almost scaling relationship between the size of recreational areas and the distance that residents are willing to travel to access them (Van Herzele et al. 2000). Francis and Hester (1995), meanwhile, highlight the benefits of individual gardens. Syme et al. (2001) showed that these benefits are valid even for small gardens, although residents then tend to make more frequent use of public spaces.

The presence of green spaces in the residential environment contributes to improving the well-being of residents and has a positive effect on their health (Jensen et al. 2000; Ellaway et al. 2001; Evans 2003; Latkin and Curry 2003). The view of green countryside improves the ability to cope with stress (Kaplan 1995). Referring to the World Health Organization (WHO Regional Office for Europe 1999), Braubach (2007) shows that health is directly linked to the quality of the residential environment. He stresses the importance of the health effects of the physical qualities of the residential environment, such as traffic nuisances, density, greenery, cleanliness of public spaces, etc. Added to this are the effects of social context, such as the presence of incivility and antisocial behaviors that influence the perception of security. These factors may result in physical and nonphysical health effects that determine residential satisfaction and quality of life in a residential area.

All of these results and observations explain the preference of many households to settle in suburban, green and low-density areas. For example, a survey conducted in the United Kingdom showed that 68% of residents in low-density areas were very satisfied with their place of residence compared to 37% of residents in very dense areas (Hedges et al. 1994).

9.1.2. *Social demand: expectations and needs for individual mobility*

In line with the objectives of sustainable development, our reflections aim to limit excessive energy-consuming and pollution-emitting trips, which is why we are considering the aspects of social demand likely to generate trips. Indeed, it seems illogical to suggest principles of urban planning without looking at the social demand for mobility. In the development plans of metropolitan areas, in particular, it seems essential to aim at reducing the use of cars, but at the same time, it is necessary to ensure good accessibility to places that meet residents' needs.

Expectations and needs in terms of mobility depend primarily on cultural context and lifestyle. For example, although in France the bakery is a business frequented almost daily, this is not the case in other countries. Tobacconists are also a French specialty. Criteria or norms for accessibility may also depend on the local context and vary significantly between a densely populated metropolitan area and a low-density rural area. In order to develop a general approach that can easily be adapted to a specific cultural or local context, we propose to mobilize a general theory of needs, rooted in the foundational works of Maslow (1954) and developed by Max-Neef (1991). This theory identifies the needs of individuals from an anthropological point of view and clarifies the links between amenities and services, on the one hand, and the basic needs of individuals, on the other hand. This makes it possible to associate different needs with amenities (stores, services, hospitals, leisure areas, etc.), many of which will be located differently depending on the cultural context. On this basis, a hierarchical typology of amenities can be established. A good match between the functional hierarchy of the territory and the distribution of amenities makes it possible to best meet the needs of its inhabitants by limiting excessive movement. This approach is in many ways similar to the concept of *capability* developed by Sen (2010) within a social justice framework and deepened by Nussbaum (2012): a development plan must give inhabitants the ability to satisfy their possible mobility needs, giving them the greatest autonomy and as wide a range of choices as possible. We thus lay the theoretical foundations of *Fractalopolis* aiming to produce an accessible city, whose benefits are obvious to all, as reflected in the contemporary idea of the "15-minute city" promoted today by public authorities.

9.1.3. *From the hierarchy of individual needs to the hierarchy of central places*

When considering different types of amenities, their location and accessibility, it is also crucial to distinguish the frequency with which individuals use them or need to use them. For example, a general practitioner or a pharmacy must be locally accessible, as must a primary school or green space with a playground. The same goes for stores that cater to daily needs. Large specialized institutions such as hospitals, universities or shopping centers are designed to accommodate large numbers of people and, because they are frequented less regularly or by a smaller part of the population, acceptable distances to access them may be greater. Just as there are different average frequencies of use of amenities depending on their nature – few people go to an embassy frequently, for example – there is also a spatial hierarchy, often based on an economic logic, or at least on the catchment area of the population concerned.

This logic is reminiscent of central place theory (Christaller 1933). The interest of establishing a functional hierarchy of relatively autonomous sub-centers to limit travel has often been advocated (Fouchier 1995) and is part of current planning strategies. This central place theory was used to design the master plans for the metropolitan region of Paris. In German-speaking countries, territorial planning also refers directly to central place theory, generally introducing four types of central places providing goods and services at different levels depending on their frequency of use and catchment area: higher centers (*Oberzentren*), middle centers (*Mittelzentren*), lower centers (*Unterzentren*) and minor centers (*Kleinzentren*) (Figure 9.1).

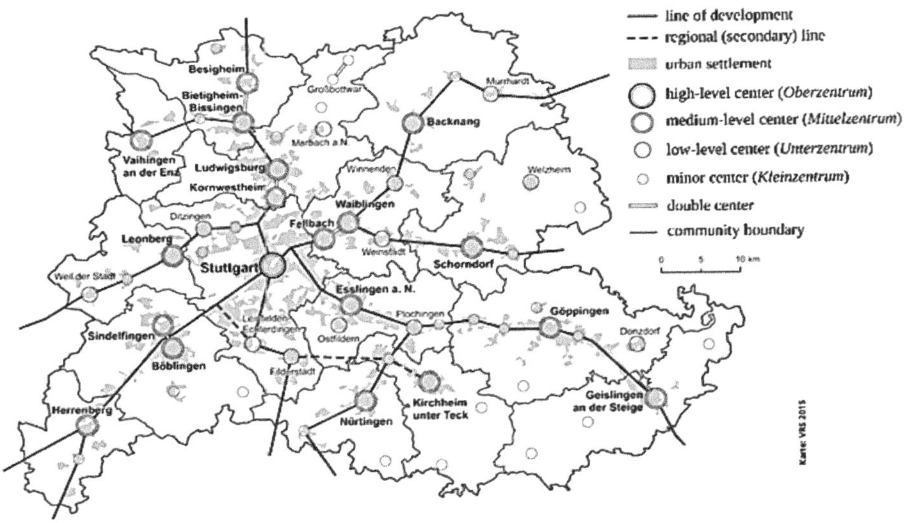

Figure 9.1. *The hierarchy of central places of the master plan for the Stuttgart metropolitan area (2009). The black lines represent the lines of development, which correspond to the RER network. Source: Regionalplanung in der Region Stuttgart, (p. 7, see: www.region-stuttgart.org). For a color version of this figure, see www.iste.co.uk/tannier/fractal.zip*

At the theoretical level, Fujita and Ogawa (1982) showed that when the urban population increases, sub-centers appear everywhere as soon as transport costs exceed a certain threshold.

Several researchers have criticized Christaller's central place theory with regard to the mathematical properties of the underlying geometric concept (Michalakis and Nicolas 1986). Berry and Garrison (1958) tried unsuccessfully to produce an economic explanation more consistent with the geometric model. The new economic geography has considered central place theory to be a descriptive schema without a micro-economic basis (Krugmann 1995). Adopting a hedonistic economic approach, Fittkau (2004) argued that, in reality, substitutability of goods is limited and that consumers may have particular preferences and are willing to travel to greater distances to access goods. Indeed, surveys have shown that consumers do not always choose the nearest shopping centers (Clark 1968). It is clear that an overly rigorous approach to the functional hierarchy of space will only be subverted by inhabitants, who retain their free will, just as they partly avoid the logic of strict concentration of the compact dense city.

At present, new types of logistic strategies, such as just-in-time delivery, drive-thru shopping, Internet shopping or teleworking, may undermine the spatial hierarchy logic of central places. Nevertheless, a good supply of local stores and services to meet the daily needs of individuals remains important, at least for perishable goods and essential services.

9.2. Integrating green and blue infrastructure into a planning process

9.2.1. *Green infrastructure to reduce heat islands*

One problem of densely built-up areas is linked to climate change, which results in an increase in urban heat islands. This leads to risks to the health of residents and, in addition, encourages increased use of air conditioners. However, it turns out that intra-urban vegetation improves the local climate as it contributes significantly to nighttime cooling. Green spaces connected to each other and linked to the open countryside (peripheral green and natural spaces), thus creating ventilation corridors, are much more efficient than isolated urban parks (Kuttler 2011). Analyzing the impact of several green infrastructures in the Cologne region, Kuttler et al. (1998) found that the flow of cold air along the ventilation corridors is superimposed on the micro-circulation of cold air in the ventilation zones, which allows horizontal transfer of air from the surrounding countryside to the city. These corridors seem far more efficient than individual rows or isolated groups of trees (Spronken-Smith and Oke 1999). Based on simulations, this was confirmed by Sachsen et al. (2013) for a green valley in the Aachen region (Germany). Frey (1999) cites another positive example the green infrastructure separating axes of development along suburban railway lines in Copenhagen's finger-plan (Figure 9.8(b)). The preservation of green infrastructure in urban areas is part of planning legislation in Germany (Figure 9.2).

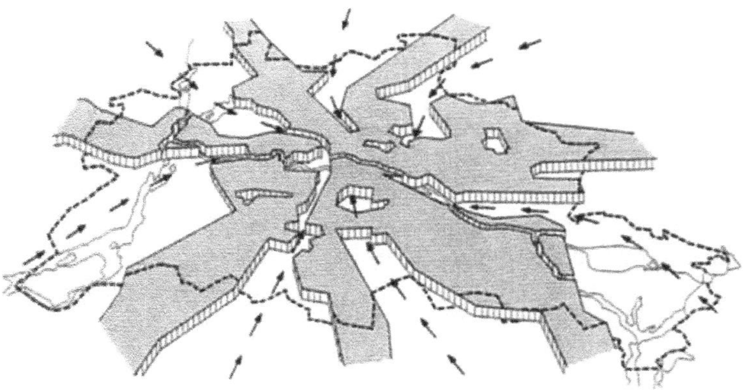

Figure 9.2. *The "ventilation corridors" scheme for Berlin. Source: BERLIN. Landschaftsprogramm Artenschutzprogramm Senatsverwaltung für Stadtentwicklung und Umweltschutz [Online]. Accessed: November 9 2022.*

9.2.2. *Green infrastructure to protect biodiversity*

So far, we have essentially considered the well-being of the resident population, and therefore the social aspect of sustainable development, even though consideration of the climate has already led us to mention the link with the environment. We will now go deeper into this environmental aspect of sustainable development and more particularly the preservation of biodiversity, which is obviously fundamental for the survival of humanity. Massive urbanization in recent decades has critically affected the natural environment. It is essentially the construction of new roads that is the problem. This concerns not only the consumption of space, but above all the fragmentation of natural spaces which reduces possibilities of travel for animals. These are crucial in the life cycle of many species. Indeed, fragmentation of ecological habitats considerably weakens biological diversity (Linehan et al. 1995). It must be considered as one of the main reasons for the disappearance of fauna and flora species (Wilcox and Murphy 1985). Green and natural spaces that are too small are generally ecologically inefficient

(Figure 9.3). It is also essential to note that distances between spaces represent a less significant barrier effect than that caused by a major road with heavy traffic (Forman 1995; Collinge 1996). Jongman et al. (2004) emphasize that biological connectivity requires landscape connectivity to preserve biodiversity. Meanwhile, Bryant (2006) emphasizes that a "piecemeal" strategy without more general coherence is ineffective.

Although biodiversity is greater in large areas compared to small areas, animal and plant populations are qualitatively different (Forman 1995). Large habitats host species that live within forest massifs, while verges are more suitable for generalist species. Thus, a mixture of habitats of different sizes seems the best guarantee for the preservation of biodiversity. Krummel et al. (1987) showed that there is a scaling relationship between the surface area of habitats and their perimeter. The edge of small habitats is less tortuous than that of habitats larger than 60 or 70 hectares. It is possible that the smooth, non-tortuous perimeters of small natural areas are the result of urban planners' tendency to smooth out urban borders.

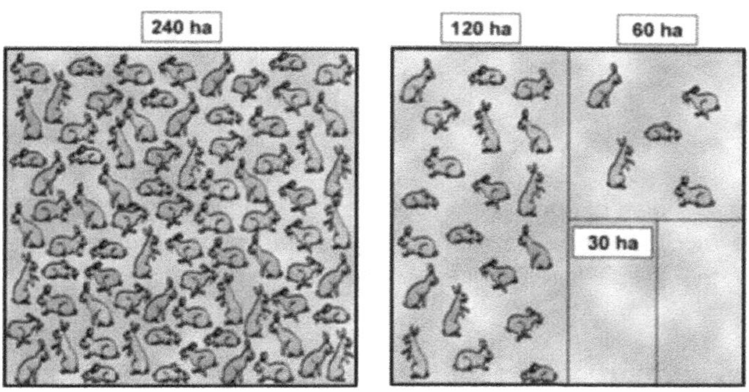

Figure 9.3. *Influence of the size of hare habitats on their population density (source: Anderegg, R. (1984). Journée "Route et faune sauvage". Organized by Office fédéral des forêts, Suisse). For a color version of this figure, see www.iste.co.uk/tannier/fractal.zip*

The United States was certainly the first country to introduce the concept of green and blue infrastructure. This goes back to Olmstedt and Eliot whose first project was the Boston Park System in the late 19th century (Fábos 2004). This project used coastal rivers to connect green infrastructure. This approach subsequently became a key idea in US planning strategy (Figure 9.4). Whyte (1961) introduced the term "greenway" in his book *Securing Open Space for Urban America*. The inclusion of greenways in the *President's Commission Report* has attracted the attention of US politicians and developers. In other countries, strategies to interconnect ecological habitats with the aim of preserving biodiversity are more recent. In Germany, for example, the preservation of green infrastructure had long been promoted for reasons of residential quality (see the next section), but its ecological function was not really considered until the 1980s (Mader 1979, 1981). In France, it was not until the Grenelle II law of 2010 that this concept was introduced into spatial planning strategies.

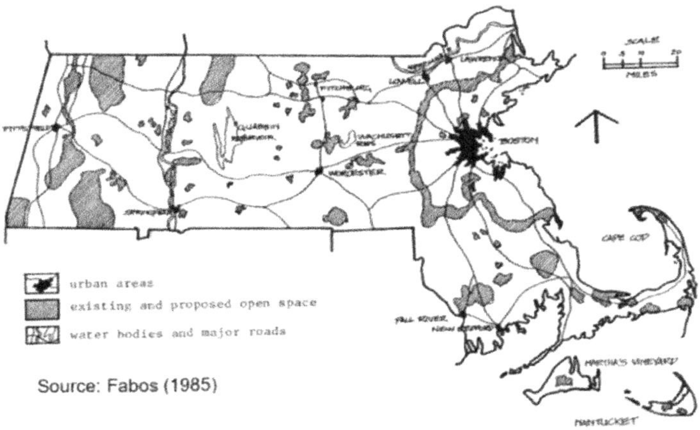

Source: Fabos (1985)

Figure 9.4. *A plan of open spaces for the Commonwealth of Massachusetts, by Eliot's son (late 19th century). It has since become the basic reference for "open space planning"*

9.2.3. *Green infrastructure as a planning concept*

The history of urban design shows that the idea of integrating green spaces into cities is quite ancient. The Renaissance took a particular interest in nature and, at least since the Baroque period, many city plans included green or natural areas as in Versailles, for example, or as shown in the fan plan of Karlsruhe (Figure 9.5). These areas were either gardens or parks reserved for the nobility, or forest massifs serving as hunting reserves. However, although these plans deal equally with gardens or natural spaces and areas to be urbanized, the two remain poorly connected. The massive urbanization of the 19th century, following industrialization and its disastrous consequences for public health, gave rise to many reflections on the integration of green areas into the urban fabric. Many cities by then had public parks, and the concept of allotment gardens appeared at this time.

Figure 9.5. *The ideal plan of Karlsruhe, in large part realized*[4]*. For a color version of this figure, see www.iste.co.uk/tannier/fractal.zip*

(a) (b)

Figure 9.6. *(a) Schematic of Howard's Garden City (1898). (b) London's green belt. Source: The Telegraph (2012)*[5]*. For a color version of this figure, see www.iste.co.uk/tannier/fractal.zip*

4. See: http://www.karlsruhe-antiquarisch.de/Plane_/Heinrich_Schwartz_koloriert/heinrich_schwarz_koloriert.html. Accessed: June 28, 2022.

5. See: http://www.telegraph.co.uk/earth/greenpolitics/planning/9708387/Interactive-map-Englands-green-belt.html. Accessed: July 4, 2022.

Apart from urban parks integrated into urban fabrics, two other planning concepts exist and relate to the scale of cities and their peripheries. These are green belts and green penetrants. The concept of green belts mainly exists in England, and refers to Howard's model of the Garden City (Figure 9.6(a)). The city is encircled by a green belt which constitutes a kind of rampart intended to contain the city and avoid sprawl (Figure 9.6(b)). The risk of green belts is that by wanting at all costs to contain the city within strict limits, suburbanization is shifted beyond the green belt, thus significantly lengthening trips to access the center. In the United Kingdom, this concept has been criticized with an emphasis on the importance of preserving green infrastructure, relatively neglected at the end of the 20th century (Turner 2006). The development plans of certain Nordic cities also participate in the spirit of urban hygiene, referring to Howard's Garden City, but are opposed to the concept of English-style green belts, advocating on the contrary the idea of green penetrants. In an urban design competition on the future development of Berlin, Eberstadt, Möhring and Petersen (1910) harshly criticized the green belt strategy. They proposed a completely different strategy, specifying that they were inspired by the reality of Berlin, which at that time was experiencing significant growth, due to its new function as the German capital after 1870, and axial sprawl along the major routes of the S-Bahn (Figure 9.7(a)). Eberstadt, Möhring and Petersen proposed concentrating urban development along the axes of this regional rail network and interposing non-urbanized green areas between the circular sectors of development (Figure 9.7(b)). The aim was to create recreational spaces near residential areas.

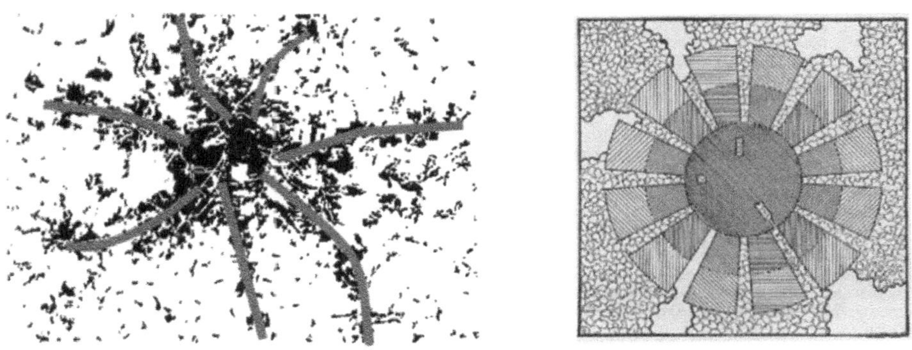

(a) (b)

Figure 9.7. *(a) The built-up area of Berlin 1945, showing the concentration of urbanized areas along the axes of the RER (in red). Source: Städtebaulisches Institut, Universität Stuttgart. (b) Eberstadt, Möhring and Petersen's plan (1910) for the development of Berlin. For a color version of this figure, see www.iste.co.uk/tannier/fractal.zip*

Another example of the application of this concept is Schumacher's plan (Güldner 1968), drawn up in the 1920s for Hamburg (Figure 9.8(a)), but the best-known example is without doubt that of Copenhagen's finger-plan (Figure 9.8(b)). Frey (1999) describes it as follows: "a considerable advantage of such a plan is that the open countryside extends to the center of the city, which makes it possible to bring fresh, clean air into the center; another is that access to the open countryside from the urbanized areas is excellent; yet another is that wood and fresh food resources are readily available near built-up areas".

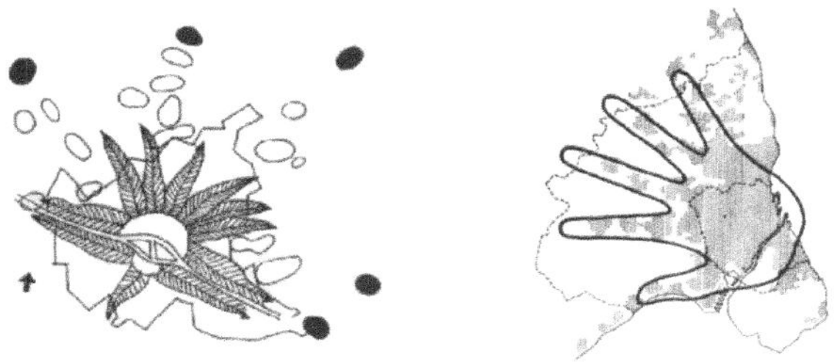

Source : Güldner (1968)

(a)

Source : Städtebauliches Institut, Universität Stuttgart

(b)

Figure 9.8. *(a) Schumacher's plan to channel the sprawl of Hamburg. (b) Copenhagen's finger-plan*

In the same vein, Schmidt proposed in 1912 a concept of development for the Ruhr separating the different cities with a network of green zones (von Haaren and Reich 2006). Some recent planning examples include the idea of inserting green zones between built-up areas to improve residents' quality of life. In the Ruhr, for example, Schmidt's idea from 1912 was taken up again in the conversion of the former heavy industrial area of the Ruhr as part of the "Emscher Park" project (Dettmar and Ganser 1999). Green zones were created between cities. These green zones are interconnected by the blue infrastructure of the Emscher and penetrate into the heart of urbanized areas (Figure 9.9(a)). They offer recreational activities integrating former industrial zones and protected natural areas. At a more local scale, the same principle was applied when setting up the Vauban eco-district in Fribourg in Brisgau by allowing penetration of the built-up area by green alleys connected to blue infrastructure (Figure 9.9(b)). In the same vein, the architect G. Schöfl (1986) proposed connections between the open countryside and the built environment at the local level (Figure 9.10(a)), opposing the widespread tendency in planning to smooth out urban borders (Figure 9.10(b)).

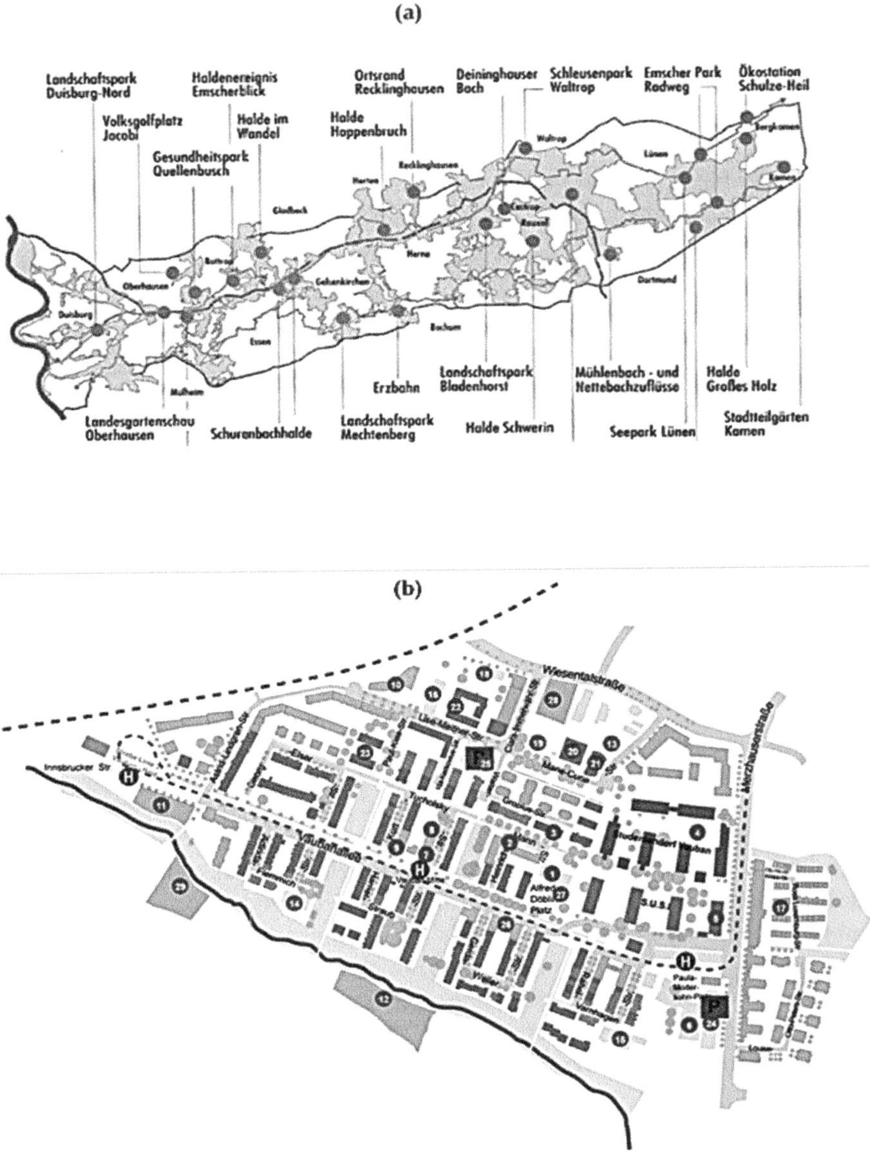

Figure 9.9. *(a) Green infrastructure separating cities in the Ruhr. Source: Art Industriel, IBA Emscher Park / ACCP Blog. (b) Urban green infrastructure in Fribourg in Brisgau (Vauban district) connected to blue infrastructure. Source: City of Fribourg (see: https://stadtteil-vauban.de/fr/vauban/). For a color version of this figure, see www.iste.co.uk/tannier/fractal.zip*

(a) (b)

Figure 9.10. *(a) Schöfl's proposal (1986) to connect the built environment with the open countryside. (b) The "built-up front" in a suburban municipality of Besançon (eastern France). Photograph by P. Frankhauser 2015. For a color version of this figure, see www.iste.co.uk/tannier/fractal.zip*

We can raise the question of the existence of conflicts of use in green infrastructure, between its ecological function for animals and plants and its recreational function for humans. According to Von Haaren and Reich (2006), these potential conflicts of use do not lead to significant reductions in the ecological, social and psychological benefits of green infrastructure. Following their evaluation of the Kronsberg project in Hanover (Germany), which aims both at improving the urban climate and creating recreational areas, they conclude that, despite the disruptions created by walkers, cars and agricultural machinery, green infrastructure completely fulfills its multiple functions. According to von Haaren and Reich (2006), the need to identify criteria for assessing risks to biodiversity has encouraged research in this area. However, they regret that this knowledge is not yet sufficient for the relevant assessment of development projects.

9.3. *Fractalopolis*: a multi-scale development model

Rather than systematically fighting against any form of extensive urban growth, it seems more useful to develop planning strategies whose aim is to channel urban development (Beaucire et al. 1999). Transit-Oriented Development (TOD) is an example of such a strategy and serves as a starting point for the presentation of the *Fractalopolis* model.

9.3.1. *A precursor: TOD, itself inspired by the Garden City*

Transit-oriented development (TOD), developed by Calthorpe (1993), places accessibility by public transport at the heart of the spatial organization of cities (Figure 9.11). TOD assigns a predominant function to transit nodes by concentrating all kinds of infrastructure around them, including green infrastructure. TOD distinguishes centers at higher and lower levels in relation to their provision of private and public services according to a hierarchical logic reminiscent of that of central place theory. This also applies to the public transport system. The density in each city decreases from the center to the periphery.

TOD has sometimes been criticized for promoting compactness, which does not guarantee a reduction in nuisances caused by car traffic (Neuman 2005). However, TOD provides for less dense land use in the secondary centers than in the main centers, and even the construction of individual housing on the periphery. In the Netherlands, a similar principle has been introduced in the planning of some new towns that incorporate old villages. Almere, which today has a population of 196,290, is one of these cities belonging to the Randstad in Holland. The urbanized area consists of three principal centers. As in TOD planning, the railway station is centrally located and is the starting point for a pedestrian boulevard with shops and restaurants at street level and apartments and multifamily accommodation above. The city benefits from a bus system with a high level of service and dedicated bus corridors as well as a network of cycle paths. These dedicated routes usually pass under roads and are therefore very efficient. A network of green spaces is accessible to residents within a radius of 500 m. In addition, large natural areas are easy to access by bike in a few minutes (Beatley 2000). The built-up area combines individual and collective housing. This concept is in line with the principle, widely discussed in the Netherlands, of "deconcentrated concentration" (Portschy 2016; Hall and Tewdwr-Jones 2019).

In many aspects, TOD is interesting but limited to the consideration of two scales (the city and the system of cities, see Figure 9.11) and is more interested in the design of intra-urban spaces and architectural issues than in the overall development of a metropolitan area. *Fractalopolis* assumes the basic principles of TOD by putting the structuring of space at the scale of metropolitan areas at the heart of its objectives. It aims to respond to the residents' needs as well as to different environmental requirements. The model incorporates the following principles:

– respecting the hierarchy of individual needs, which induces a hierarchical spatial organization of "urban" amenities;

– setting up a spatial system of diversified and hierarchical green zones, accessible to individuals according to their potential frequency of use;

– avoiding the fragmentation of green spaces and the disintegration of urbanized areas to preserve biodiversity.

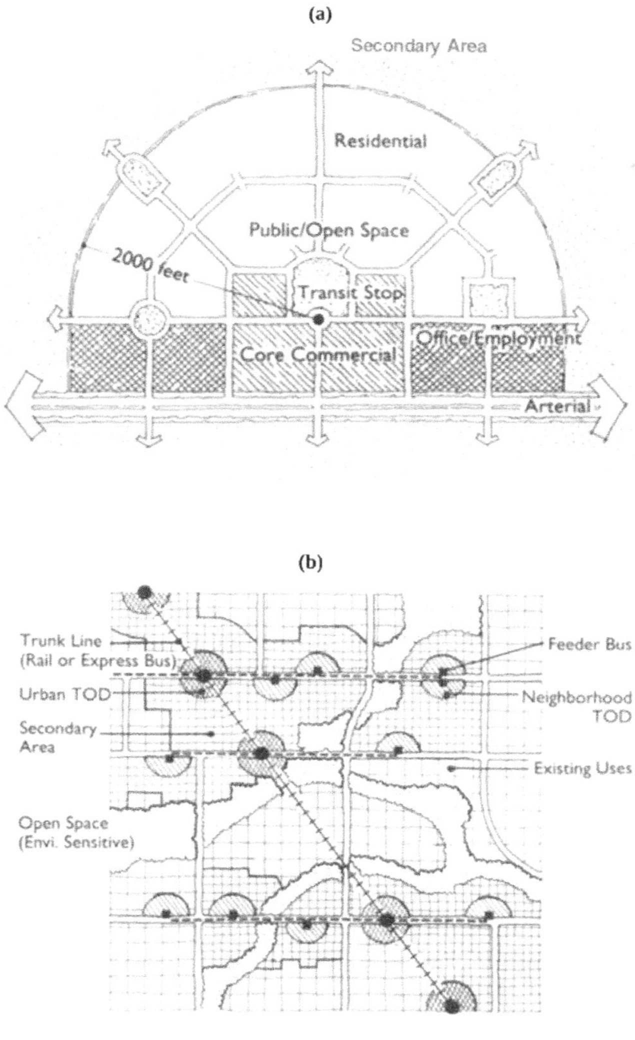

Figure 9.11. *(a) Calthorpe's schema (1993, p. 56) at the scale of an urban center: the heart of the city is the public transport stop. (b) Calthorpe's schema (1993, p. 62) for a regional city system: a two-tier public transport system*

It is clear that accessibility to the most basic amenities must be good, both to better meet the daily needs of residents and to avoid long and frequent, and therefore polluting, trips. For smaller or more rarely used amenities, a greater distance may be acceptable. These may be heavy and expensive resources, whose radius of attraction is necessarily wider. In addition, a hierarchical structure of connected recreational areas should allow these areas to be sized in accordance with their frequency of use.

The properties of fractal geometry make it possible, in our opinion, to meet these requirements. Since Mandelbrot's work, the question has been raised as to what extent a fractal structure has advantages over other geometric configurations. It has been pointed out that many systems in biology are organized according to fractal logic when it comes to efficiently connecting several spatial systems. This is the case, for example, for the lungs, the circulatory system or for trees which must allow light to penetrate within their crowns. In these systems, it is assumed that the mechanisms of genetic selection have favored the emergence of fractal structures.

9.3.2. *Fractal geometry as a basis for reflections on planning*

A survey of residential satisfaction in Wallonia showed that the degree of dissatisfaction was the highest in municipalities where the fractal dimensions of the built-up fabrics were close to $D = 2$, while the largest percentage of satisfied households was in municipalities with fractal dimensions between $D = 0.5$ and $D = 1.8$ (Thomas et al. 2008). The lowest fractal dimensions corresponded to suburban and rural municipalities. This result is consistent with the observations described above concerning the criteria of residential choice and satisfaction, although it is obvious that satisfaction is conditioned by a large number of criteria that are unrelated to urban morphology. The paradox that follows is that suburbanization jeopardizes, at the scale of a metropolitan area, what residents look for at the local level, namely, connections between green and natural areas and built-up areas. However, at the local level, planners try to smooth out urban borders, which reduces or even prevents any connection of green and natural areas with built-up areas.

Tentacular sprawl along the transport axes lengthens the distances to be covered to access different places. One solution is to create lateral branches of urbanization that are grafted onto a network of preexisting main axes. It is in this respect that we see the interest of using a concept of planning based on fractal geometry. For reasons highlighted above, we wish to promote the non-fragmentation of built-up areas and green areas and their connection (interpenetration). This objective can be achieved with the Nordic city planning concept and some recent urban planning projects. However, we want to go further by maximizing residents' access to green and natural spaces and ensuring good ventilation of the city at the level of urban heat islands. The model must therefore offer amenities at different scales to take account of the hierarchy of needs and the resulting diversified use of urban and green amenities. It must also provide for different levels of built density.

Through cross-scale linking of urbanized and green spaces and multi-scale modeling of local built-up densities, fractal geometry makes it possible to simultaneously respond to these requirements. It allows cross-scale connections between the three spatial systems that interest us (Frankhauser 2008, 2021; Frankhauser et al. 2018):

– the built-up zones we designate as development zones, which will be the elements of the fractal figure at the base of the model;

– the public transport networks that make the development zones accessible;

– the preserved zones for which low development, freezing of development or possible deconstruction are recommended.

9.3.2.1. *The box fractal, the basic figure of planning with* Fractalopolis

The basic planning figure with *Fractalopolis* is a box fractal, also called a Vicsek fractal (Frankhauser 1988, p. 68; Vicsek 1989, p. 12). Its multifractal version allows the introduction of both a hierarchy of central places and a hierarchical organization of green spaces and transport networks (Figure 9.12). It also ensures good accessibility to various amenities according to their hierarchical level.

The construction of a fractal figure is based on iteration, that is, on the step-by-step repetition of the same geometric operation, in principle until infinity. First, a square area corresponding to the study area is defined (Figure 9.12(a)). The first step of the iteration, which corresponds to defining a generator, is crucial. Indeed, the generator defines how the initial square is duplicated in the form of several reduced copies of itself, positioned in a certain way with respect to each other. In Figure 9.12(b), for example, the initial square is reduced by two reduction factors r_1 and r_0, thus defining the size ratio between a central zone corresponding to the main center (order 1) and a number $N = 4$ of sub-centers of order 2. At the next step, this operation is applied to each newly created gray square (sub-center). The process of creating the fractal figure can be continued down to the scale of urban blocks. For medium-sized cities, this corresponds to three or four steps of iteration. For a large metropolitan area such as Paris, six or seven steps of iteration are required. As the number of steps is limited, the geometric figure thus created is, in Mandelbrot's terminology, a "prefractal".

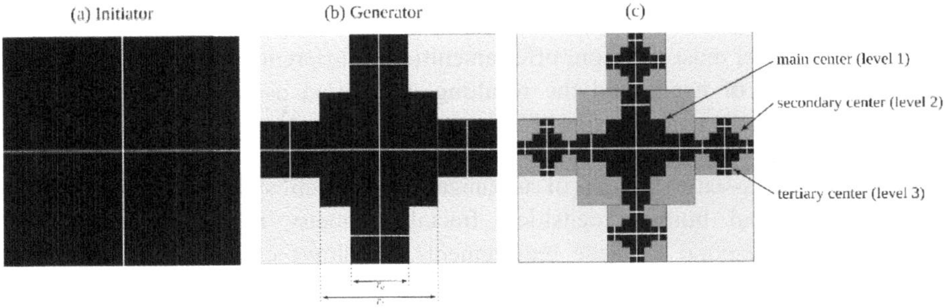

Figure 9.12. *Construction of a theoretical development plan by iterations based on a multifractal box fractal. In white, the main axes of the associated public transport system. Unbuilt (green) areas are created as the iterations progress*

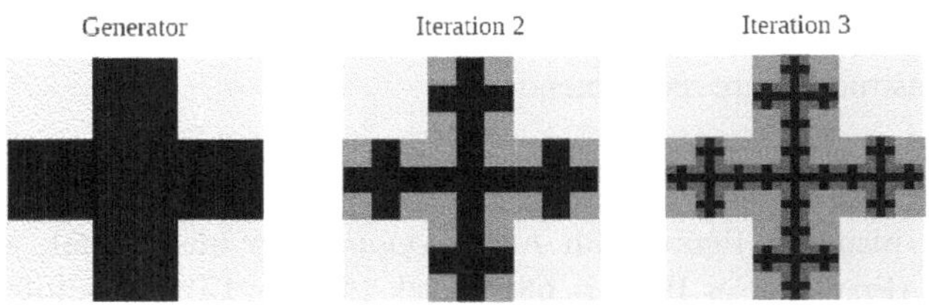

Figure 9.13. *First three steps of building a box fractal. The hierarchical structure of the gaps is shown in darker or less dark grays*

The box fractal shows an over-proportional elongation of its border (i.e. of its perimeter) in relation to its surface area as the steps of its creation progress. It is this property that allows the interpenetration of green and natural spaces (areas to be preserved) and built spaces (development zones). Indeed, the areas not occupied by the fractal (gaps) are of very different sizes. The relationship between their size and their number follows a hyperbolic law, that is, there are very few large gaps and an increasing number of small gaps (Figure 9.13).

9.3.2.2. *Properties of a transport network associated with the box fractal*

Other particular properties of the box fractal appear when considering the properties of the associated transport network. Figure 9.14(d) shows this network as well as three non-fractal branching networks (Figures 9.14(a)–(c)). The total length of the network is the same in all cases. Networks (a) and (b) each comprise two main axes around which are deployed secondary axes, all of the same length and fairly short. In doing so, residents' access to unbuilt spaces of different sizes is limited. Branches of the network (c) are of the same length and cover the space in a fairly uniform manner, leaving no very large empty spaces. In the case of fractal network (d), the boundary separating the space served by the network and the empty space is longer than in the case of the simple branching networks (a) and (b). Indeed, the fractal network shows much more local penetration of the empty space, and the contact zone between the network and the empty space is longer. These properties are related to the fact that fractal geometry allows generation of non-uniform figures, which are intermediate between lines and two-dimensional shapes in the way that they fill space.

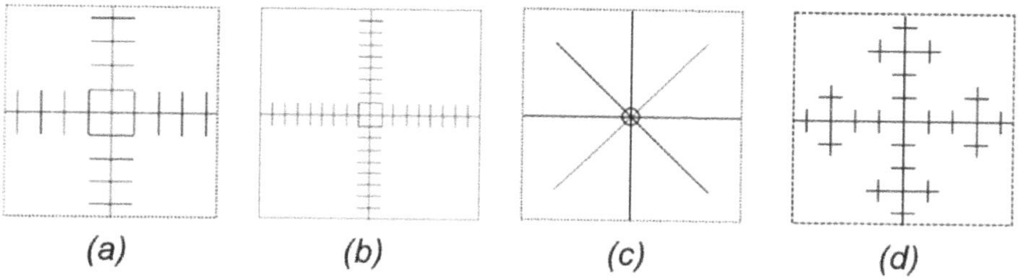

Figure 9.14. *Three non-fractal networks (a, b and c) and a prefractal network corresponding to a box fractal (d). Sources: (a), (b), (d): Frankhauser (2021); and (c): Frankhauser (2017)*

One of the arguments put forward to defend the dense compact city model is that distances would be reduced. Let us now consider two theoretical cities: a square compact city and a hierarchical axial city (box fractal). In both cases, the cities are surrounded by green areas. Starting from an initial figure representing an urban core, we imagine that the two figures grow in stages according to a general mechanism preserving the initial structure of each core (Figure 9.15). For the fractal city, each iteration enlarges the structure, leading to a branching shape (Figure 9.15(a)). The generator consists of $N = 9$ elementary squares of side L. The next step assembles nine figures identical to the generator following the position of the generator elements. The fractal city then comprises $N^2 = 81$ squares of side L. The compact city consists of an initial square urban core made up of $N = 9$ elementary squares identical to those of the fractal generator. At the next step, $N^2 = 81$ squares are arranged to form a new square of side $9L$. The number of elementary squares is therefore identical between the fractal city and the compact city; the cities have the same area but not the same shape. Each city has a main center and secondary centers at a lower hierarchical level.

For each city, it is possible to calculate the average distance to reach the center of the city as well as the average distance to leave the city. For this purpose, we use the Manhattan distance rather than the Euclidean distance in order to account for the fact that, in an urban space, travel takes place on the network of streets and axes of communication (Table 9.1).

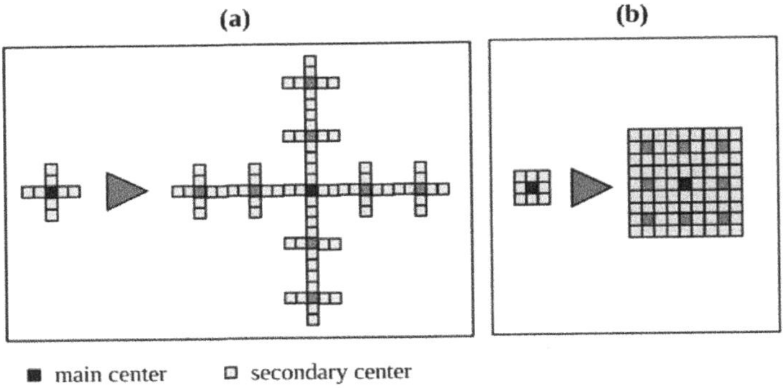

Figure 9.15. *Two steps to create a monofractal city model (box fractal) (a) and a square compact city model (b), both having an identical urbanized area*

	First step (generator)		Second step		
	Average distance to the outside of the city	Average distance to the center	Average distance to the outside of the city	Average distance to the main center	Average distance to the nearest secondary center
Theoretical monofractal city	~0.26 L	~1.61 L	~0.27 L	~7.66 L	~1.61 L
Theoretical compact city	~0.5 L	~1.5 L	~1.91 L	4.5 L	~1.5 L
Ratio of distances of compact city to fractal city	~1.9	~0.93	~7.07	0.58	~0.93

Table 9.1. *Distances to the outside of the city and to centers of different hierarchical levels for the two theoretical cities shown in Figure 9.15*

We therefore find, in the second step of iteration, that the average distance to the outside of the city is about seven times longer for the compact city than for the fractal city, while the average distance to the main center is only lengthened by a factor of 1.72 for the fractal city over the compact city. The distances to the nearest secondary center are comparable for the two models.

Let us now calculate a distance budget d for both models by introducing a number of trips and a distance to reach the nearest secondary center (respectively f_0 and $\overline{d_0}$), a number of trips and a distance to reach the main center (respectively f_1 and $\overline{d_1}$), and a number of trips and a distance to leave the city (respectively f_b and $\overline{d_b}$). The distance budget d is defined as follows:

$$f_0\overline{d_0} + f_1\overline{d_1} + f_b\overline{d_b} = d \quad [9.1]$$

Figure 9.16 shows the number of possible trips to the nearest secondary center, the main center and outside the city given the distances calculated for each city (Table 9.1). All of the points on the gray and black planes (triangles) represent numbers of trips which, multiplied by the associated distances, correspond to a weekly travel budget of 100 km. The compact city (black triangle) allows about 50 trips per week outside the city, provided that travel is limited to this reason alone, while the fractal city allows 370. The number of possible trips each week to the main center, if the trips are limited to this reason alone, is 13 for the fractal city and 22 for the compact city, respectively. The number of possible trips to the nearest secondary center, if travel is limited to that reason alone, is equivalent for both city models.

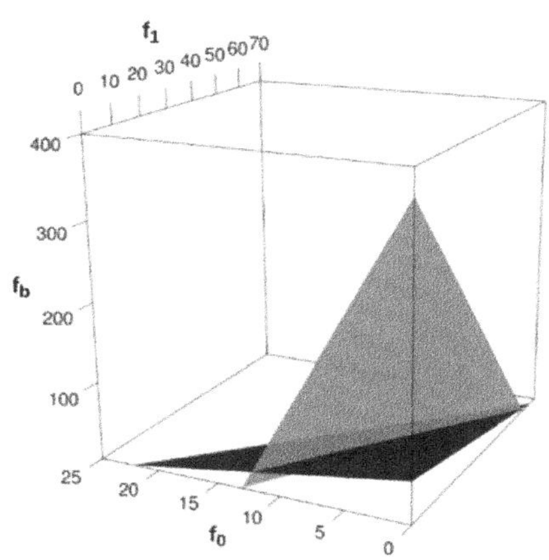

Figure 9.16. *Number of possible trips to the nearest secondary center (axis f_0), towards the main center (axis f_1) and outside the city (axis f_b) for the two city models shown in Figure 9.15, conditional on a cumulative distance of 100 km per week. The gray triangle represents possible trips within the fractal city and the black triangle represents possible trips within the compact city*

In the light of these results, we may consider that the fractal city presents a form of optimality in how it fills space and that it offers certain advantages compared to more traditional city forms. This does not mean that each fractal figure meets all of the criteria (here, for example, good access to the main center, the nearest secondary center and outside the city) as there is an infinity of fractal figures which are distinguished from each other by their topological properties even though they may have the same fractal dimension (see Chapter 3). When designing development plans with *Fractalopolis*, it is therefore a question of choosing the fractal figure that best meets the objectives.

9.3.3. *Hierarchy of central places linked by green and blue infrastructure*

In *Fractalopolis*, at each step of creating a fractal development plan, more details are added to the geometry of the development zones (centers and sub-centers in black in Figure 9.12). In doing so, new and increasingly smaller preserved zones are created at the same time, corresponding to "gaps" in the reference fractal figure (gray zones in Figure 9.12). Thus, at each step of design of the plan, the quantity of preserved areas increases while the extent of development zones is increasingly reduced. Such a principle corresponds to a "concentration–deconcentration" logic of urban development. Different types of preserved zones are distinguished based on the step of iteration at which they were created. In these zones, small-scale urban development may be permitted. It may be a little more intense in the preserved zones closest to the development zones (dark gray areas in Figure 9.12).

The preserved zones form a connected system of green infrastructure that not only allows good accessibility to green and natural spaces for residents, but also avoids their fragmentation and preserves large and small green spaces. This is consistent with the recommendations of Borgström et al. (2006), which emphasize that "ecosystem management used for the protection, maintenance and restoration of ecosystem services requires not only planning at multiple scales, but also crossing and linking between hierarchical scales in order to recognize ecosystem dynamics". This system of green infrastructure ensures not only ecological functions but also good ventilation of urbanized areas, thus reducing the risk of urban heat islands. It should be noted that the model is also clearly distinguished, at local level, from the tendency to smooth out urban borders, due to the interpenetration of built-up areas and the open countryside that it foresees. This recalls Sieverts' (1997) reflections on the *Zwischenstadt* or those of Dubois-Taine and Chalas (1997) on the "emerging city".

The centers and sub-centers at different levels form a hierarchical system of central places (Figure 9.17). As in Christaller's (1933) model, the assumption in *Fractalopolis* is that ordinary stores, services and amenities are present in all centers, but centers at higher hierarchical levels also offer amenities that are rarely frequented. Only the main center has rare stores, services and amenities, such as high-end stores and specific cultural offerings. The distance to be traveled by residents to access the different types of stores, services and amenities is thus consistent with their frequency of use. Such a principle may also be found in Calthorpe (1993). Some authors emphasize that consumers do not necessarily choose to visit the amenities closest to their place of residence (Clark 1968). Overdevelopment of the road network favors such "commercial tourism". Hence, the public transport network that we propose in Figure 9.17 has no circular branches to avoid competition between centers at the same level – we argue instead for internal competition within each center and do not favor bypasses for local use. Good commercial infrastructure to meet daily needs will also help to maintain a certain economic dynamism at the local level.

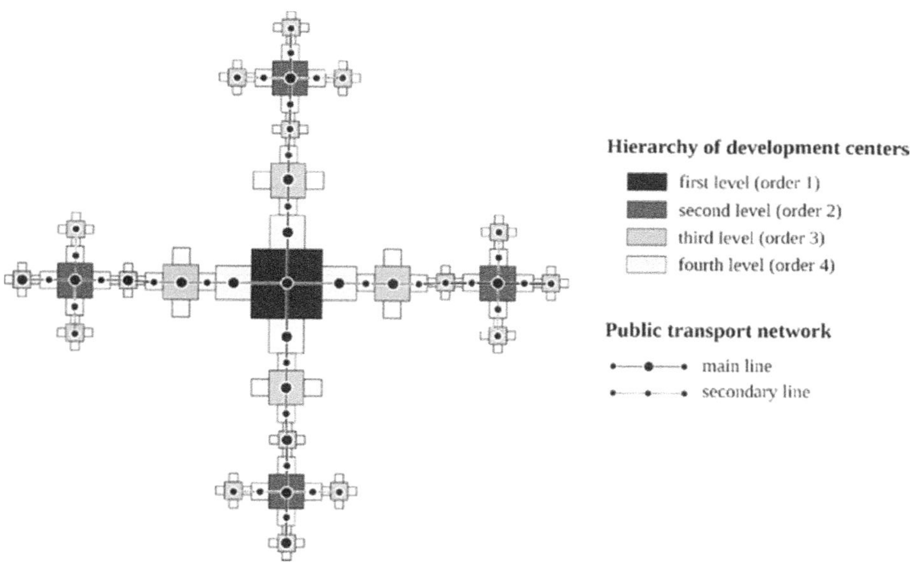

Figure 9.17. *Hierarchy of central places and main transport axes in a theoretical development plan based on a multifractal box fractal*

Contrary to Christaller's (1933) central place theory, centers are not uniformly distributed in space; they are located on nodes of a hierarchical system, which are generally desired to coincide with the public transport system. Centers at the same hierarchical level do not necessarily have the same surface area. They are larger if they are located near a major center and smaller if they are farther away from it (Figure 9.17).

	Principles of Fractalopolis		
Goals of planning	*Functional urban hierarchy*	*Contiguous axial development*	*Connected and hierarchical green infrastructure*
Avoiding urban sprawl		×	
Avoiding fragmentation of green and natural spaces (biodiversity)		×	×
Reducing distances to access stores and services	×	×	
Assuring good access to green and leisure areas	×		×
Promoting the use of public transport and soft modes	×	×	
Promoting short supply chains, (sub)urban agriculture			×
Promoting the infiltration of rainwater			×
Avoiding heat islands			×
Integrating planning levels (block, district, city, metropolitan area)	×		×

Table 9.2. *The three principles underlying the design of a development plan with* Fractalopolis, *responding to the goals of sustainable development*

Finally, the three principles underlying the design of *Fractalopolis*, namely the introduction of a functional hierarchy of central places, the concentration of development zones along public transport axes and the preservation of green infrastructure, make it possible to meet the requirements for sustainable development of metropolitan areas as summarized in Table 9.2.

9.3.4. *Formalization of accessibility*

We adopt the definition of accessibility suggested by Hansen (1959). Accessibility from a given place is defined as the total of the destinations of interest accessible from this place, each amenity being negatively weighted by its ease of access measured in the form of a distance, a travel time or a generalized transport cost. Planning and transport strategies are often evaluated in terms of accessibility (Geurs and van Wee 2004).

The great advantage of this approach to accessibility is that it does not focus solely on home-to-work journeys; although certainly these largely determine road and public transport flows for the morning rush hour, they do not reveal much about the functioning and potential of an area, its attractiveness and the reasons that lead people and businesses to settle there. Accessibility represents a potential rather than a demand for mobility observed through the spatial practices of households. Evaluation of accessibility also corresponds quite well to the arguments of real estate developers, who in their communications praise proximity to nature, schools, stores and public transport.

Working on accessibility requires answers to three questions:

– which choice of destinations to consider;

– which standards for acceptable distances and travel times;

– how to calibrate the cost functions for access to destinations.

9.3.4.1. *From needs to satisfiers*

In his book *Human Scale Development*, Max-Neef (1991) clearly identifies three concepts that underpin our approach to choosing destinations for accessibility calculations. The first concept is the universal one of need, which is shared by all human beings. For Max-Neef, the basic needs are as follows: *subsistence, protection, affection, understanding, participation, idleness, creativity, identity* and *freedom*. Each universal need can then be more precisely subdivided. Empirical studies have shown the merits of a needs-based approach (Ryff and Keyes 1995; Tay and Diener 2011).

The second concept is that of *satisfiers*. This concept is very powerful for Max-Neef, who distinguishes *satisfiers*, *pseudo-satisfiers*, *inhibiting satisfiers*, *singular satisfiers* and *synergetic satisfiers*. These satisfiers depend on the cultural context, and for some require access to goods or services located in the territory, but are not to be confused with the goods themselves. Thus, food is not seen as a need; it is a satisfier of the basic need of *subsistence*. Another example is education, which corresponds to the need for *understanding*. The health care system is a satisfier of the need for *protection*. Max-Neef emphasizes that satisfiers are "forms of organization, political structures, social practices, subjective conditions, values and norms, spaces, contexts, modes, types of behavior and attitudes". For him, goods are the means by which individuals will empower the satisfiers to meet their needs.

The third concept is that of physical goods and services, which must be accessible to play the role of satisfiers. It is usually this dimension that is addressed in accessibility studies, without explicitly referring to needs and satisfiers. In this chapter, we look at satisfiers and consider their importance, location and accessibility. This will then allow us to define a degree of satisfaction among a place's residents, considering the global cultural context and local specificities (Figure 9.18).

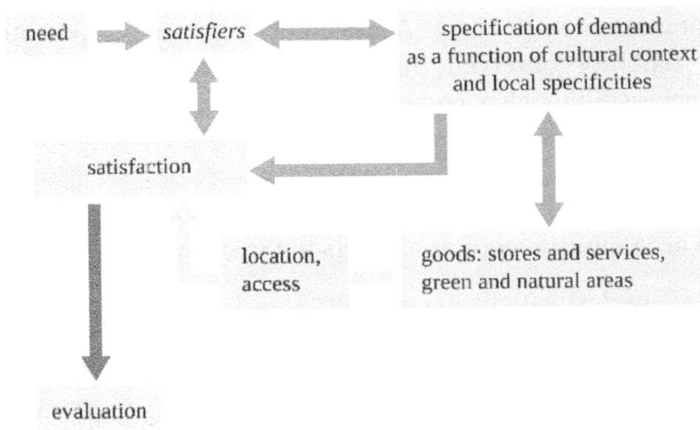

Figure 9.18. *The transition from needs to goods that are the materializations of satisfiers*

9.3.4.2. *Identification of destinations (stores, services, green and natural spaces) and acceptable distance thresholds*

Based on Max-Neef's (1991) theory of needs, it is possible to select, for a given territory in a given cultural context, a list of stores and services whose proximity ensures the well-being of the inhabitants (whether they actually frequent them or not). The facilities taken into account are "physical places that involve travel outside the home, which are open to the public, and in which it is possible to access services" (A'urba, urban planning agency of Bordeaux Aquitaine 2017). This definition is deliberately inclusive in order to refer to all the physical sites which the population may visit, whatever the type of activity (commercial, health, school, cultural, etc.). Services which only involve travel by the suppliers (electricians, plumbers, etc.) have been excluded. Here, we consider the European, or even French, cultural context, but the types of facilities considered may also be found in other contexts.

Level	Potential frequency of use	Type of facility	Associated needs	Acceptable distance (in meters)	*	Sources
1	Rare	Department stores	Subsistence	Not considered	70%	8
		Retail sale of leather goods and travel goods	Subsistence		20%	8
		Retail sale of timepieces and jewelry in specialized stores	Subsistence, idleness		20%	8
		Cinema	Idleness understanding		74%	1
		Hospital activity	Protection		54%	3
2	Occasional	Retail sale of frozen products	Subsistence	3,000–6,000	48%	1
		Hypermarkets	Subsistence		97%	1
		Retail sale of fruits and vegetables in specialized stores	Subsistence		63%	1
		Retail sale of hardware, paints and glassware in big-box stores (400 m^2 and more)	Subsistence		77%	1
		Distance selling from general catalog (drive-through)	Subsistence		17%	1
		Retail sale of rugs, carpets and wall and floor coverings in specialized stores	Subsistence		77%	1
3	Frequent	Convenience stores	Subsistence	800–1,600	74%	1
		Supermarkets	Subsistence		97%	1
		Retail sale of fuel in specialized stores	Subsistence		85%	1
		Retail sale of hardware, paints and glassware in small stores (less than 400 m^2)	Subsistence		77%	1
		Retail sale of clothing in specialized stores	Subsistence		88%	1
		Retail sale of pharmaceutical products in specialized stores	Protection		93%	1
		Post office	Protection		90%	1
		Traditional dining	Affection, idleness		92%	1
		Fast food	Subsistence, affection, idleness		92%	1
		Bank, savings bank	Protection		97%	1
		General practitioners	Protection		80%	1

4	Proximity	Butchery-charcuterie	Subsistence	300–600	56%	1
		Bakery and pastry	Subsistence		94%	1
		General food sales	Subsistence		74%	1
		Retail sale of newspapers and stationery in specialized stores	Subsistence, idleness		80%	1
		Retail sale of meat and meat products in specialized stores	Subsistence		57%	1
		Kindergarten	Understanding, participation		15%	1
		Primary school	Understanding, participation		56%	1

*Percentage of residents using the facilities.
Sources: 1: A'urba, agence d'urbanisme Bordeaux Aquitaine (2017); 3: Agence Technique de l'Information sur l'Hospitalisation (2019), key figures on hospitalization; 8: Estimates.

Table 9.3. *Stores and services taken into account for evaluating the accessibility of centers of development in* Fractalopolis

In order to identify the facilities associated with different satisfiers, we relied on scientific knowledge in the field and also on surveys conducted in France, such as household travel surveys or the INSEE (National Institute of Statistics and Economic Studies) use of time survey, and many French planning and development documents. More specifically, the Territorial Coherence Schemes (SCoT) provide information on the characteristics and effects of stores and services according to the frequency of their use and their levels of function in the urban framework. Indeed, the challenge for the public authorities is to try to include an offer well suited to their needs, both to meet public demand, and also in a logic of complementarity and balance of provision over the territory. For this they use planning tools such as commercial planning documents (DAC) or orientation and objectives documents (DOO) that recommend a set of principles and strategies for commercial development, adapted to the characteristics of the territory of application. In addition to the information provided by the urban design documents, we took into account other classification proposals such as the INSEE 2019 permanent database of facilities, the consumption and lifestyle surveys of CREDOC (Research Center for the Study and Observation of Living Conditions) and the exploratory approach to the hierarchy of facilities in Île-de-France by the Institut Paris Région (Albe-Tersiguel et al. 2008). This review, carried out by Xavier Lehman (research at the ThéMA laboratory, Besançon, and LVMT, Champs-sur-Marne, France), chose a set of businesses and services as shown in Table 9.3, which are associated with acceptable distances to access them.

Like stores and services, green, natural and leisure spaces have been the subject of numerous studies dealing with the hierarchy between them and acceptable distances to access them. Accessibility standards for green spaces depend on their attractiveness, which in planning practice is often measured in terms of surface area (hectares or square meters) (Ståhle 2010). Several studies show that the catchment area of a park is proportional to its size (Liu et al. 2010; Feng et al. 2019). Surface area is therefore the indicator most used by urban designers to build hierarchical typologies of green spaces. For example, the *Accessible Natural Greenspace Standards* (ANGSt) propose a hierarchy of satisfactory distances to access green spaces in the UK based on their surface area (Handley et al. 2006):

– a 2 ha site less than 300 m from the home;

– a 20 ha site less than 2 km;

– a 100 ha site less than 5 km;

– a 500 ha site less than 10 km.

The attractiveness of a green space can also be associated with other factors such as types of use and the diversity of activities available to visitors (Ballion and Grandjean 1982; Bussey 1996; Sugiyama et al. 2010). A classification that combines usage and surface area has been implemented in the *National Open Space Guidelines* (Boverkets publikationsservice 1999, Gröna områden I planeringen) recommended by the Swedish government.

The urban green spaces usually considered are public squares and gardens, as well as parks and wooded and natural areas. However, it is possible to adopt a broader definition that includes private or semi-private places that may accommodate a smaller public. These include family gardens and shared gardens to which demand for proximity is high in dense urban environments, although they are often considered inaccessible by residents (A'urba, Bordeaux Aquitaine 2017 urban planning agency). On this basis, we have identified different types of green amenities as shown in Table 9.4.

In Tables 9.3 and 9.4, we have distinguished four hierarchical levels of facilities. These levels were defined accounting for various criteria such as frequency of use, the demand for proximity and the constraints of location in the urban setting.

Level 4 includes facilities with the highest demand for proximity. Their frequency of use is often daily. The trips made by users are not necessarily long, but numerous and frequent. Level 3 facilities are also associated with questions of proximity, but involve facilities with a wider catchment area. In the case of health services, ease of access must be guaranteed despite a potentially lower frequency of use. Level 2 includes facilities whose frequency of use is more occasional, or which are frequently used but the size of the relevant infrastructure does not allow local implantation. Finally, level 1 includes facilities with the rarest frequency and the largest catchment area. As they are generally used quite rarely by individuals, their attendance may generate long-distance travel.

Level	Potential frequency of use	Type of green amenity	Associated needs	Acceptable distance (in m)	*	Sources
1	Rare	Wooded area	Idleness, affection	0–20,000	65%	4
		Castle park	Idleness, affection		35%	8
		Inclusion garden	Idleness, affection		10%	2
		Botanical garden	Idleness, affection, understanding		35%	8
		Regional leisure center	Idleness, affection		22%	5
2	Occasional	Therapeutic garden	Idleness, affection, protection	1,500–3,000	10%	2
		Natural area of educational character	Idleness, affection, understanding		65%	4
		Leisure center	Idleness, affection, understanding, protection		22%	5
		Large urban park	Idleness, affection,		86%	4
		Shared garden	Idleness, affection, understanding		10%	2

3	Frequent	Unidentified garden (e.g. vegetable garden)	Idleness	500–1,000	10%	2
		Family garden	Idleness, affection		10%	2
		City park	Idleness		86%	4
		Other garden	Idleness		10%	2
		Water park	Idleness		86%	4
		Sports park	Idleness, protection		35%	8
4	Of proximity	Playground	Idleness, participation	150–300	35%	8
		Square and public garden (small)	Idleness, participation		54%	4
*Percentage of residents using the facilities. Sources: 2: A'urba, agence d'urbanisme Bordeaux Aquitaine (2020); 4: Boutefeu (2011); 5: Institut Régional de Développement du Sport (2013); 8: Estimates.						

Table 9.4. *Green and leisure facilities taken into account for evaluating accessibility of development centers in* Fractalopolis

9.3.4.3. *Synthetic measures of accessibility*

In *Fractalopolis*, we introduced different types of indicators providing information on household satisfaction. They synthesize the quality of access of each development center (or zone) to stores and services, on the one hand, and to green and leisure amenities, on the other hand. They are calculated at each stage of creation of a development plan. Formalization of the indicators is based on the theory of fuzzy sets and is inspired by the rules set out in (Tannier et al. 2012). Here, we present the indicators included in the latest version of the model, *Fractalopolis 1.2*, developed within the framework of the SOFT research project (Bonin et al. 2020; Frankhauser 2021). These indicators are simpler than those included in the previous version of the model (*Fractalopolis 1.1*, Frankhauser et al. 2018)[6].

6. *Fractalopolis 1.2* actually includes both versions of the calculations for accessibility indicators. By default, the indicators included in *Fractalopolis 1.1* and described in (Frankhauser et al. 2018) are offered to the user of *Fractalopolis 1.2*, through the *Macroscale/Assess the suitability of each planned center* menu. The user can choose to calculate the indicators described in this chapter by checking a box labeled "V2".

Evaluation of accessibility of stores and services or green and natural spaces of a given frequency of use

In *Fractalopolis*, each development zone (i.e. each center) contains buildings that themselves contain dwellings. Facilities are located in more or less close proximity to each building, inside or outside the development zone under consideration. These facilities consist of stores and services (each represented by a point) or green and leisure spaces (each represented by their entry points, i.e. the intersections of their perimeter with the network of roads and paths).

Let us first consider a type of facility a, closest to a given building i. This facility is located at a distance $r_{a,i}$ from the building i. It may be located in or outside the zone to which i belongs. Distances are calculated as the crow flies. The value of the distance $r_{a,i}$ between facility a and building i is denoted $m_a(r_{a,i})$. Its value lies between 0 if the distance $r_{a,i}$ is totally unacceptable because it is much too far, and 1 if $r_{a,i}$ is perfectly acceptable. Two thresholds of acceptable distance, r_{crit} and r_{max}, characterize the different types of facilities considered (see Tables 9.3 and 9.4). When the distance $r_{a,i}$ is greater than an initial acceptability threshold denoted $r_{a(crit)}$, the value of $m_a(r_{a,i})$ decreases linearly until $m_a(r_{a,i}) = 0$ is reached when $r_{a,i}$ lies beyond a second acceptable distance threshold $r_{a(max)}$ (Figure 9.19).

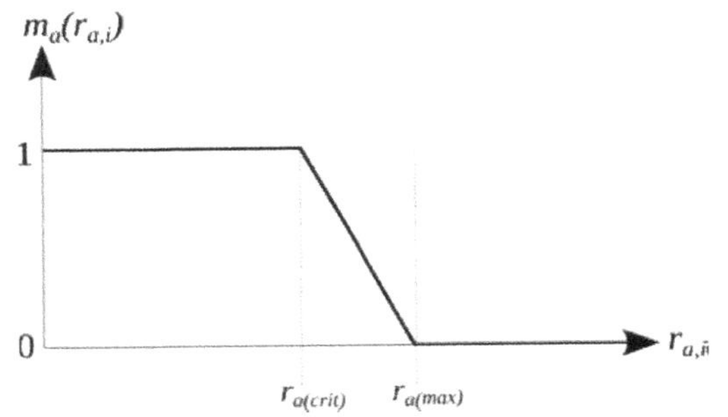

Figure 9.19. *Evaluation $m_a(r_{a,i})$ of the distance $r_{a,i}$ between a facility of type a and a building* i

The evaluation S_a of the overall accessibility of a development zone considering all buildings i with regard to a given type of facility a is calculated as follows:

$$S_a = \frac{1}{N}\sum_i n_i m_a(r_{a,i}) \qquad [9.2]$$

where N represents the total number of dwellings in the area and n_i represents the number of dwellings contained in each building i.

Synthetic evaluation of accessibility of all types of facility present in a development zone

Residential satisfaction of individuals is greater if they have good accessibility to several types of facilities (Nelson 1959; Brown 1989; Hotelling 1990; Jones and Simmons 1995). As indicated in Tables 9.3 and 9.4, we distinguish eight categories of facility $A^{c,l}$ according to the level of facility (levels 1 to 4) and whether the facility is a store or a service (category s) or a green and leisure amenity (category g). For example, local stores and services (level 4) fall under category $A^{s,4}$. The accessibility of all facilities of category $A^{c,i}$ in a development zone is calculated as follows:

$$S^{A_{c,l}} = \frac{1}{\sum_a w_a} \sum_{a \in A_{c,l}} w_a S_a \qquad [9.3]$$

where w_a represents the proportion of the population that uses type a facilities (see Tables 9.3 and 9.4). The term w_a represents, in a way, the importance of facilities of type a to the population in the sense that facilities frequented by a large part of the population provide more satisfaction, at the scale of the zone, than facilities that are only rarely used. The value $w_a = 1$ corresponds to the case where the type of facility under consideration is used by the whole population.

The different categories of facility $A^{c,l}$ are each of different importance for individuals, particularly to the extent that they use it more or less often. This differentiated importance, denoted $f_{c,l}$ ($f_{s,1}, f_{s,2}, f_{s,3}, f_{s,4}, f_{g,1}, f_{g,2}, f_{g,3}, f_{g,4}$), is included in the calculation of the synthetic accessibility S^{tot} in a development zone of all categories of facility:

$$S^{tot} = \frac{1}{F}\left(\sum_{s,l} S^{A_{s,l}} f_{s,l} + \sum_{g,l} S^{A_{g,l}} f_{g,l}\right) \qquad [9.4]$$

where F is the number of categories $A^{c,l}$ of facility and $\sum_1^F f_{c,l} = 1$.

9.4. Example of application: from Greater Paris to the *Est Ensemble* sector

We present here an example of the application of *Fractalopolis* to the area of the territorial authority of *Est Ensemble* (Bonin et al. 2020; Frankhauser 2021), carried out as part of the SOFT project funded by ADEME. *Est Ensemble* is an inter-municipal structure created in 2016. It includes the municipalities of Bagnolet, Bobigny, Bondy, Le Pré-Saint-Gervais, Les Lilas, Montreuil, Noisy-le-Sec, Pantin and Romainville, and has a total population of 408,000 inhabitants. It is a very high-density area with 10,877 inhabitants/km^2. *Est Ensemble* is indeed part of the inner ring of

Paris where the average population density amounts to 8,234 inhabitants/km^2. The density of the Île-de-France region is approximately 1,000 inhabitants/km^2.

The data used for this application come from the BD Topo 2018 (French National Geographic Institute) for buildings and from the 2018 population census files of INSEE (French Institute of Statistics and Economic Studies) for dwellings. The sources of data on facilities have been presented in section 9.3.4.2.

9.4.1. *Multifractal development planning*

9.4.1.1. *Design of a multifractal development plan for Greater Paris*

We consider the *Est Ensemble* sector, not in an isolated way, but as integrated into the metropolitan area of Paris (Greater Paris). This allows us to reflect on the planning of the development of the study area in a broad context, which aims to rebalance the population within the Paris metropolitan area through the development of sub-centers located on its periphery. The area of Greater Paris, as defined for this project (Figure 9.20), incorporates five peripheral cities of historical importance that do not necessarily play as significant a role as they could (Figure 9.21): *Mantes la Jolie* (01), *Rambouillet* (02), *Montereau-Fault-Yonne* (04), *Coulommiers* (05) and *Meaux* (03).

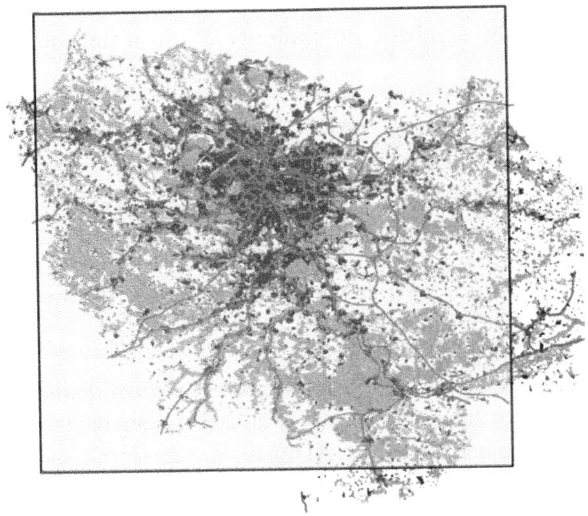

Figure 9.20. *The study area of Greater Paris. The red lines represent the RER network and the Île-de-France train network. For a color version of this figure, see www.iste.co.uk/tannier/fractal.zip*

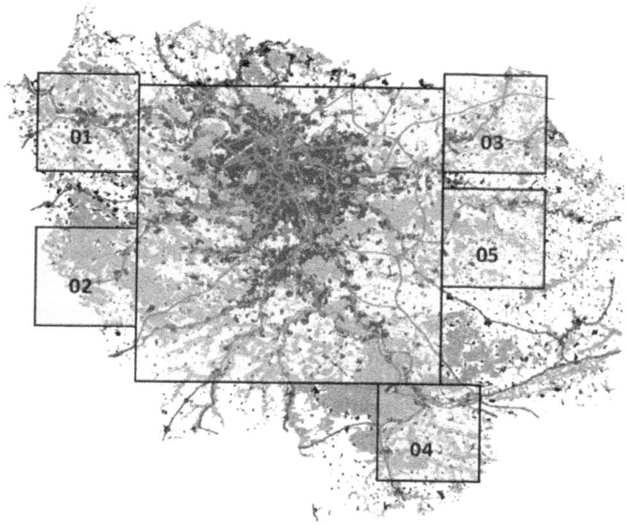

Figure 9.21. *Generator of the multifractal development plan for Greater Paris. The five highest ranked peripheral centers in the urban hierarchy are numbered from 01 to 05. For a color version of this figure, see www.iste.co.uk/tannier/fractal.zip*

Figure 9.21 shows the generator designed in *Fractalopolis*, which will be applied iteratively, step by step, to identify both sub-centers (development zones) and preserved zones at different hierarchical levels. This generator has the following characteristics. The five peripheral cities (sub-centers at the first hierarchical level) are connected to the capital by public transport axes, represented in red in Figure 9.21. The large existing forest areas, which must be preserved, can be part of development zones because these latter are very large in the early stages of designing a plan with *Fractalopolis*. Then, as we progress through the steps of creating the plan, the development zones will become smaller and smaller while additional preserved zones, of smaller and smaller size, will be created. Small quantities of housing construction may be permitted in preserved zones. In addition, it is possible to plan for reducing the density of certain development zones (via the deconstruction of buildings) with the aim of integrating them into green or blue infrastructure in the medium or long term.

At the second and third steps of creating the development plan of Greater Paris, the generator presented in Figure 9.21 is applied iteratively in each development zone (i.e. the large square and the five smaller squares in Figure 9.21). The new development zones (new centers) created by application of the generator in each of the initial squares must, in some cases, be moved to correspond to the location of the existing urban and suburban centers. This displacement from their position as initially defined by the generator is performed manually in *Fractalopolis*. Nevertheless, the computer application helps to maintain a certain number of rules governing their movement:

– In order to maintain the statistically self-similar character of the multifractal plan, the squares representing each zone of development may not overlap. Nor may they overlap the preserved zones created in the previous steps of iteration. This ensures both the preservation of interconnected preserved zones (green and blue infrastructure) and the penetration of unbuilt spaces into built spaces at different scales.

– Good access to facilities (stores and services, green and leisure amenities) must be ensured. To do this, the accessibility of each center is represented using a color code in *Fractalopolis* (see section 9.4.2). Each time a square is moved, all of its accessibility indicators are automatically recalculated, allowing a choice of location of the squares that ensures the best possible access to the different types of facility.

– In the early steps of creating a plan, since the squares representing the development zones are very large, they can overlay preserved areas. Then, as the iterations progress, the squares become smaller and it is possible to position them in such a way as to preserve green alleys or corridors, as well as other protected zones.

Figure 9.22. *Third step of iteration of the multifractal plan proposed for Greater Paris. The red square delimits the Paris Center sector. For a color version of this figure, see www.iste.co.uk/tannier/fractal.zip*

The third step in creating the development plan for Greater Paris is shown in Figure 9.22. Development zones are concentrated along the axes of public transport. Preserved zones (green and natural spaces) penetrate into the heart of built spaces.

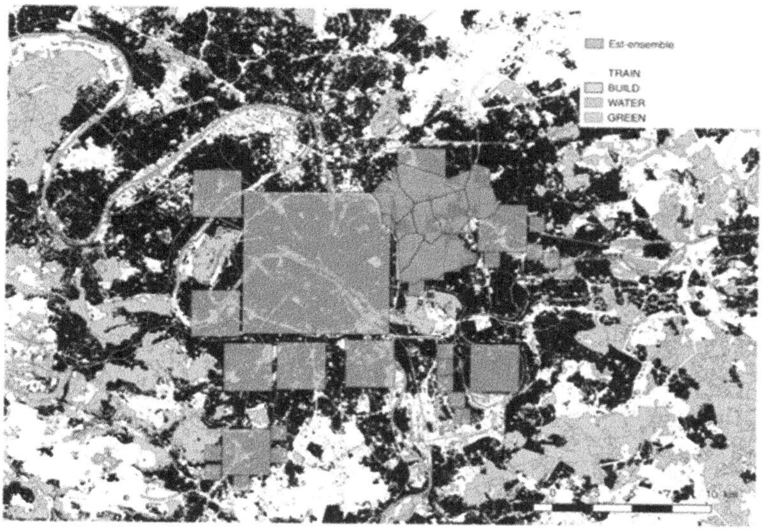

Figure 9.23. *Second step of creating the multifractal plan of the Paris Center sector obtained by applying the generator shown in Figure 9.21. The* Est Ensemble *sector is represented by a solid orange color. For a color version of this figure, see www.iste.co.uk/tannier/fractal.zip*

9.4.1.2. *Design of a multifractal development plan for the Paris Center sector*

Starting from the third step of the design of the development plan for Greater Paris, we created a *Fractalopolis* project specific to the *Paris Center* sector, shown with a red outline in Figure 9.22. The *Paris Center* sector contains the city of Paris as well as the part of the inner ring of suburbs bordering Paris whose operation is strongly conditioned by its significant interactions with Paris. The *Est Ensemble* sector is part of this. The development plan created with *Fractalopolis* for *Paris Center* uses the same generator as that defined for Greater Paris, namely, a center of greater importance and five sub-centers of secondary importance with a certain size ratio between them (Figure 9.23). In this way, the proposed development plan is spatially coherent from the global scale of Greater Paris to the very local one of the *Est Ensemble* sector.

Figure 9.24. *Third step in the creation of the multifractal plan of Paris Center: a zoom-in on the* Est Ensemble *sector represented by a solid orange color. The pink squares represent the planned development zones. The red lines represent the Greater Paris public transport network project*[7]. *For a color version of this figure, see www.iste.co.uk/tannier/fractal.zip*

The multifractal planning implemented at this local scale favors the conservation and establishment of green and blue infrastructure more than access to public transport. Indeed, in this over-densified space, the presence of ventilation corridors to reduce the risk of heat islands and the presence of recreational green zones seemed to us to be a priority. We have thus developed a plan that respects the territory's green aeration zones, but is not fully in line with the public transport network project at the intra-urban scale of the area under consideration (Figure 9.24).

Figure 9.25 shows another view of the third step of creating the multifractal plan of *Paris Center*. The preserved zones are represented in darker green where they were created during the first or second design step of the plan. In the preserved

7. Source: APUR (Parisian urban design workshop). See: https://opendata.apur.org. Data downloaded in 2019.

zones in darker green, the degree of concentration of housing allowed is generally lower than in preserved areas near the lighter-colored development zones. We identified three existing examples of green and blue infrastructure, to be preserved or developed through de-densification of the preserved areas.

Figure 9.25. *Third stage of design of the Paris Center multifractal plan: a zoom-in on the* Est Ensemble *sector. Development zones are shown in blue-purple. Preserved zones are represented in darker green the further they are from the development zones. The proposed green and blue infrastructure is shown in fluorescent green. For a color version of this figure, see www.iste.co.uk/tannier/fractal.zip*

9.4.2. *Assessment of potential residential satisfaction*

The *Fractalopolis* software calculates potential residential satisfaction for each development zone, considering the accessibility of different types of facilities (stores and services, green and leisure amenities) according to the logic described in section 9.3.4. Accessibility indicators are calculated for different frequencies of use of facilities (proximity, frequent, occasional and rare). They may be visualized through cartographic representations. A color code indicates the quality of accessibility in each development zone. The darker the color, the better the accessibility.

Figure 9.26 shows the accessibility values of the development zones of the *Est Ensemble* sector (Bobigny, Villemomble and Montreuil zones numbered, respectively, (1), (2) and (3) in the figure) at the first step of creating the *Paris Center* development plan for level 4 facilities (proximity use). We find, on this scale, that the accessibility of stores and services is generally quite good, but that accessibility of green and leisure spaces is poor. This trend is the same for all frequencies of use, except for level 1 (rare use) relating to green spaces (Table 9.5). Recall here that accessibility of level 1 facilities only concerns the presence of facilities in the metropolitan area, not their distances to places of residence.

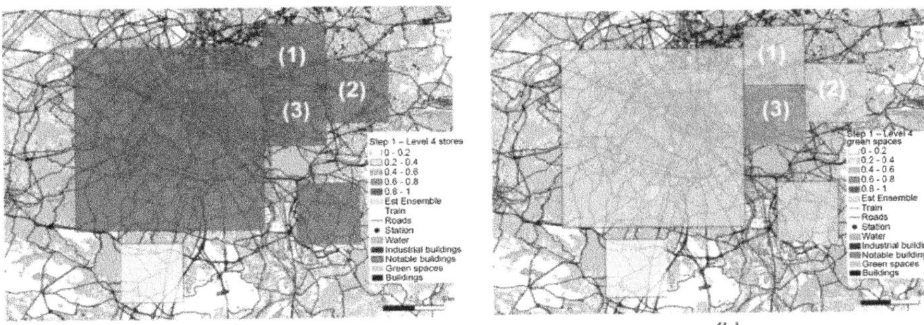

Figure 9.26. *First step in the design of the Paris Center development plan: assessment of accessibility to stores and services (a) and green spaces (b) at level 4 (proximity frequency of use). (1) Bobigny, (2) Villemomble and (3) Montreuil. For a color version of this figure, see www.iste.co.uk/tannier/fractal.zip*

	Stores and services				Green and leisure areas				Overall satisfaction
Level	4	3	2	1	4	3	2	1	
Bobigny	0.73	0.97	0.99	0.88	0.30	0.20	0.43	0.78	0.66
Villemomble	0.64	0.93	1.00	0.92	0.26	0.30	0.11	0.80	0.62
Montreuil	0.79	0.99	1.00	0.94	0.42	0.27	0.51	0.83	0.72

Table 9.5. *Evaluation of accessibility of the development zones of the* Est Ensemble *sector at the first design step of the Paris Center development plan*

At the third design step of the *Paris Center* plan, evaluation of the accessibility of the development zones of the *Est Ensemble* sector is much more nuanced. In general, accessibility is not excellent for local stores and services (Figure 9.27(a)), except in a few dark blue areas. Accessibility is much better for services and shops of weekly use (level 3), for which acceptable access distances are greater. This also applies to services and shops of monthly use (level 2) (Figure 9.27(b)).

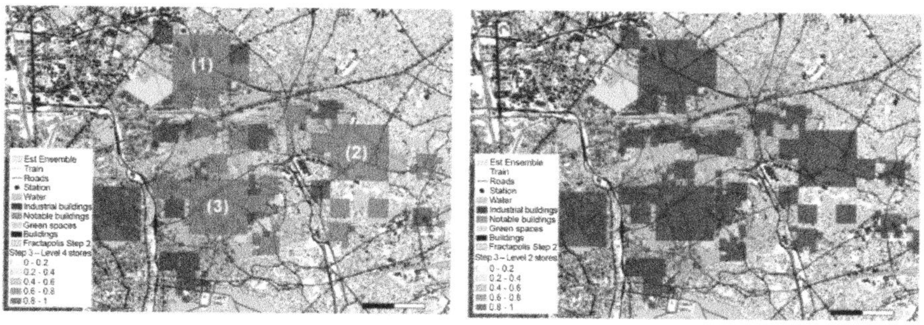

(a) (b)

Figure 9.27. *Evaluation of accessibility to stores and services in the development zones of the* Est Ensemble *sector at the third design step of the Paris Center development plan: (a) level 4 (proximity use), (b) level 2 (occasional use). (1) Bobigny-center, (2) Villemomble-center and (3) Montreuil-center. For a color version of this figure, see www.iste.co.uk/tannier/fractal.zip*

For services and shops of rare use (level 1), the situation is more differentiated (Table 9.6). There is indeed a lack of local stores compared to establishments such as supermarkets or shopping centers. Regarding green spaces, accessibility is generally poor for local facilities (Figure 9.28(a)), which can be explained by the over-densification of this sector of the Paris metropolitan area. The situation is even worse when it comes to green spaces at level 3 (frequent use) (Table 9.6). Regarding level 2 facilities (occasional use, Figure 9.28(b)), the situation is much better in the Montreuil area due to the presence of the Bois de Vincennes.

Thus, studying the accessibility of development zones at different scales, that is, for different steps of designing a development plan with *Fractalopolis*, allows a nuanced reading. Accessibilities may be acceptable on an aggregate scale but not on a fine scale and vice versa. In addition, on a fine scale, large disparities between areas may appear.

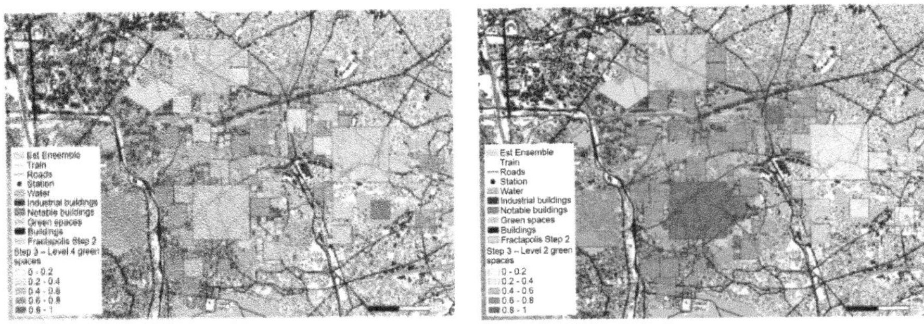

(a) (b)

Figure 9.28. *Evaluation of the accessibility of green and leisure spaces in the development areas of the* Est Ensemble *sector at the third design step of the Paris Center development plan: (a) level 4 (proximity use) and (b) level 2 (occasional use). For a color version of this figure, see www.iste.co.uk/tannier/fractal.zip*

	Stores and services				Green and leisure areas				Synthesis
Level	4	3	2	1	4	3	2	1	
Bobigny-center	0.7204	0.9813	0.9995	0.8657	0.2517	0.2194	0.3937	0.7750	0.6516
Villemomble-center	0.7120	0.9775	1.0000	0.9365	0.2389	0.2034	0.0706	0.8184	0.6196
Montreuil-center	0.7724	0.9958	1.0000	0.9450	0.3377	0.2530	0.6831	0.8180	0.7256

Table 9.6. *Evaluation of accessibility of the development zones of the* Est Ensemble *sector at the third design step of the Paris Center development plan*

9.4.3. Scenarios for spatial distribution of dwellings

We begin by considering the number of dwellings present in each development zone and each preserved zone of the development plans for Greater Paris and *Paris Center*[8]. These dwelling numbers are obtained from data from the 2018 census of INSEE (French Institute of Statistics and Economic Studies). On this basis, the

8. The total number of dwellings in 2018 was 5,679,788 in Greater Paris and 3,384,039 in the *Paris Center* area.

proportion of dwellings present in the development and preserved zones at each hierarchical level is calculated. At the first step of creating the Greater Paris plan, it is noted that 97.31% of dwellings are located in a development zone (parameter $\alpha \times 100$ of Table 9.7, column "Initial situation"). The remaining dwellings are located in a preserved zone. Of the dwellings located in a development zone at the first stage of plan creation, 93.31% ($\beta \times 100$ in Table 9.7) are still located in a development zone at the second stage of plan creation. The remaining dwellings are located in the preserved zones created at this second stage. In the end, at the third stage of creation of the development plans of Greater Paris and *Paris Center*, respectively, the share of dwellings located in the development zones is $\alpha \times \beta \times \gamma \times 100$.

The same logic is applied to calculate the proportion of dwellings present in the development zones at each hierarchical level. At the first step of creation of the Greater Paris plan, 95.50% of the dwellings located in a development zone are located in the main center (parameter a_1, Table 9.8), the remaining percentage of dwellings being distributed among the five peripheral sub-centers (a_0) (see Figure 9.21 for the location of the main center and the five sub-centers). By definition, $a_1 + a_0 = 1$. At the second step, 90.20% of the dwellings in a development zone created in the previous step are located in a higher-level development zone (parameter b_1, Table 9.8), with the remaining percentage of dwellings distributed among the five lower-level zones (b_0). Following this logic, at the third step of creation of the *Paris Center* development plan, the share of dwellings in each development zone at the second hierarchical level, including Bobigny-center, Villemomble-center and Montreuil-center (see Figure 9.27 for their location), is $a_0 \times b_1 \times c_1$.

			Share of dwellings located in development zones	
	Step of plan creation		Initial situation	Scenario 2
Greater Paris	1	α	0.9731	0.950
	2	β	0.9331	0.933
	3	γ	0.8212	0.830
Paris Center	1	α	0.8181	0.830
	2	β	0.7325	0.750
	3	γ	0.5817	0.650

Table 9.7. *Concentration of dwellings in development zones created at the first, second and third steps of the development plans, respectively*

	Step of plan creation		Initial situation	Scenario 2
			Share of dwellings located in development zones at a higher hierarchical level	
Greater Paris	1	a_1	0.9550	0.9500
	2	b_1	0.9020	0.9100
	3	c_1	0.8940	0.9000
Paris Center	1	a_1	0.8267	0.8300
	2	b_1	0.7737	0.8000
	3	c_1	0.6048	0.6500

Table 9.8. *Concentration of dwellings in development zones at a higher hierarchical level created at the first, second and third steps of the development plans, respectively*

We now propose two alternative scenarios for the distribution of dwellings in the multifractal development plans of Greater Paris and *Paris Center*. These two scenarios are based on a logic of "deconcentrated concentration" of housing unit construction. The idea is to spatially redistribute existing and future dwellings between the denser development zones and the less dense preserved zones. The scenarios include changing the number of dwellings within development zones, with reference to the urban functions they contain, and changing the number of dwellings within preserved zones according to their remoteness from development zones.

To create these two scenarios, we used the *Fractalopolis* spatial distribution model, which is described in detail in the paper by Frankhauser et al. (2018). This model allows us to set new values for the housing unit concentration parameters α, β, γ and a_1, b_1, c_1, a_0, b_0, c_0, in accordance with the hierarchical organizational logic of a given multifractal plan. In *Fractalopolis*, the values of the housing unit concentration parameters can vary from one design step to another, which is why the spatial distribution model of dwellings does not fit into a strict fractal logic. If it is desired to model a spatial distribution of dwellings that is relatively monocentric, the values of the parameters must meet the following conditions:

$$b_1 > c_1 \wedge a_1 > b_1 \wedge a_1 > c_1 \quad [9.5]$$

On the contrary, if we desire to model a strictly hierarchical spatial distribution of dwellings, the values of the parameters must comply with the following conditions:

$$a_1>a_0 \wedge b_1 > a_1 \wedge b_1>b_0 \wedge c_1 > b_1 \qquad [9.6]$$

The first scenario created with *Fractalopolis* (scenario 1) aims to obtain a spatially balanced distribution of dwellings in the development and preserved zones at each hierarchical level, without modifying the overall distribution of existing dwellings in 2018 described in Tables 9.7 and 9.8. The underlying assumption of scenario 1 is that a spatially balanced distribution of dwellings corresponds to the configuration in which each development zone and each preserved zone has a number of dwellings equal to the average number of dwellings in the development areas or preserved areas at the same hierarchical level. For example, the total number of dwellings in the five peripheral development zones (second hierarchical level) in Figure 9.21 is 248,940. The number of dwellings in each zone, according to scenario 1, should be 248,940 divided by five, or 49,788. Reading Table 9.9, we can see that there is a certain disparity between the five zones. The Mantes-la-Jolie area appears to be the most oversupplied with housing compared to the average (objective of scenario 1) while the Rambouillet and Coulommiers areas are undersupplied.

Code*	Peripheral development zones	Number of dwellings in 2018	Average number of dwellings in each area (scenario 1)	Difference between the initial number and the average number of dwellings predicted by scenario 1
01	Mantes-la Jolie	81,338	49,788	+31,584
02	Rambouillet	24,225		-25,563
03	Meaux	50,842		+ 1,054
04	Montereau-Fault-Yonne	64,150		+14,362
05	Coulommiers	28,386		-21,402

*According to Figure 9.21.

Table 9.9. *Number of dwellings in 2018 and the number targeted by scenario 1 in each peripheral development zone identified by the generator of the Greater Paris development plan shown in Figure 9.21*

Figure 9.29. *Surpluses and deficits of dwellings compared to the average number calculated for each hierarchical level (scenario 1) at the third step of creation of the multifractal development plan for Greater Paris. The brighter the color, the greater the surplus. For a color version of this figure, see www.iste.co.uk/tannier/fractal.zip*

At the third step of creation of the development plan for Greater Paris, we note an overall deficit of housing in the peripheral zones compared to the numbers forecast in scenario 1 (dark colored squares in Figure 9.29).

Scenario 2 envisages a change in the overall spatial distribution of dwellings, keeping the same number of dwellings as in the initial situation and scenario 1. The choice of parameter values for housing unit concentration relative to the parameters estimated from the initial situation is shown in Tables 9.7 and 9.8. At the scale of the Greater Paris region, scenario 2 envisages a slight reduction in the share of housing in the development zones created during the first two steps of creation of the plan (reduction of the values of the parameters α, β), which implies a slight densification of the preserved zones furthest from Paris compared to the situation in 2018. However, in preserved zones located near development zones, the degree of housing unit concentration (parameter γ) is increased in order to free up more local green spaces. Concerning the distribution of dwellings between development zones (i.e. the centers) at different hierarchical levels, the slight decrease in the value of parameter a_l compared to the initial situation implies a loss (deconstruction) of 132,993 dwellings for the *Paris Center* sector. However, the small increase in the values of the parameters b_l and c_l produces a slight strengthening of the secondary centers. At the scale of *Paris Center*, the values chosen for α, β and γ increase the concentration of housing units in the development zones and reduce it in the preserved zones (de-densification) in order to open up the green and blue infrastructure. At the same time, the net increase in the values of the parameters a_l, b_l and c_l leads to a strengthening of the urban hierarchy by increasing the density of dwellings in the largest centers and by decreasing that of the smallest centers.

Table 9.10 shows the changes in the number and density of dwellings projected in scenario 2 for the three main development and preserved zones of the *Est Ensemble* sector at the second step of creation of the *Paris Center* plan. Only the Montreuil development zone is impacted by deconstruction of dwellings, while constructions of dwellings are envisaged in the Bobigny and Villemomble development zones. Thousands of housing deconstructions are planned in the preserved zones located near these two areas, while massive deconstructions are planned in the preserved zones located near Montreuil.

	Development zones			Preserved zones		
	Change in the number of dwellings	Initial density*	Density target*	Change in the number of dwellings	Initial density*	Density target*
Bobigny	+24,630	48	86	-8,775	42	30
Villemomble	+26,574	45	87	-5,430	36	29
Montreuil	-2,352	70	41	-40,086	76	28

*Measured in dwellings/ha of buildable area.

Table 9.10. *Evolution of the number and density of dwellings forecast by scenario 2 in the zones of the Paris Center development plan defined at the second step of creation of the plan and located in the* Est Ensemble *sector*

It is possible to imagine many other housing distribution scenarios based on the proposed development plans for Greater Paris and *Paris Center*, notably by varying the total number of housing units present in the study area. This is likely to increase from 5,679,788 in 2018 to about 7,327,000 by 2050 based on INSEE population growth forecasts (Omphale 2017 model). In order to avoid a significant increase in the number of dwellings in the already over-dense central areas, one option could be to increase the concentration of dwellings in peripheral areas, where large wooded areas are located, and to increase the share of dwellings allocated to centers of secondary importance. One scenario tested with *Fractalopolis* by applying these principles shows that the factor of increase in the number of dwellings would be 1.289 in the central zone (*Paris Center* sector). However, on the periphery, for example, at Meaux (zone 03 in Figure 9.21), the factor would be 2.349.

9.5. Conclusion

Territorial planning consists of trying to influence the trajectory of development of territories through regulation, incentives or urban projects driven by public authorities, in a direction that seems useful for the needs of both individuals and the community, respecting the environment and for the good of future generations. A very striking feature of cities, regardless of their size, history and geographical location, is that they present physical infrastructures (buildings, roads, networks) that are of great spatial extension, relatively enduring, and of complex form in terms of filling space.

Addressed through spatiality, that is, the shape of cities, the question of their sustainability and their possible evolution presupposes, on the one hand, proposing a conceptual or theoretical model of their current organization and desirable changes in these forms, and on the other hand, having tools to measure, evaluate and build scenarios. As there is no privileged or natural scale, from that of the individual, of the building, to that of the world, passing through neighborhoods and blocks, districts, boroughs, cities, conurbations, metropolitan areas, etc., it is desirable that these concepts and tools can be used at different scales or be perfectly multi-scalar (or even cross-scalar).

The model proposed in this chapter, *Fractalopolis*, as well as the associated software, make it possible to apply in a coherent way, at several scales, a concept of development aimed at synthesizing different approaches observed in the history of urban planning, from the Garden City to TOD, passing through the finger-plan city and deconcentrated concentration. The basic principle of *Fractalopolis* is to create a functional hierarchy of centers (or development zones) compatible with a public transport network and making available, at each scale, a diversity of densities and built forms. A development plan designed with *Fractalopolis* must allow the penetration of green and blue infrastructure into the center of the metropolitan area at different scales. When designing a development plan, it is possible to consider the whole of a metropolitan area and also to choose within it a particular area, for example, a city or a district, which will be studied up to the fine scale of the urban block, while being consistent with the plan drawn up at the aggregate scale of the metropolitan area.

Fractal geometry has original characteristics in the way it fills space, in its similarity between scales and in its ability to resist description by the usual geometric primitives. Hence, it is able to provide geometric figures that are able to support the concept of the *Fractalopolis* model, notably through the figure of the box fractal and its different variations. It is possible to adapt the geometry of the figure of the box fractal to many real territories and to use the components of this figure to evaluate, on the one hand, a theoretical model of housing unit distribution that reflects a use of space in accordance with the intensity targeted by the model, and on the other hand, accessibility to different services and amenities for different spatial scales and different frequencies of use. Analysis of the differences between this theoretical model and reality then makes it possible to suggest planning initiatives aimed at strengthening green and blue infrastructure, shortening lengths of trips, and ensuring overall a correct level of satisfaction in the resident population.

Although the *Fractalopolis* software is operational, its implementation remains for the moment within the research domain. It requires good knowledge of the underlying planning concept, and the design of a plan often goes through several attempts before achieving a satisfactory proposal. However, there are many avenues that could facilitate or even automate its implementation. In particular, analyses of curves with scaling behaviors of radial fractal analyses of the distribution of buildings would make it possible to highlight the fractal dimensions that dominate in different places, for example, the characteristic dimensions of centers, suburbs and peripheral areas. It would be possible to use this information to specify the form of the generator of a development plan. In addition, morphological analyses of urban fabrics could help to determine a priori the number of sub-centers to consider, thus further constraining the fractal form of the generator, the basis for creation of a development plan. Finally, a lot of information associated with buildings, such as the numbers of residents, jobs, stores and services, could be analyzed by multivariate multifractal analysis tools in order to further help design a generator adapted to each study area.

Whether cities are really fractal or not, fractal geometry as used in *Fractalopolis* makes it possible to reconcile objectives that are a priori incompatible, and to address, in a coherent theoretical framework, the question of forms, distribution of dwellings, accessibility to stores and services, all while respecting green and blue infrastructure. Rather than serving as a principle of absolute optimization, fractals can be used as tools to formally integrate different planning principles into a multi-scale operational framework, thus offering a form of multi-objective optimization to meet planners' specific expectations.

9.6. References

Albe-Tersiguel, S., Blum, E., de Berny Riche, C., Delaporte, C., Mangeney, C., Parnaix, A., Peuvergne, C. (2008). Démarche exploratoire pour une hiérarchisation des équipements en Île-de-France. Institut Paris Région.

Antoni, J.-P., Bonin, O., Frankhauser, P., Houot, H., Nicolas, J.-P., Tomasoni, L., Toilier, F., Vuidel, G. (2014). VILMODes, Ville et mobilité durables, évaluation par la simulation. Report, PREDIT, ThéMA, LVMT and LET.

A'urba (Agence d'urbanisme Bordeaux Aquitaine) (2017). L'accès aux commerces et équipements de la vie quotidienne : résultats d'une enquête quantitative sur les usages et attentes des habitants dans l'aire métropolitaine de Bordeaux. Report.

A'urba (Agence d'urbanisme Bordeaux Aquitaine) (2020). Enquête sur les usages des espaces de nature : analyse et interprétation des résultats. Report.

Ballion, R. and Grandjean, A. (1982). Aire d'attraction des espaces verts publics urbains de la région d'Île-de-France. Report, Étude pour l'agence des espaces verts de la région d'Île-de-France.

Bandemer, H. (2005). *Mathematics of Uncertainty: Ideas, Methods, Application Problems*. Springer, London.

Beatley, T. (2000). *Green Urbanism: Learning from European Cities*. Island Press, Washington, DC.

Beaucire, F., Rosales-Montano, S., Duflos, E., Turchetti, I. (1999). Les outils de planification urbaine au service de la relation urbanisme/transport : approche dans la perspective du développement durable. Research summary, project DRAST/ PREDIT 98MT115. Fédération Nationale des Agences d'Urbanisme, Paris.

Berry, B.J.L. and Garrison, W.L. (1958). The functional bases of the central place hierarchy. *Economic Geography*, 34, 145–154.

Bolitzer, B. and Netusil, R. (2000). The impact of open spaces on property values in Portland, Oregon. *Journal of Environmental Management*, 59, 185–193.

Bonin, O., Bonneau, P., Clerc, M., Cousin, J., Frankhauser, P., de Gouvello, B., Haffner, M., Lehmann, X., Pioli, R., Poirel, M. et al. (2020). Projet SOFT : sobriété énergétique par les formes urbaines et le transport. Research report, ThéMA laboratory, Institut pour la Transition Energétique, Efficacity, LVMT.

Borgström, S.T., Elmqvist, T., Angelstam, P., Alfsen-Norodom, C. (2006). Scale mismatches in management of urban landscapes. *Ecology and Society*, 11(2), 16 [Online]. Available at: http://www.ecologyandsociety.org/vol11/iss2/art16/.

Botkin, D.B. and Beveridge, C.E. (1997). Cities as environments. *Urban Ecosystems*, 1, 3–19.

Boutefeu, E. (2011). Les pratiques citadines de la nature en ville. In *Villes et sciences sociales*, MRSH. Centre d'étude sur les réseaux et les transports urbains (CERTU), Caen.

Braubach, M. (2007). Residential conditions and their impact on residential environment satisfaction and health: Results of the WHO large analysis and review of European housing and health status (LARES) study. *International Journal of Environment and Pollution*, 30, 384–403.

Brown, S. (1989). Retail location theory: The legacy of Harold Hotelling. *Journal of Retailing*, 65, 450–470.

Brun, J. and Fagnani, J. (1994). Lifestyles and locational choices – Trade-offs and compromises: A case-study of middle-class couples living in the Île-de-France Region. *Urban Studies*, 31, 921–934.

Bryant, M.M. (2006). Urban landscape conservation and the role of ecological greenways at local and metropolitan scales. *Landscape and Urban Planning*, 76, 23–44.

Bussey, S.C. (1996). Public uses, preferences and perceptions of urban woodlands in Redditch. PhD Thesis, University of Central England, Birmingham.

Calthorpe, P. (1993). *The Next American Metropolis: Ecology, Community, and the American Dream*. Princeton Architectural Press, Princeton.

Christaller, W. (1980). *Die zentralen Orte in Süddeutschland. Eine ökonomisch-geographische Untersuchung über die Gesetzmäßigkeit der Verbreitung und Entwicklung der Siedlungen mit städtischer Funktion*. Wissenschaftliche Buchgesellschaft, Darmstadt 1980.

Clark, W.A.V. (1968). Consumer travel patterns and the concept of range. *Annals of the Association of American Geographers*, 58, 386–396.

Collinge, S.K. (1996). Ecological consequences of habitat fragmentation: Implications for landscape architecture and planning. *Landscape and Urban Planning*, 36, 59–77.

Czerkauer-Yamu, C. (2012). Strategic planning for developing sustainable metropolitan areas with a multiscale decision support system. The Vienna case. PhD Thesis, Université de Franche-Comté, Besançon.

Dettmar, J. and Ganser, K. (1999). *IndustrieNatur, Ökologie und Gartenkunst im Emscher Park*. Verlag Eugen Ulmer, Stuttgart.

Dubois-Taine, G. and Chalas, Y. (1997). *La ville émergente*. Éditions de l'Aube, La Tour d'Aigues.

Eberstadt, R., Möhring, B., Petersen, R. (1910). *Gross-Berlin. Ein Programm für die Planung der neuzeitlichen Grosstadt*. Wasmuth, Berlin.

Ellaway, A., Macintyre, S., Kearns, A. (2001). Perceptions of place and health in socially contrasting neighbourhoods. *Urban Studies*, 38, 2299–2316.

Evans, G.W. (2003). The built environment and mental health. *Journal of Urban Health*, 80, 536–555.

Fábos, J.G. (2004). Greenway planning in the United States: Its origins and recent case studies. *Landscape and Urban Planning*, 68, 321–342.

Feng, S., Chen, L., Sun, R., Feng, Z., Li, J., Khan, M.S., Jing, Y. (2019). The distribution and accessibility of urban parks in Beijing, China: Implications of social equity. *International Journal of Environmental Research and Public Health*, 16, 4894.

Fittkau, D. (2004). Beeinflussung regionaler Kaufkraftströme durch den Autobahnlückenschluß der A 49 Kassel-Gießen zur empirischen Relevanz der "New Economic Geography" in wirtschaftsgeographischen Fragestellungen. Georg-August-University, Göttingen.

Forman, R. (1995). *Land Mosaics, the Ecology of Landscapes and Regions*. Cambridge University Press, Cambridge.

Fouchier, V. (1995). La densification : une comparaison internationale entre politiques contrastées. *Les annales de la recherche urbaine*, 67, 94–101.

Francis, M. and Hester, R.T. (eds) (1995). *The Meaning of Gardens: Idea, Place, and Action*. MIT Press, Cambridge, MA.

Frankhauser, P. (1988). Fractal aspects of urban systems. *Sonderforschungsbereich 230 "Natürliche Konstruktionen"*, 1, 67–76.

Frankhauser, P. (2008). Fractal geometry for measuring and modeling urban patterns. In *The Dynamics of Complex Urban Systems – An Interdisciplinary Approach*, Albeverio, S., Andrey, D., Giordano, P., Vancheri, A. (eds). Springer Physica Verlag, Heidelberg.

Frankhauser, P. (2012). The *Fractalopolis* model – A sustainable approach for a central place system [Online]. Available at: https://hal.archives-ouvertes.fr/hal-00758864.

Frankhauser, P. (2015). From fractal urban pattern analysis to fractal urban planning concepts. In *Computational Approaches for Urban Environments, Geotechnologies and the Environment*, Helbich, M., Jokar Arsanjani, J., Leitner, M. (eds). Springer International Publishing, Cham.

Frankhauser, P. (2017). Between self-organization and planning: Cities and the fractal order principle. In *The Virtual and the Real in Planning and Urban Design*, Yamu, C., Poplin, A., Devisch, O., de Roo, G. (eds). Routledge Abingdon.

Frankhauser, P. (2021). *Fractalopolis* – A fractal concept for the sustainable development of metropolitean areas. In *Complex Systems, Smart Territories and Mobility*, Patricia, S. and Cyrille, B. (eds). Springer International Publishing, Cham.

Frankhauser, P., Tannier, C., Vuidel, G., Houot, H. (2018). An integrated multifractal modelling to urban and regional planning. *Computers, Environment and Urban Systems*, 67, 132–146.

Frey, H. (1999). *Designing the City: Towards a More Sustainable Urban Form*. Taylor & Francis, Abingdon.

Fujita, M. and Ogawa, H. (1982). Multiple equilibria and structural transition of non-monocentric urban configurations. *Regional Science and Urban Economics*, 12, 161–196.

Gérard, M., Hauser, M., Belaghene, Y. (2021). Les Français et les sorties culturelles post-crise. Report, Harris interactive pour le ministère de la culture.

Geurs, K.T. and van Wee, B. (2004). Accessibility evaluation of land-use and transport strategies: Review and research directions. *Journal of Transport Geography*, 12, 127–140.

Güldner, H. (1968). *Unsere Stadt Tragödie einer Spätkultur*. Pinneberg, Beig.

Guo, J. and Bhat, C. (2002). Residential location choice modeling: Accommodating sociodemographic, school quality and accessibility effects. Research paper, University of Texas, Austin.

von Haaren, C. and Reich, M. (2006). The German way to greenways and habitat networks. *Landscape and Urban Planning*, 76, 7–22.

Hall, P. and Tewdwr-Jones, M. (2019). *Urban and Regional Planning*, 6th edition. Routledge, Abingdon.

Handley, J., Pauleit, S., Slinn, P., Ling, C., Lindley, S. (2006). *Accessible Natural Green Space Standards in Towns and Cities: A Review and Toolkit for their Implementation*. Report, School of planning and landscape, University of Manchester [Online]. Available at: https://www.blaenau-gwent.gov.uk/media/epkbgstp/sd106.pdf [Accessed October 2022].

Hansen, W.G. (1959). How accessibility shapes land use. *Journal of the American Institute of Planners*, 25, 73–76.

Hedges, B. and Clemens, S. (1994). Housing attitudes survey. Social and community planning research (Great Britain). Survey, Department of the Environment, H.M.S.O., London.

Hobden, D.W., Laughton, G.E., Morgan, K.E. (2004). Green space borders. A tangible benefit? Evidence from four neighbourhoods in Surrey, British Columbia. *Land Use Policy*, 21, 129–138.

Hotelling, H. (1990). Stability in competition. In *The Collected Economics Articles of Harold Hotelling*. Springer, New York.

Howard, E. (1898). *To-Morrow: A Peaceful Path to Real Reform*. Swan Sonnenschein & Co, London.

Institut Régional de Développement du Sport (IRDS) (2013). Notoriété et fréquentation des bases de loisirs en Île-de-France. Report, 23.

Jensen, M.B., Persson, B., Guldager, S., Reeh, U., Nilsson, K. (2000). Green structure and sustainability – Developing a tool for local planning. *Landscape and Urban Planning*, 52, 117–133.

Jones, K. and Simmons, J. (1995). *The Retail Environment*. Routledge, London and New York.

Jongman, R.H., Külvik, M., Kristiansen, I. (2004). European ecological networks and greenways. *Landscape and Urban Planning*, 68, 305–319.

Kaplan, S. (1995). The restorative benefits of nature: Toward an integrative framework. *Journal of Environmental Psychology*, 15, 169–182.

Kestens, Y., Thériault, M., Des Rosiers, F. (2004). The impact of surrounding land use and vegetation on single-family house prices. *Environment and Planning B: Planning and Design*, 31, 539–567.

Krugmann, P.R. (1995). *Development, Geography and Economic Theory*. MIT Press, Cambridge, MA.

Krummel, J.R., Gardner, R.H., Sugihara, G., O'Neill, R.V., Coleman, P.R. (1987). Landscape patterns in a disturbed environment. *Oikos*, 48, 321–324.

Kuttler, W. (2011). Climate change in urban areas. Part 2, Measures. *Environmental Sciences Europe*, 23, 21.

Kuttler, W., Dütemeyer, D., Barlag, A.-B. (1998). Influence of regional and local winds on urban ventilation in Cologne, Germany. *Meteorologische Zeitschrift*, 77–87.

Lamb, R.J. and Purcell, A.T. (1990). Perception of naturalness in landscape and its relationship to vegetation structure. *Landscape and Urban Planning*, 19(4), 333–352.

Latkin, C.A. and Curry, A.D. (2003). Stressful neighborhoods and depression: A prospective study of the impact of neighborhood disorder. *Journal of Health and Social Behavior*, 44, 34–44.

Lerman, S.R. (1976). Location, housing, automobile ownership, and mode to work: A joint choice model. *Transportation Research Record*, 610, 5–11.

Linehan, J., Gross, M., Finn, J. (1995). Greenway planning: Developing a landscape ecological network approach. *Landscape and Urban Planning*, 33, 179–193.

Liu, C.F., Li, X.M., Han, D. (2010). Accessibility analysis of urban parks: Methods and key issues. *Acta Ecologica Sinica*, 30, 5381–5390.

Luttik, J. (2000). The value of trees, water and open space as reflected by house prices in the Netherlands. *Landscape and Urban Planning*, 48, 161–167.

Mader, H.J. (1979). *Die Isolationswirkung von Verkehrsstrassen auf Tierpopulationen untersucht am Beispiel von Arthropoden und Kleinsaeugern der Waldbiozoenose*. Bundesforschungsanstalt für Naturschutz und Landschaftsökologie Landwirtschaftsverlag. Bonn-Bad Godesberg. Münster-Hiltrup.

Mader, H.J. (1981). Untersuchungen zum Einfluß der Flächengröße von Inselbiotopen auf deren Funktion als Trittstein oder Refugium. *Natur und Landschaft*, 56, 235–242.

Malaure, J. (2020). Goût du chlore : la piscine, une passion française. *Le Point* [Online]. Available at: https://www.lepoint.fr/sport/gout-du-chlore-la-piscine-une-passion-francaise-21-06-2020-2381073_26.php#11.

Maslow, A.H. (1954). *Motivation and Personality*. Harper & Row, New York.

Max-Neef, M.A. (1991). *Human Scale Development: Conception, Application and Further Reflections*. The Apex Press, New York and London.

Michalakis, M. and Nicolas, G. (1986). Le cadavre exquis de la centralité : l'adieu à l'hexagone régulier. *Ératosthène*, 1, 38–87.

Nelson, R.L. (1959). *The Selection of Retail Locations*. F.W. Dodge Corp, New York.

Neuman, M. (2005). The compact city fallacy. *Journal of Planning Education and Research*, 25, 11–26.

Nussbaum, M. (2012). *Capabilités – Comment créer les conditions d'un monde plus juste ?* Éditions Climats, Paris.

Portschy, S. (2016). Community participation in sustainable urban growth, case study of Almere, the Netherlands. *Pollack Periodica*, 11, 145–155.

Rieniets, T. (2009). Shrinking cities: Causes and effects of urban population losses in the 20 Century. *Nature and Culture*, 4, 231–254.

Ryff, C.D., and Keyes, C.L.M. (1995). The structure of psychological well-being revisited. *Journal of Personality and Social Psychology*, 69, 719–727.

Sachsen, T., Ketzler, G., Knörchen, A., Schneider, C. (2013). Past and future evolution of nighttime urban cooling by suburban cold air drainage in Aachen. *Die Erde – Journal of the Geographical Society of Berlin*, 144, 274–289.

Schöfl, G. (1986). *Minimalnetze*. ARCUS, Alaska.

Schwanen, T., Dijst, M., Dieleman, F.M. (2004). Policies for urban form and their impact on travel: The Netherlands experience. *Urban Studies*, 41, 579–603.

Sen, A. (2010). *L'idée de justice*. Le Seuil, Paris.

Sieverts, T. (1997). *Zwischenstadt – zwischen Ort und Welt, Raum und Zeit, Stadt und Land*. Vieweg+Teubner Verlag, Wiesbaden.

Spronken-Smith, R.A. and Oke, T.R. (1999). Scale Modelling of nocturnal cooling in urban parks. *Boundary-Layer Meteorology*, 93, 287–312.

Ståhle, A. (2010). More green space in a denser city: Critical relations between user experience and urban form. *Urban Design International*, 15, 47–67.

Sugiyama, T., Francis, J., Middleton, N.J., Owen, N., Giles-Corti, B. (2010). Associations between recreational walking and attractiveness, size, and proximity of neighborhood open spaces. *American Journal of Public Health*, 100, 1752–1757.

Syme, G.J., Fenton, D.M., Coakes, S. (2001). Lot size, garden satisfaction and local park and wetland visitation. *Landscape and Urban Planning*, 56, 161–170.

Tannier, C., Vuidel, G., Houot, H., Frankhauser, P. (2012). Spatial accessibility to amenities in fractal and nonfractal urban patterns. *Environment and Planning B*, 39, 801–819.

Tay, L. and Diener, E. (2011). Needs and subjective well-being around the world. *Journal of Personality and Social Psychology*, 101, 354–365.

Thomas, I., Tannier, C., Frankhauser, P. (2008). Is there a link between fractal dimensions and other indicators of the built-up environment at a regional level? *Cybergeo: European Journal of Geography*, 413 [Online]. Available at: https://www.researchgate.net/publication/41717494_Is_there_a_link_between_fractal_dimensions_and_other_indicators_of_the_built-up_environment_at_a_regional_level.

Turner, T. (2006). Greenway planning in Britain: Recent work and future plans. *Landscape and Urban Planning*, 76, 240–251.

Van Herzele, A., Wiedemann, T., Van Overmeiren, M. (2000). Environment and nature report for flanders MIRA-S 2000, chap. Stedelijk Milieu (Urban Environment). Report, Vlaamse Milieumaatschappij & Garant Uitgevers NV, Leuven/Apeldoorn.

Vicsek, T. (1989). *Fractal Growth Phenomena*. World Scientific Publishers, Singapore.

Whyte, W.H.J. (1961). Securing open space for urban America: Conservation easements. *Soil Science*, 92, 153.

Wilcox, B.A. and Murphy, D.D. (1985). Conservation strategy: The effects of fragmentation on extinction. *The American Naturalist*, 125(6), 879–887.

Yamu, C. and Frankhauser, P. (2015). Spatial accessibility to amenities, natural areas and urban green spaces: Using a multiscale, multifractal simulation model for managing urban sprawl. *Environment and Planning B: Planning and Design*, 42, 1054–1078.

Conclusion

Isabelle THOMAS
Centre for Operational Research and Econometrics (CORE), Université catholique de Louvain-la Neuve, Belgium

How can we conclude nine extremely pedagogical chapters devoted to fractal logics in geography, and more particularly to fractal tools and models applied to the study of the spatial distribution of human settlements and to the management of urban territories? The task is not an easy one, but one thing is certain: thanks to this book, our readers will no longer be confused when using fractal tools, especially those developed and made available as open source by the ThéMA laboratory in Besançon (France). The organization of the chapters is straightforward, the language is clear, and the examples and many illustrations are undeniable assets. This book represents a valuable guide for measuring the spatial organization of built urban realities, for interpreting and understanding the measures and fractal indicators, and for planning/modeling space across scales. This guide should help researchers who wish to perform informed and relevant fractal analyses, and planners who wish to meet current and especially future urban challenges. With this work, fractal geometry becomes an accessible starting point both for the measurement of the spatial organization of human settlements (buildings, intersections of roads) and for the extraction and characterization of urban borders.

As a reminder, this book, which is part of the ISTE SCIENCES series, is organized into three distinct parts: the first part clearly deals with the basic notions for understanding and performing fractal analyses of urbanized fabrics. The second part presents the principles of four major families of methods of analysis (point patterns, morphological delimitation, multifractal analyses and cross-scale signatures). Theoretical and empirical fractal shapes form the basis for illustrating

Fractal Geometry in Human Geography and Planning,
coordinated by Cécile TANNIER.

the applicability of the measures. Finally, the third part shows the usefulness of two fractal geographic models of urban and regional planning: *MUP-city* and *Fractalopolis*. The first model allows the simulation of fractal residential development under accessibility constraints, and the second model is devoted to the interactive design of multifractal urban and regional development plans. The simulations identify cells that are potentially valuable for urbanization and urban or suburban centers in which new housing units could be built.

Each chapter may be read independently, which certainly sometimes involves a little repetition when reading the book in its entirety, but these repetitions are far from embarrassing. Quite the contrary, they make it possible to master the material. Moreover, fractal notions follow one another in a very pedagogical manner, from the simplest to the most complex, to hybrid fractals and multifractals. Beyond the technical terms, the reader will become familiar with various theoretical figures (Sierpinski carpets and gaskets, Fournier dusts, box fractals, etc.), and will perceive in a concrete way how the study of these structures across scales allows for a better understanding of the measurement and planning of real urban structures.

We know that fractals are still frightening due to their strong mathematical connotation, but this work should be able to reassure more than one skeptic. Indeed, although the important mathematical formulations are included in the text, the intention of the authors is plain throughout the work to make fractal tools usable and practical in both human geography and spatial planning. The mathematical language is very clear, accessible, limited to the essentials, and also is especially accompanied by numerous examples and counter-examples in France, Belgium and elsewhere in the world.

We should recognize here that the scientific literature on fractals in general is already very abundant and touches on many disciplines. Fractals in human geography are no exception, and the various chapters of this work make extensive reference to them. Each chapter is accompanied by several pages of bibliographic references from the international literature. This literature review is not only a source of inspiration for future research, but also provides a fairly complete overview of the situation since Mandelbrot and the founding works on urban geography of Batty and Longley (1994) and Frankhauser (1994). Cécile Tannier, the scientific editor of this book, complements this general framework with a presentation of the principles of fractal planning and urbanism (particularly in Chapter 8), where the fractality of gaps (green zones in cities) proves to be fascinating in the context of greening the city of tomorrow.

We also particularly appreciate the anchoring of the fractal approach at the heart of geography by regular appeals to well-known notions of quantitative geography such as the MAUP (*Modifiable Areal Unit Problem*), scale, density, the shape of open-field or bocage landscapes, etc., as well as appeals to basic models in human geography such as those of von Thünen, Burgess or Christaller. These reminders thus make it possible to introduce the fractal tool well and truly into the toolbox of the quantitative geographer and the urban modeler.

The two models presented in Part 3 (*MUP-city* and *Fractalopolis*) are the result of a long research journey on the part of the authors and are promising in terms of their potential in urban planning. Through examples of application in France to the cities of Paris and Besançon, these two fractal simulation models are a major step forward in understanding and planning urban complexity. They deserve still greater development. Due to their geometric nature and intrinsic scale dependence, they account for both spatial and functional hierarchies. They provide a better understanding of how built and unbuilt spaces are organized across scales and how it is possible to modify this spatial organization. The caution of the scenarios and the nuanced nature of the conclusions drawn from them will be appreciated.

This book is useful but above all inspiring for the researcher, whether they work in an academic or operational context, at this time of crucial environmental, climatic or societal changes. Built and green space (mentioned throughout the chapters) must certainly be organized inside cities, but we will also consider various networks for the servicing of land such as electricity networks, drainage networks, roads or simply the height of buildings in the third dimension in later versions of the models. We might also think of combining other functions of the built-up area, since urban extension is not just residential satisfaction expressed in terms of housing and accessibility. The quality of urban life is more complex than ever. The current work opens many doors with a modern synthesis of practices, tools and knowledge.

To conclude, this book constitutes a significant methodological asset. It should convince more than one quantitative geographer of the usefulness of fractal analysis and modeling for measurements of spatial reality, as well as and above all contribute to a better understanding of space and its organization across scales and thus help to better plan the urban space of tomorrow, as demonstrated in the last two chapters.

References

Batty, M. and Longley, P.A. (1994). *Fractal Cities: A Geometry of Form and Function.* Academic Press, San Diego and London.

Frankhauser, P. (1994). *La fractalité des structures urbaines.* Anthropos, Paris.

List of Authors

Olivier BONIN
Laboratoire Ville Mobilité Transport (LVMT)
École des Ponts ParisTech (ENPC)
Gustave Eiffel University
Champs-sur-Marne
France

Pierre FRANKHAUSER
Laboratoire ThéMA
CNRS-Université de Franche-Comté
Besançon
France

Gaëtan MONTERO
Centre for Operational Research and Econometrics (CORE)
Université catholique de Louvain-la-Neuve
Belgium

François SÉMÉCURBE
Laboratoire ThéMA
CNRS-Université de Franche-Comté
Besançon
France

Cécile TANNIER
Laboratoire ThéMA
CNRS-Université de Franche-Comté
Besançon
France

Isabelle THOMAS
Centre for Operational Research and Econometrics (CORE)
Université catholique de Louvain-la-Neuve
Belgium

Index

F, G, H

I, L

M, N

O, P

R, S

T, U

Printed and bound by CPI Group (UK) Ltd, Croydon, CR0 4YY

13/09/2026

4961069-0007